# Physical Design Automation of VLSI Systems

*Edited by Bryan T. Preas and Michael J. Lorenzetti*

**Bryan D. Ackland**
*AT&T Bell Laboratories*

**D. Scott Baeder**
*Digital Equipment Corporation*

**William E. Donath**
*IBM T. J. Watson Research Center*

**Alfred E. Dunlop**
*AT&T Bell Laboratories*

**Daniel M. Gajski**
*University of California, Irvine*

**Donald M. Hanson**
*Integrated CMOS Systems*

**Patrick G. Karger**
*Tektronix/CAE Systems Division*

**Thomas Lengauer**
*University of Paderborn*

**Y-L. Steve Lin**
*Tsing Hua University*

**Michael J. Lorenzetti**
*Microelectronics Center of North Carolina*

**Bryan T. Preas**
*Xerox Palo Alto Research Center*

**Thomas G. Szymanski**
*AT&T Bell Laboratories*

**Christopher J. Van Wyk**
*AT&T Bell Laboratories*

**Wayne H. Wolf**
*AT&T Bell Laboratories*

*Sponsoring Editor:* **Alan Apt**
*Production Supervisor:* **Karen Gulliver**
*Production Coordinator:* **Sharon Montooth**

LIBRARY OF CONGRESS
Library of Congress Cataloging-in-Publication Data


Physical design automation of electronic systems / edited by Bryan T.
   Preas and Michael J. Lorenzetti ; Bryan D. Ackland ... [et al.].
      p.    cm.
   Includes index.
   ISBN (invalid) 0-8053-0412-9
   1. Integrated circuits--Design and construction--Data processing.
2. Computer-aided design.   I. Preas, Bryan T.   II. Lorenzetti,
Michael J.   III. Ackland, Bryan D.
   TK7874.P47  1988
   621.381'73--dc19                                    88-14479
                                                          CIP

ISBN 0-8053-0142-9

ABCDEFGHIJ-HA-898

**The Benjamin/Cummings Publishing Company, Inc.**
*2727 Sand Hill Road*
*Menlo Park, California 94025*

# Preface

The field of design automation has received increasing attention and commercialization in the last two decades but has produced relatively few books. A body of knowledge exists in journals and conference proceedings, but only a few comprehensive (and now out-of-date) books have been produced. As a result, it is difficult for people new to the field to know what has happened before and to avoid reinventing old ideas and duplicating old results. Several universities offer graduate courses in design automation (DA) for VLSI, but until now no suitable textbook for physical design existed. This book fills that void.

**Goals of the Book**

This book is intended as a textbook for an upper division or first-year graduate course in VLSI physical design automation or as a reference book for the practicing professional. The important aspects are covered: synthesis, analysis and verification, knowledge-based techniques, and combinatorial complexity. We provide a review of existing techniques, a guide to related literature, and an indepth discussion of the state of the art. Each chapter covers a particular aspect of physical design automation. The individual chapters can be studied independently or as part of a larger, more comprehensive course that includes all of physical design automation. Advances are put into historical and taxonomic perspective. We have collected and described algorithms to give the reader a perspective on the state of the art in each area. We have avoided detailed descriptions, data structures, and implementation details in favor of concepts and algorithms. Each chapter discusses applications of algorithms; the examples are drawn from MOS ICs.

Taken as a whole, this book represents a survey of electronic system, physical design methods for a wide range of design styles. To the greatest possible extent, a technology-independent presentation has been maintained so that the algorithms and techniques are applicable to a wide range of fabrication processes.

The area of physical design automation is too large to be covered by a single author to the depth necessary for a book such as this. Unifying a book by many authors is always a challenge, and we are indebted to the authors for the many changes necessary to produce a consistent presentation. The authors are recognized experts in their individual fields. Their opinions are sometimes subjective but well informed.

We cover *physical* design automation in detail but do not address other aspects of VLSI design automation such as design capture, simulation and test. This is not a book about "advances" but rather a review of the current state of the art. Neither is this a cookbook. This book is not an implementation guide for layout systems; rather, it is a means to understanding existing techniques and systems. The many references will permit readers to use this book as a guide to and review of the existing literature.

## Use of the Book and the Exercises

When used as a textbook, we suggest that students be familiar with digital design and have a strong computer science background. The course, as we envision it, would be tutorial in nature. This text should be supplemented with some of the references when a more in-depth study of a particular topic is desired. For such a course, this book provides definitions of problems, a taxonomy of the solution space, and a perspective on previous solutions.

Some of the exercises at the end of each chapter test understanding while others extend the concepts presented in the chapters. The exercises are divided into three levels of difficulty.

• Questions to test basic understanding of the material covered in the chapter. These exercises have no special marking.

• Problems of intermediate difficulty. These problems are preceded by a dagger (†).

• Suggested projects which take a week or more to complete. These projects are preceded by a double dagger (‡).

When used as a reference by the practicing professional, this book provides a survey of the state of the art with editorial comments and an ample list of references for further research into individual topics.

## Acknowledgements

The editors wish to acknowledge the Xerox Palo Alto Research Center (PARC) and the Microelectronics Center of North Carolina (MCNC) for the use of their equipment in the preparation of this book and for allowing us the time required to complete it. Without their continued support, this book could not have been written. In particular, Bob Ritchie provided enthusiastic encouragement and both tangible and intangible support.

The Association for Computing Machinery Special Interest Group on Design Automation (ACM SIGDA) recognizes the need for up-to-date texts and references in design automation and provided funds to improve the quality of this book. We used the funds for editing, formatting, checking references, and proofreading the manuscript. Chuck Shaw's support is appreciated.

Wayne Wolf and Lisa Kelly did an excellent job of unifying the text and presentation across the chapters. Special thanks go to Lorna Fear of Xerox PARC for her tireless efforts and long hours in managing the production of the book, as well as providing camera-ready copy. Additional thanks go to Jeri Williams of MCNC and Kathy Preas for their time and effort in editing and reformatting the individual chapters. Jeri Williams also checked each of the references. We are also grateful to Giordano Beretta for producing the color plates and the cover art, to Maureen Stone for color consultation, to Peter Kessler for help with the typesetting programs, to Tim Diebert for help with the printers and typesetters, and to Bridget Scamporrino for the typesetting.

Many of our colleagues have assisted us by reviewing and commenting on early drafts; we are especially indebted to William Dees of Automated Design Enterprises and Jim Cohoon of the University of Virginia for the time that they devoted to the project. Jim Cohoon did an excellent job of unifying the exercises. Lissy Bland, Ed Dorsey, Louis Monier and Wayne Stewart all made helpful comments.

*Bryan T. Preas*  
*Palo Alto, California*

*Michael J. Lorenzetti*  
*Raleigh, North Carolina*

# Contents

# Chapter 1

# Introduction to
# Physical Design Automation

**Bryan T. Preas**

*Xerox Palo Alto Research Center*
*Palo Alto, California*

## 1.1  Introduction

This book focuses on the automation of the physical design process for electronic circuits and describes physical design methods for a wide range of design styles. In this chapter, physical design automation is introduced by defining automation and the physical design process within the context of electronic system design. This establishes a framework for the remaining chapters in the book.

Section 1.2 provides an overview of electronic system design by describing the design process and discussing broad categories of design automation tools. Section 1.3 further places physical design within the context of electronic system design by providing a prescription for the design process. Section 1.4 focuses on physical design through a brief introduction to fabrication methods and provides an overview of the synthesis, analysis, and verification methods for physical design. Design styles, the focus of Section 1.5, are a restriction to a particular class of circuit structures and are defined by a combination of the fabrication methods and physical design methods described in Section 1.4. The physical design methods are further illustrated through an example circuit produced by a modern design automation system; this example is presented in Section 1.6. The organization of this book is the subject of Section 1.7.

## 1.2  Automating the Design Process

The creation of large, complex electronic systems has grown beyond the capabilities of any number of people without computer support; successful completion of large design projects requires that computers be used in virtually all aspects of the design. This trend toward automation will accelerate as improved circuit fabrication technologies permit higher levels of integration and as more powerful computers allow more sophisticated tools.

While we cannot yet specify a large, high-performance computer system and automatically produce its design, more restricted goals are well within the state of the art. Computer-based, design automation (DA) tools enable designs that are too large or complex to undertake otherwise, shorten design time, improve product quality (performance and reliability), and reduce product costs. To understand the impact of these tools, let's first look at the elements in the design process.

The design process can be viewed as a sequence of transformations on behavioral, structural and physical design representations, at various levels of abstractions. For example, *functional design* defines the behavior of system outputs as a function of the inputs; *logic design* manipulates structural descriptions (schematic or logic diagrams) while preserving functionality. *Physical design* involves either transformations on or transformations into information used in the manufacture of physical systems.

The DA tools used in each of these transformations can be placed into the following general categories: management of design information and flow; synthesis; analysis; or verification and validation. The first category, *management of design information and flow*, provides the foundation on which DA systems are built. This foundation must be flexible and adaptable to new tools, design styles, and fabrication technologies. A design management system and paradigm for accessing design data and maintaining consistency are described in [Kat85]. The latter three categories transform, analyze or verify the design representations.

*Synthesis* creates new representations, or provides refinements to existing representations, for objects being designed. For example, automatic placement and routing systems create new representations of layouts from structural representations such as lists of components that comprise the design and lists of pins on the components to be connected. As an alternative to creating new representations, refinement operations involve optimizations to improve quality or reduce cost. These optimizations may be either *combinatorial* (for example, cell placement improvement or wiring compaction) or *parametric* (for example, sizing transistors to improve performance).

*Analysis* operations evaluate the consistency or correctness of design representations. An example of an analysis tool is a design rule checker; it checks physical layout representations for geometric rule violations.

*Verification* operations provide a formal process for demonstrating the equivalence of two design representations under all specified conditions, while validation is an informal, less rigorous correctness check.

## 1.3  Electronic System Design

This section places physical design within the larger context of system design by providing an overview of the design process for electronic systems. Automation of the design process is enabled by extensive use of computer design tools. Effective use of these programs requires a structured approach; otherwise, the synthesis, analysis and verification tools would be unmanageable. From the perspective of physical design, the important aspects of electronic design are the following:

- Design representations.
- The design process — to define the limits and interfaces of physical design.
- The degree of automation in the design process.
- A structured design method.

Each of these elements is described in the remainder of this section.

### Design Representations

During the design process several different *representations*, or *views*, are used to show different aspects of the system under design. These views are classified as behavioral, structural and physical, and represent various levels of abstraction [New81, Gaj83 and Chapter 7].

*Behavioral representations* describe a circuit's function. Procedural descriptions (for example, **if** *clock = high* **then** *counter := counter + 1*) and Boolean expressions (for example, *out := a\*b + c*) are behavioral representations; they say nothing about implementation. Some hardware description languages are purely behavioral representations while others include several views.

*Structural representations* describe the composition of circuits in terms of *cell*s (abstractions of circuit element definitions) and *component*s (abstractions of instances of circuit elements) and interconnections among the components. Such descriptions are usually hierarchical; a component at one level may be decomposed into constituent components until elementary (primitive or leaf) components are reached. Examples of structural descriptions are block diagrams, schematic drawings, and net lists of logic gates. Structural representations say nothing about functionality of a circuit except what can be inferred from the behavior of the components given the structure.

*Physical representations* are characterized by information used in the manufacture or fabrication of physical systems. These representations only implicitly describe how a circuit behaves. Physical information may be either a geometric layout (for example, location of transistors or wiring on a silicon surface) or a topological constraint (for example, a higher order cell of a counter should be placed to the left of a lower order cell).

## The Electronic Design Process

Electronic design is carried out in many ways by various designers for a variety of purposes; it is impossible to describe one methodology that applies in all cases. Instead of attempting to provide a comprehensive description, this section provides a prescription for the design process that places physical design in perspective and defines the interfaces of physical design with other design phases. This general method is adopted with variations by different designers.

A top-down design methodology [Lat81] divides the design process into phases. As shown in Figure 1.1, the phases are design specification, functional design, logic design, circuit design, and physical design. Each design phase is further divided into three steps consisting of synthesis, analysis, and verification. The generic design steps within the phases are discussed next; this is followed by a discussion of the design phases.

**Steps Within the Design Phases.** Each design phase is characterized by synthesis, analysis and verification steps as shown in Figure 1.2. *Synthesis* derives a new or improved representation based on the representation derived in the previous phase. At lower levels of abstraction, synthesis is typically automatic; at higher levels of abstraction, synthesis is a topic of intense research, but computer-aided synthesis is still the order of the day.

*Analysis* follows synthesis in each design phase and generally takes two forms. First, a design representation must be evaluated against its requirements. For very large-scale integrated (VLSI) circuits the requirements are usually specified in terms of die size, performance, and power consumption. A design must also be evaluated for behavioral, structural and physical correctness and completeness. Demonstrating that a design representation meets these criteria can be very difficult. The criteria are analyzed many times and the process is often forced to backtrack when constraints cannot be met.

*Verification*, the final step within a design phase, demonstrates that the synthesized representation is equivalent under all conditions of interest to another representation. An example of verification is the exhaustive comparison of two representations such as a schematic diagram and the physical layout synthesized from the schematic. In some cases verification is not possible, and validation, a semiformal process, is substituted. *Validation* demonstrates the equivalence of representations, in behavior or structure, under restricted conditions. For example, a logic model is validated against a functional model if both produce the same sequence of outputs for the same sequence of inputs. Validation may also involve test cases or worst case models that stress a design representation to its worst case limits.

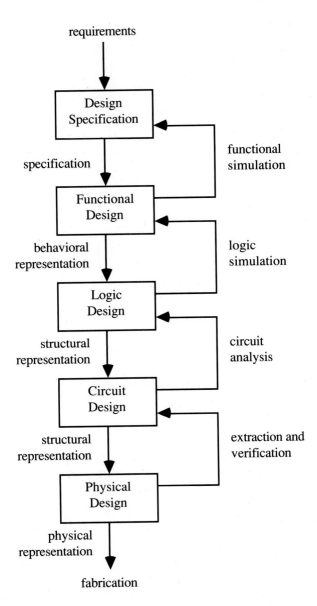

**Figure 1.1**    The phases of electronic system design for a top-down design methodology. This prescription provides a structure for, and defines the boundaries of, physical design automation.

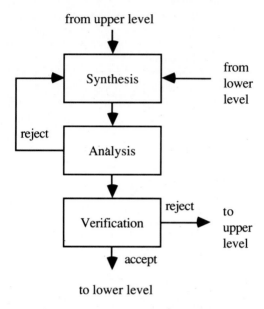

**Figure 1.2**    Each of the design phases of Figure 1.1 is composed of synthesis, analysis and verification steps. Sometimes verification is too difficult and validation is substituted.

**The Phases of Electronic Design.** Now that the steps within the design phases have been described, the design phases of the top-down process are described. The first design phase, *specification*, is a time-consuming, normally manual step. Important factors to be considered are the application of the system, the performance required to meet the application, the architecture of the system, the external interfaces and protocols, the competitive market, the product cost targets, the choice of manufacturing technology, and the design tools that are available. Particular attention must be paid to the design methodology, its impact on the cost of the design, and the time required to complete the design.

*Functional design* follows specification. In this phase, a functional behavior is synthesized to meet the specifications. The result may be a purely behavioral representation (for example, an instruction set description or a timing diagram), or it may include structural aspects by partitioning functionality into components. Behavioral simulation is the normal method of analysis.

*Logic design*, the next phase, concerns the logic structure that implements the functional design. The design representations may be either a textual, register transfer level (RTL) description or a graphic, schematic description. For analysis, these representations are simulated at the transistor, gate, or register level. The logic design is validated by comparing the results from the logic level and behavioral level simulations.

The *circuit design* phase concerns the electrical laws that govern the detailed behavior of the basic circuit elements such as transistors, resistors, capacitors and inductors. In this phase, transistors are sized to meet signal delay requirements. Automatic synthesis tools are available to determine the transistor sizes. Analysis is performed using circuit and timing simulations. Timing verification tools can determine if signal delay specifications are met.

In the *physical design* phase, the behavioral or structural representations from the previous phases are transformed into the geometric shapes that are used in the fabrication of the system. Automatic synthesis, analysis and verification tools for this phase are widely available. Physical design is discussed in greater depth in Section 1.4, and on a larger scale, it is the subject of this book.

## Automation in the Design Process

Computers play a major role in electronic design, but the degree of automation varies widely. The extremes in automation are at one end of the spectrum, handcrafted design (where the designer supplies the intellectual content or inspiration of the design process), and at the other end of the spectrum, completely automated design (where a computer supplies the inspiration as well as the perspiration of the design process). Computer-aided design (CAD) falls between these extremes. The following descriptions do not represent discrete classifications but, instead, represent points on a continuous spectrum.

**Handcrafted Design.**  Handcrafted design describes a less constrained method of design. This is the oldest method and is rarely used today to create large designs; its use is relegated to creating small cells or to finishing any incomplete wiring left by automatic routing. Even though the term implies manual operation, it also applies to the intellectual content of the design process. Generally speaking, the term is applied to the design method where a computer is used as a high-speed drafting machine and the designer provides all the creativity.

**Computer-Aided Design.**  Computer-aided design systems provide tools to synthesize, analyze and verify portions of the design but require active participation of the designer. The designer directs and sequences the tools to complete the design. He or she must supply a degree of creativity, but the CAD system frees him or her from tedious tasks.

**Design Automation.**    Given an abstract specification of an object to be designed, a design automation system generates the physical design automatically and verifies that the design conforms to the specification. Currently, most of the individual areas of physical design are understood, but no complete design automation system exists. Even though much research remains, promising results are available. Design automation also encompasses correct-by-construction concepts or verification of correctness of the design. Design automation and computer-aided design differ in the degree to which the designer supplies the intellectual content of the design process.

## A Structured Design Method

A structured design method is of major importance for automating the design process. The design method discussed in this section permits the design to evolve rationally as a modular system. Describing the design as a formal hierarchy leads to well-defined interactions among the components of the design system, and produces a simplified design. From the perspective of physical design, hierarchical design provides the framework that makes automation possible and enables design styles, which are the subject of Section 1.5.

A structured design method enables consistent descriptions in the three design representations (behavioral, structural and physical) at all relevant levels of abstraction. Consistent descriptions are imperative for design automation systems to work. Furthermore, because the design of large electronic systems is so complicated, a structured design method provides the simplification necessary to complete the design activity. Good methods for simplification are abstraction, regularity and standardization. These simplification methods are described below. Through standardization, developers of design automation systems can create efficient tools; abstraction and regularity allow system designers to reduce complexity by working with simpler objects. More comprehensive discussions of managing the complexity of large electronic system designs are available [Seq83, Nie83].

**Abstraction.**    Abstraction replaces an object with a simplified model that defines the interaction of the object with its environment. Any details of internal organization or implementation are deleted. Abstraction reduces the amount of data needed to describe an object. However, for large systems, one level of abstraction is insufficient to reduce objects to manageable sizes. Thus, hierarchical abstraction or *hierarchy* is required. Hierarchy involves decomposing a system into a set of components. These components are recursively decomposed into subcomponents until all components are small enough to manipulate. In order for a structural, hierarchical decomposition to be of greatest value, the abstraction process must follow certain rules [Seq83]. Two of the most important of these rules are modularity and locality.

*Modularity.*  Modularity implies a well-defined, unambiguous functional interface. In the physical domain, modularity dictates clear specifications of connection points. Modularity allows the designer to understand his design and permits a design system to verify the attributes of an electrical component in its environment.

*Locality.*  Locality implies that the details outside a component are not important while the interior of a component is being considered. In the physical domain, locality is clearly related to area. One method is to allocate to a component a rectangular area and require that any internal physical implementation be design rule correct with respect to the boundary of the area. Of course, more relaxed (and less wasteful of area) locality rules can also be specified at the cost of more complex rules and algorithms. Thus, locality is a form of information hiding that reduces complexity.

**Regularity.**  Regularity is often used to reduce complexity. One- or two-dimensional arrays of random access memory (RAM) or read only memory (ROM) cells can be formed by iteration. Less regular forms are apparent in datapaths where the iterated components may be slightly different. Hierarchy amplifies regularity because iterated components may exist at any level of abstraction.

**Standardization.**  Standards that restrict the design domain can be applied at all levels of abstraction to simplify components. Restriction to a few well-understood constraints has many advantages. At the lower levels of abstraction for the physical domain, design rules [Gla85, Wes86] and meta-design rules [Seq81] have been standardized. Another example of low-level standardization is the restriction to Manhattan geometry containing only rectilinear features. Higher levels of abstraction require that more features be standardized. Cell-based design (where the definitions of the primitive circuit elements have been previously proven to work and are now stored in a library) makes a system more modular and reduces the possibility of errors. Some physical design styles are defined by restrictions on power distribution and on sizes and shapes of components. For example, one cell-based design style, standard cells, provides a further degree of standardization: cell heights and power distribution through the cells are common for all cells. This allows the cells to be placed neatly in rows and greatly simplifies the layout problem.

## 1.4  The Physical Design Process

To understand the physical design process it is necessary to be acquainted with the methods used to fabricate physical systems and to understand the important physical design methods. This section provides a brief introduction to fabrication methods of modern electronic systems: silicon integrated circuits, hybrid integrated circuits, and circuit boards. The remainder of the section discusses the physical design methods: full custom design, cell-based design, symbolic design, procedural design, analysis, and verification.

## Fabrication Methods

A wide variety of methods are used to manufacture circuit assemblies for modern electronic systems. The most important categories are shown in Figure 1.3. Certainly, many more methods exist, but these are the most important from the perspective of physical design automation. A detailed discussion of the fabrication processes is beyond the scope of this book, but good descriptions are available [Gla85, Wes86]. For this discussion it is only necessary to abstract the features of the processes that are important to physical design automation.

> Integrated Circuits
> > Programmable Arrays
> > Full Custom Fabrication
> Hybrid Integrated Circuits
> > Thin Film
> > Thick Film
> > Silicon-on-Silicon
> Circuit Boards
> > Discrete Wiring
> > Printed Circuits

**Figure 1.3**    These fabrication methods for electronic systems are important for physical design automation.

Integrated circuit fabrication technology has provided the impetus for the revolution in the electronics industry. It is now possible, in batch fabrication processes, to manufacture individual chips with millions of interconnected transistors. Since modern fabrication processes are so complicated, an abstraction called design rules captures the elements that are important to the physical design process. *Design rules* are constraints imposed on the geometry or topology of layouts and are derived from two sources:

- Basic physics of circuit operation such as electromigration, current carrying capacity, junction breakdown, or punch through.

- Limitations in the fabrication process such as minimum widths and spacings that can be exposed on a single layer, misalignment among layers, or edge shifts of the materials during processing.

A basic distinction of fabrication methods is between programmable arrays and full custom processing. A programmable array does not implement any specific function when it is initially manufactured. It is fabricated as a one- or two-dimensional array of repeated cells. Generic features such as empty memory cells or unconnected transistors are fabricated; the array is later customized to obtain the specified function. Wafers are stockpiled by the manufacturer in a partially processed state. A particular circuit function is implemented by custom processing for only those mask layers that are unique to the particular circuit. Examples of programmable arrays are ROMs, programmable logic arrays (PLAs), and gate arrays. In ROMs and PLAs only the transistor gate is customized, while in gate arrays the metal interconnection layers and via layers between the interconnection layers are customized to produce a specific function. Fewer unique masks for programmable arrays (typically only one to four layers) leads to reduced tooling costs and shorter fabrication times.

In contrast to the programmable array method, a full custom chip is designed and manufactured for a specific function. The design automation system has the additional freedom to optimize all of the mask layers of the physical design for the particular circuit being designed. This extra freedom results in increased tooling cost and fabrication time. However, if the part is manufactured in high volumes, full custom processing is more economical because the chips are usually smaller. The increased yield of the smaller chips more than compensates for the increased cost of processing all of the mask layers.

A large number of board-level fabrication methods are available to electronic system designers. The two primary methods involve either discrete, point-to-point wiring or printing conductive material (usually copper) onto a fiberglass-epoxy substrate. The most important point-to-point wiring method for modern systems is *wire-wrapping*. With this method, the electrical connection points for the components (usually integrated circuits) are connected by individual wires wrapped around metal posts called *pin*s. Wiring is usually performed by machine [Gri61] but may be done manually. A survey of this and other discrete wiring techniques is found in [Lym80a]. The second board-level fabrication method, *printed circuit*s, involves printing one or more layers of metal interconnection on a substrate [Ein77]. Except at *via* locations, wiring on one layer is insulated from wiring on other layers.

Hybrid integrated circuit fabrication provides a technology that is intermediate between that of integrated circuits and circuit boards. Hybrid processes can be divided into three categories: thin film, thick film, and silicon-on-silicon. Thick film hybrids are built by screening and firing alternate conductive, resistive and dielectric pastes onto a ceramic substrate. Thin film hybrids have resistors and conductors deposited in a vacuum onto a ceramic substrate [Lym80b]. A newer hybrid packaging technique mounts silicon integrated circuits directly on a wafer sized silicon substrate. Interconnection wiring is deposited on the substrate using conventional silicon metallization processing. This technique allows fine (very narrow) lines and high packing densities compared to other hybrid or board-level packaging technologies. [Joh86] provides a review of the process and an excellent bibliography.

**Physical Design Methods**

The physical design phase within the system design process (as described in Section 1.3) is divided into synthesis, analysis and verification steps. This section discusses the methods used within these steps. Synthesis is further divided into full custom design, cell-based design, symbolic design, and procedural design.

**Full Custom Design.**    This method is characterized primarily by the absence of constraints on the design process. It usually requires a handcrafted level of automation since the lack of constraints makes synthesis tools difficult to develop. The designer is responsible for layout optimization and may span all levels of abstraction if he or she chooses. The only means of standardization are design rules. Depending on the ability of the designer, layout quality can be high. However, design rule checking and verification are mandatory because of inevitable errors. Full custom design is very time-consuming (five to ten devices/designer/day); thus the method is inappropriate for large circuits. However, the full custom method is widely used for smaller cells that are inputs to synthesis tools.

**Cell-Based Design.**    The notions of the structured design method, especially abstraction and standardization, discussed in Section 1.2 are directly embodied in the cell-based design method. Abstraction is implemented by composing the design from cells exhibiting modularity and locality. When the designer and the design system are operating at a particular hierarchical level, only the public interfaces of the constituent cells need to be considered. For board-level and hybrid integrated circuit design, cell-based design is the only synthesis method available. In these cases, the cells are pre-packaged silicon integrated circuits.

For the design of integrated circuits, the cells may range in complexity from logic gates with a few transistors to large subsystems with many thousands of transistors. The degree to which cells are standardized determines the design style (see Section 1.5).

Automatic placement and routing systems are the primary synthesis tools for cell-based design methods. A result of the logic design phase is a list of components and *public pin*s (interface connections) to be placed as well as net lists specifying the required connections among the components' pins. The components must be placed to minimize layout area and perhaps signal delay. All the connections must be routed within the space and wiring layers provided. Integrated circuit technology provides an extra degree of freedom: the sizes and shapes of the cells are flexible. This flexibility leads to floorplanning which determines sizes and shapes of the cells as well as their placement.

Interconnection analysis and partitioning are functions within cell-based design. Interconnection analysis determines the compatibility of a cell's design and its fabrication technology. When a design and a fabrication technology are not compatible or the design is suboptimal, partitioning can be used to modify the hierarchical assignment of cells within other cells. Partitioning is also an important aspect of some modern placement algorithms.

**Symbolic Design.**  The cell-based design method reduces the complexity of the design process through abstraction, hierarchy, and standards. Symbolic design methods when combined with compaction simplify the design process through a different kind of abstraction: hiding implementation design rules and capturing structural and physical requirements primarily as topological constraints.

Symbolic design allows the designer to think and act in a topological domain. His or her design tools can synthesize the geometric domain representation to form the smallest area, yet still honor the physical and electrical rules required. Since a given topological representation is valid over a range of physical and electrical layout rules, a symbolic design can be updated to new fabrication technologies by entering the new layout rules and resynthesizing the physical design. This updating flexibility leads to the notion that a topological representation of a design is an appropriate "assembly language."

With the symbolic method, design is simpler since design rule correctness for the layout is enforced by the synthesis tool. Layout quality can be high, but designer expertise in creating the topology still has a big impact. The symbolic method is acceptable as a primary design method for small- to medium-size physical designs and is widely used for the physical design of cells that are input to other synthesis tools.

**Procedural Design.**  The design methods discussed above are classified as data driven because the input data provides the primary source of information for the synthesis process. A second classification of design methods is called *procedural* (or *program driven*) *design*. These methods have become key components in modern physical design systems. Procedural methods may be algorithmic in nature, such as module generation or silicon compilation, or may be rule based, expert systems.

*Module Generation.*  Module generation is advantageous when a particular class of circuit structures or layout architectures is to be used, with variations, in different designs. A programming language procedure, with parameters to define the required variations, can generate the physical representation of the specified module; the procedure adapts a given module to the context in which it is used. The parameters can specify complex variations that can include repetitions and conditionals. This design method is particularly successful for regular arrays, such as RAMs, ROMs and PLAs, as well as less regular structures such as datapaths.

*Silicon Compilation.*  Silicon compilers translate high-level, behavioral or structural design representations into layouts. Here the operative term is "high-level." Silicon compilers hide intervening transformations that a designer would otherwise see. An ideal silicon compiler, which does not yet exist, would be able to take a design specification (for example, a behavior defined by a program) and produce a system design which implements that program in hardware.

*Knowledge-Based Design.*    Knowledge-based design enables system designers to explore the large number of design possibilities that must be considered in order to produce a design that meets specification. Exploring the design space is particularly difficult even for expert designers. All the information needed to make decisions is not available until the design is complete, but the complete design process takes a long time and produces large amounts of data. To avoid exhaustive search of the design space and still achieve good solutions, information concerning the problem domain must be explicitly represented in the design system. Expert systems combine the problem domain information, an inference engine, and a working memory to make informed decisions early in the design process. This allows the programs to concentrate on the parts of the design space that are likely to produce good results.

**Analysis and Verification.**    Design and fabrication times for integrated circuits are so long and debugging is so difficult that it is crucial to eliminate all errors during the design process (before the circuit is first fabricated). Modern analysis and verification tools are capable of achieving this goal; many designs work after the first fabrication. Analysis and verification for physical design can be divided into three areas:

- *Design rule checking* analyzes mask geometries to determine if they meet the size, spacing and enclosure rules specified by the fabrication technology.

- *Circuit extraction* and *connectivity verification* determine the equivalence of the physical circuit topology to some other representation. Typically, this is the representation from which the physical circuit was synthesized.

- *Parameter extraction* determines electrical parameters from layout information that can be used in simulating the timing of the signals.

## 1.5  Integrated Circuit Design Styles

Soon after computers were applied to the problem of electronic design, it became apparent that a structure was necessary in order to use both computer and human capabilities effectively. Structure can reduce complexity of both the design problem and the automation tools that determine a solution. A design style provides such a unifying structure. It is the use of a particular class of circuit structures and is defined by the combination of a fabrication method and a design method as shown in Figure 1.4. Currently, the popular integrated circuit design styles are gate arrays, standard cells, and general cells. A full custom design style also exists, but is not discussed because of declining importance and limited opportunity for automation.

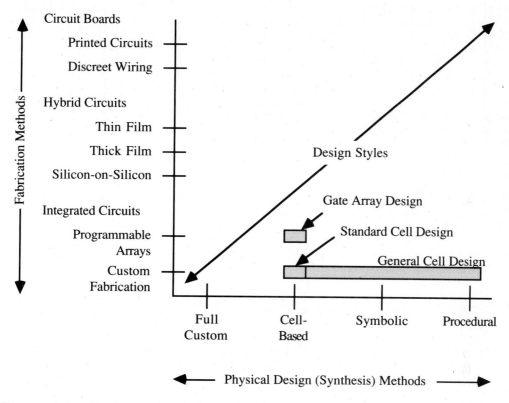

**Figure 1.4**    Design styles are a function of fabrication methods and physical design methods.

In addition to providing background and terminology of the popular design styles, this section describes how the design methods, which are the focus of this book, are used to create physical representations of electronic circuits.

### Gate Array Design

The gate array design style combines the programmable array fabrication method and the cell-based design method. In this approach, a cell-based design is mapped onto a prefabricated, two-dimensional array of uncommitted transistors.

Gate array foundries usually fabricate large numbers of wafers containing identical, unwired arrays. These arrays are then personalized for a particular circuit function by adding interconnection wires. Two layers of wiring are the

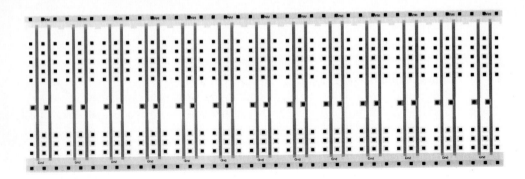

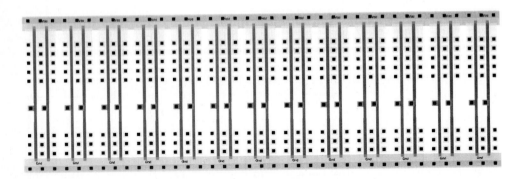

**Figure 1.5**  A drawing showing a portion of the array of uncommitted transistors of a gate array with a row-based floorplan.

most common, but one and three layers are also used. Figure 1.5 shows a drawing of the features that comprise a two-dimensional array of transistors. This array has a row-based floorplan. With this floorplan, the transistors that form the gates are arrayed in rows that are separated by horizontal routing channels. Other popular floorplans are columns, islands, and sea-of-gates. Island floorplans have routing channels running both horizontally and vertically. The cell sites form the islands. Sea-of-gates floorplans either have no routing channels or the channels are small compared to the gate sites. In this floorplan, wiring must be routed over the cells.

Although all wiring in a particular layer is deposited on the gate array surface at one time, the intra-cell wiring and inter-cell wiring are composed differently by the design system. The basic cells that comprise the design are usually defined in a library; part of the cell definition is an interconnection pattern that is imposed on a local group of transistors. Interconnection among the cells of the design is generated by automatic placement and routing. An inherent disadvantage of the gate array design style is that, in general, completion of all of the required interconnections cannot be guaranteed. The fixed-size routing areas can only hold a fixed amount of routing.

The gate array design style is very popular and many mature design systems are available. The advantage of this design style is clear for prototyping and low-volume production. Tooling costs are low because only four masks (contact or cut, metal, via or cut 2, and metal 2) are required for two interconnection layers. Applying the interconnection layers is a low-risk, inexpensive operation; yields are high and fabrication time is short. However, for higher volume production, other design styles are usually more area-efficient and less costly to manufacture.

### Standard Cell Design

Circuit design with standard cells is similar to gate array design; both use cell-based design methods where the cells are small- and medium-scale integration gates. However, full custom processing is required for standard cells. The resulting full mask set affords the circuit designer an extra degree of flexibility; circuit functions such as memory or analog elements (that cannot be mapped onto the predefined array of transistors of gate arrays) can readily be included in standard cell designs. For a given circuit, area is almost always smaller compared to gate arrays. Interconnection area is smaller since the routing and the cells can be positioned to accommodate the specific circuit being designed. Basic cells are smaller because they do not have to be mapped onto a predefined array of transistors; the layout of each cell can be customized for its specific function. Modern physical design systems can further reduce area by generating basic cells that use procedural design methods to generate whatever specialized functions are required [Nai85].

Figure 1.6 shows a line drawing of a standard cell design that has 3014 components and 3029 nets. (This example is the Primary2 design from the 1987 Physical Design Workshop [Pre87].) The placement and routing process is similar to that of gate arrays. The difference is that the positions of the cell rows as well as the positions of cells within the rows are completely determined by the layout system. When *feedthrough* cells are needed to complete a connection between two routing channels, they can be added between any two cells, as shown in Figure 1.7. Because standard cell design systems have flexibility in determining positions for rows, cells and feedthroughs, they can guarantee that all of the interconnections can be completed.

The standard cell design style is well-suited for circuits having several thousand gates and medium production volumes. Physical design is straightforward and efficient using modern design tools. The standard cell design style is also widely used to implement the "random logic" (that portion of a design that is difficult to cast into a regular, efficient layout style) of general cell designs.

### General Cell Design

It is difficult to implement a wide variety of logic functions in a single design style. The predefined array of transistors that defines gate array design prohibits structures other than standard, digital logic. Standard cell design is less restrictive than gate array design, but it is still expensive to implement memory intensive functions (for example, RAMs and ROMs) or regular functions such as datapaths. The general cell design style, a generalization of the standard cell design style, addresses these issues. The cells (available from a library or constructed as required by the design system) may be large and irregularly shaped. The cells may be customized by logic function and synthesized by any of the methods discussed in Section 1.4.

Automatic placement of general cell designs is complicated because the cells must be represented as two-dimensional objects and their sizes and shapes can vary widely. Automatic routing is also more difficult (compared to standard cells and gate arrays) since the channels may interact in complex ways. An example of a general cell design produced by a modern design automation system is presented in the following section.

## 1.6  A General Cell Design Example

This section provides a comprehensive example of a general cell IC design; it serves to illustrate the material introduced in this chapter. Plates 1 and 2 show the physical representation of a general cell design that was (completely) automatically generated by the DATools System at Xerox PARC; Plate 1 shows a photograph and Plate 2 shows a color plot of the same circuit. The generation of the physical design took 3 hours and 20 minutes on a Xerox PARC Dorado (a 2-MIPS machine). The DATools System was designed and implemented at Xerox PARC and provides most of the synthesis, analysis and verification methods discussed in this chapter.

This section first provides an overview of the example circuit and its function. Next the overall design methodology is discussed; this discussion is followed by an overview of the techniques used to produce the layout of this circuit.

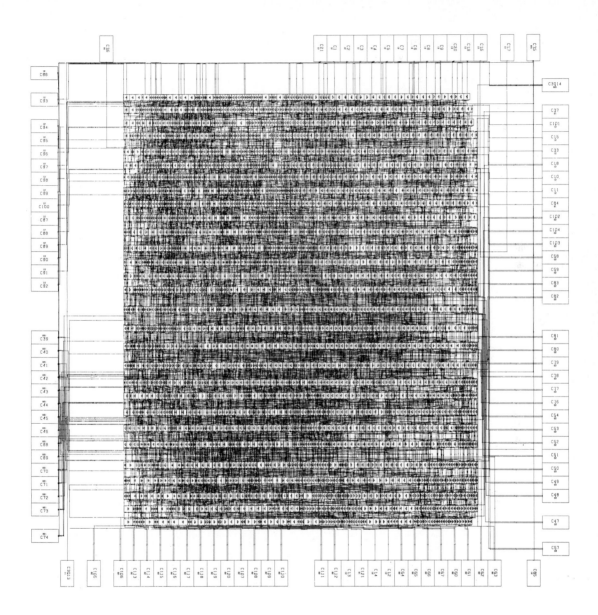

**Figure 1.6** A standard cell design with 3014 components and 3029 nets. This is the Primary2 benchmark of the Physical Design Workshop benchmarks. (Photograph provided by Sandia National Laboratories, Albuquerque, NM.)

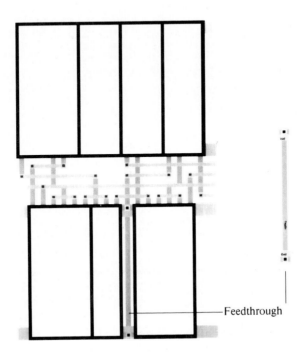

**Figure 1.7**    Feedthrough cells are used to form a connection across a row of standard cells. They are inserted by standard cell design systems as a function of the placement.

### The Circuit

The example circuit, known as the Execution Unit (EU), is a heavily pipelined, 32-bit datapath; the combination of this chip and an associated instruction-decode chip forms a high performance (5-MIPS), multiprocessing, central processing unit. The top-level architecture includes a large 3-port register file, a full carry look-ahead arithmetic-logic unit (ALU), a field unit (including a 64-bit funnel shifter and two mask generators), address decoding logic for the register file, condition-code logic, and control logic. The control logic is small because most of the normal control functions are provided by the associated instruction-decode chip. The periphery of the chip contains 89 signal pads and 54 power pads. A floorplan of the circuit (Figure 1.8) shows the top-level cells and their synthesis methods.

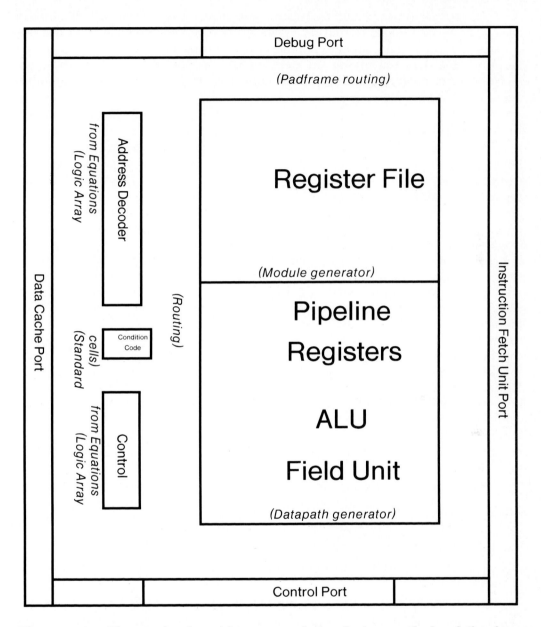

**Figure 1.8**  The top-level architecture and the design methods of the functional units for the circuit shown in Plates 1 and 2. The condition-code logic is shown in more detail in Plate 3.

The EU comprises 68,000 transistors and covers 10.892 mm by 9.092 mm in area. The large amount of routing between the core logic and the bonding pads is caused by external packaging constraints; routing on the IC is cheaper and more reliable than routing on the next higher level of packaging (a silicon-on-silicon hybrid integrated circuit). The active area of the circuit is much smaller, about 8 mm by 4.5 mm.

**Layout Methodology**

A designer enters his or her structural and behavioral specification and the physical design for any leaf cells that do not exist in a library using a graphics editor; a single editor is used for both purposes, and the two types of descriptions typically reside in the same graphic design. The schematics may carry arbitrary annotations which are expressions in the underlying Cedar programming language [Swi86]. The schematics (the structural specification) are then extracted; in addition to determining connectivity, the extractor provides a scoping model and evaluates the annotation expressions. The result of the extraction is a hierarchical netlist with procedural annotations. Most synthesis, analysis and verification tools (for example, layout generators, simulators, and design rule checkers) share this representation and use the annotations. The hierarchical description of the circuit is thus a combination of schematics and code. However, the distinction is actually fuzzy since any object in the system can be generated by program, and conversely the extraction of any object can invoke an arbitrary piece of code. Typically, description by program is reserved for large objects whose structure is best expressed algorithmically. Examples are irregular tilings of cells, such as ROMs or PLAs, or complex structured interconnection specifications, such as the carry tree of an adder. On the other hand, schematic representations are the most convenient way to capture hierarchical structure and are used extensively. A more complete description of the DATools and the layout methodology is given in [Bar88a, Bar88b, Bar88c].

For physical design, important annotations on the schematics are the procedures used to synthesize the layout and any parameter values needed to override the default values. This approach allows layout synthesis to be invoked by a single, common procedure call that takes as input a structural description and returns a physical layout. Thus, complete chips (or any simpler cells) are laid out by a single, recursive call to one procedure (called *Layout* in the remainder of this section).

The DATools System provides high-level synthesis procedures such as pad frame routing, placement and routing for cell-based design methods, module generators for datapaths, memory elements, and Boolean equations. A one-dimensional stack operator allows combination of automatic routing and module generation. Lower level synthesis procedures include getting cells from libraries

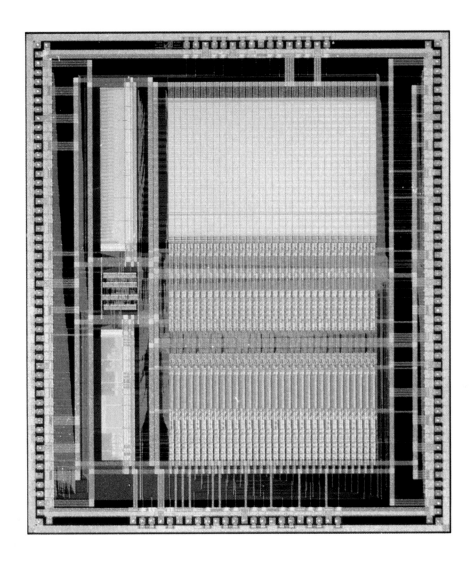

**Plate 1** A photograph of a 32-bit execution unit designed by the DATools System at Xerox PARC. This physical design was automatically constructed in 3 hours and 20 minutes. (Photograph supplied by Xerox PARC, Palo Alto, CA)

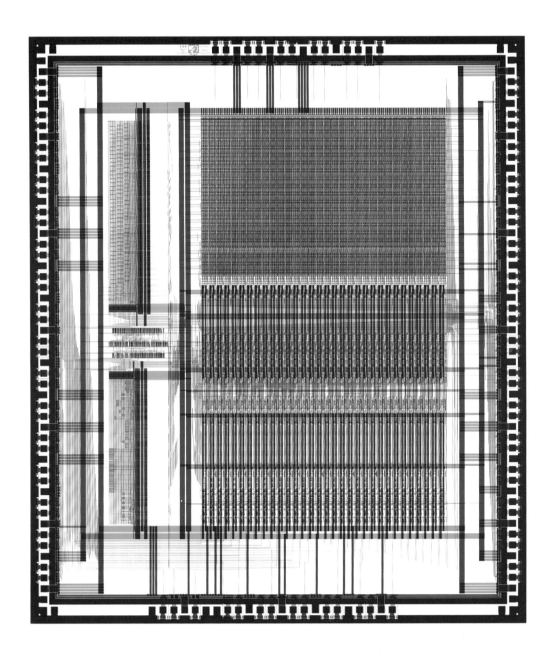

**Plate 2**    A Plot of the 32-bit Execution Unit Pictured in Plate 1. (Plot supplied by Xerox PARC, Palo Alto, CA)

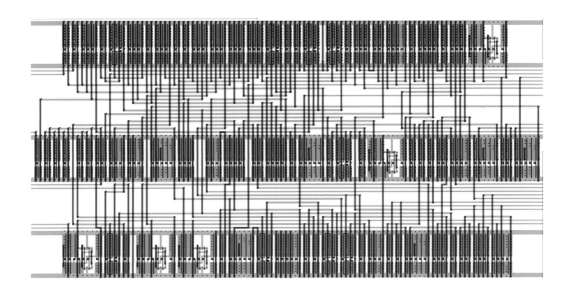

| | n-diffusion | | p-diffusion and well |
| | polysilicon | | metal |
| | overglass | | metal 2 |
| | cut | | cut 2 |

**Plate 3**   A plot of the condition-code logic of the 32-bit execution unit pictured in Plate 1. This is a standard cell implementation. (Plot supplied by Xerox PARC, Palo Alto, CA)

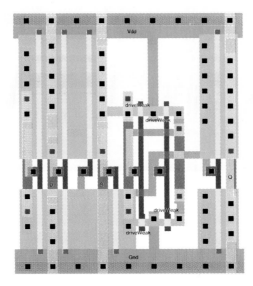

(a)  A plot of a D latch. This leaf cell is one of the standard cells used to construct the standard cell assembly shown in Plate 3.

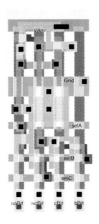

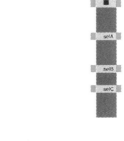

(b)  A plot showing *RamCell*. This is the memory cell used to construct the register array.

(c)  A plot showing *RamStitch*. This leaf cell is used to "stitch" the register array together.

**Plate 4**    This shows three of the leaf cells used to construct the integrated circuit shown in Plates 1 and 2. (Plots supplied by Xerox PARC, Palo Alto, CA)

(*Get*), abutting cells (*Abut*), iterating cells (*Array*), and manipulating cells'
orientations (*Flip* and *Rotate*). New layout synthesis procedures can be added
by a simple registration mechanism.

### Physical Synthesis of the Execution Unit

**The Register File.** At the top of the data path is the register file. The following description of its assembly illustrates cell construction by a module generator. As expected in densely packed areas, the main layout functions are the
low-level assembly operations.

The basic cell is a triple-ported *RamCell*, shown in Plate 4b. The
corresponding schematics with several annotations are shown in Figure 1.9.
The lengths and widths of transistors are typical annotations (only the width is
shown here); this information is used later by the schematics *vs.* layout comparator, the static, electrical checker, and various simulators. An interesting
annotation is *Layout*. Its value is a registered layout generator. In the case of
*RamCell*, the value is *Get*; this means that the layout is available from a library
as a user-drawn cell. Computing the *Layout* of *Get* simply returns the layout of
the requested cell, after checking that the schematics and layout are representations of the same cell and the representations have compatible interfaces. A
small cell called *RamStitch* distributes power through the array and is described
similarly and shown in Figure 1.9 and Plate 4c; both cells have a corresponding
iconic form, shown on the right of the schematics in Figure 1.9.

The schematic description needed to construct the register array is shown
in Figure 1.10. The next level of assembly forms a quad-word; the *RamCell* is
sequenced; the graphic slash symbol distinguishes the wires to be sequenced.
The resulting cell is composed with *RamStitch*, forming a *RamQuad* cell. The
layout annotation *AbutX* indicates that cells must be abutted to compose the
layout. The order of cells is derived from the schematics. This *RamQuad* is
then sequenced to form *RamRow*. Notice that the repetition factor is a parameter; it could in fact be any expression of the underlying Cedar programming
language that evaluates to a discrete numeric type. This evaluation is performed during the extraction to produce a fully expanded netlist.

A symmetric version of the row is produced using the *FlipY* annotation;
the rows are abutted and sequenced vertically to form the entire array. This
array is parameterized by the number of quads per words and the number of
words; in order to generate the layout of the register file, the specific values are
attached to an instance of the icon of *RamArray*. At the next higher level of
hierarchy, periphery cells for pre-charging and I/O are added to the array to
make the full register file.

**The Pipeline.** The pipeline is built from a small library of cells by a datapath
generator. This generator parses a schematics description to infer the placement of rows, the need for inter-row routing channels (for example, the ALU
carry tree is automatically routed), and the location of public pins. Explicit

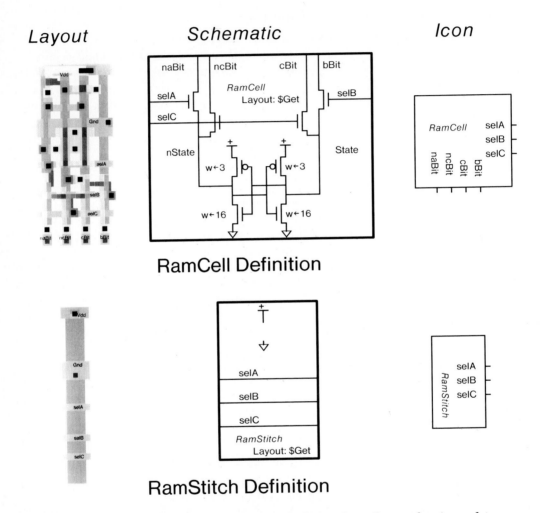

**Figure 1.9**    These schematic diagrams and icons describe, and are used to construct, a storage cell of register file. The primitive cells that correspond to these diagrams and icons (the *RamCell* and *RowStitch*) are shown in Plates 4b and 4c. Figure 1.10 shows how these cells are combined to construct the register array.

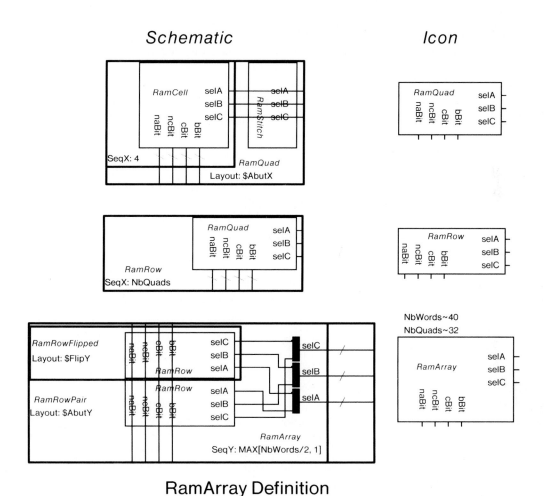

**RamArray Definition**

**Figure 1.10** These schematic diagrams and icons describe the construction of the register file array. The inputs are the single storage cell shown in Figure 1.9 and Plates 4b and 4c. The completed register array is generated from this information and shown in Plates 1 and 2.

parameters include the pitch and number of internal buses (six buses are used in this case) and inter-leaving. During the layout phase, the generator determines power routing and decides whether to insert routing channels or simply to abut adjacent rows. The register file and pipeline are then stacked to form the data portion of the core of the circuit.

**Control and Address Decode Logic.**    Most of the address decode logic for the register file and control logic for the pipeline is compiled from equations into a PLA-like structure which uses cascode logic to provide a fully static logic array [Ser87]. The upper cell is the address decoder for the register file, and the lower cell performs the opcode decoding and register control functions.

**The Condition-Code Logic.**    The remaining random logic is collected into one cell which is described with heavily parameterized library elements. The resulting layout is a small block of standard cells. The sides on which signals exit the cell are inferred from the schematics, while the number of rows is an explicit parameter. One of the standard cells, a D-latch, is shown in Plate 4a.

**Final Assembly.**    Automatic routing was used to connect the functional blocks and the core to the pad ring. Each row of pads is constructed from a schematics description. The pad ring is a combination of the four pad sides (positioned and oriented as specified by the schematics) and four corner cells. The pad frame router assembles nine blocks (the core of the circuit, the four sides, the four corners) into a complete circuit. Default wire widths can be modified by annotations on the nets in the schematic; power bus widths were so specified. Plates 1 and 2 show the result.

## 1.7  Organization of the Book

To understand physical design automation, it is necessary to understand the design methods that are introduced in Section 1.4. Each of the remaining chapters in this book describes, in detail, the concepts and algorithms that are required to automate a physical design method or a major aspect thereof. The state of the art in all major areas is described in detail and placed in historic and taxonomic perspective. In most cases, the concepts and algorithms are presented independently of fabrication technology, and, therefore, are interesting to a wide range of design automation specialists and electronic system designers.

The field of physical design automation is very broad; it is not possible to cover all aspects in a single volume. In order to concentrate on the main subject, many other important topics are omitted: integration of the functions into a system, management of design information, underlying data structures, and implementation techniques.

The cell-based design method is discussed in Chapters 2 through 5. Interconnection analysis (Chapter 2) determines the compatibility of a structural design representation and its fabrication technology. Partitioning, covered in

Chapter 3, divides logic cells into smaller logic cells while maintaining restrictions on cell sizes or on the number of external communication paths. Chapter 4 describes placement, pin assignment, gate assignment, and floorplanning. These functions are related; they all address some aspect of positioning circuit elements on a two-dimensional layout surface. Automatic routing of interconnection wiring is the topic of Chapter 5.

Symbolic layout and compaction, two related design methods, are covered in Chapter 6. Symbolic layout and compaction help automate physical design by synthesizing or optimizing geometrical representations from topological specifications.

Two chapters, 7 and 9, are devoted to procedural-based design. Chapter 7 covers the algorithmic methods: module generation and silicon compilation. Chapter 9 discusses knowledge-based physical design through expert systems.

Chapter 8 covers analysis and verification, the techniques used to demonstrate that physical representations of the design are correct.

Serving as an underlying framework for the previous chapters, Chapter 10 focuses on the computational complexity of physical design algorithms. This complexity is a subject of increasing importance given the steady growth in the size of VLSI circuits and in the systems in which these circuits are embedded.

## Acknowledgements

Louis Monier was responsible for the design of the integrated circuit described in Section 1.6 and contributed the material for that section. The DATools System was developed by the following members of the Design and Architecture Area at Xerox PARC: Rick Barth, Giordano Beretta, Don Curry, Jean-Marc Frailong, Jim Gasbarro, Neil Gunther, Christian Jacobi, Christian LeCocq, Ed McCreight, Louis Monier, Bertrand Serlet, Pradeep Sindhu, and Mike Spreitzer. Carl Diegert of Sandia National Laboratories provided the standard cell layout shown in Figure 1.6.

## Exercises

**1.1**  Discuss the synthesis, analysis and verification steps for your physical design system.

**1.2**  In the prescription for circuit design presented in this chapter, the design phases are design specification, functional design, logical design, circuit design, and physical design. Define these design phases and discuss the design representations that the phases produce for your design system.

**1.3**  What are the different floorplans for gate arrays? What are the advantages and disadvantages of each?

**1.4**  What properties differentiate full custom design, cell-based design, symbolic design, and procedural design?

**1.5†**  Why cannot we yet fully automate the physical design of VLSI circuits? What innovations are necessary before we can fully automate the physical design of large, high-performance circuits?

# References

[Bar88a]  Barth, R., and B. Serlet, "A structural representation for VLSI design," in *Proc. of the 25th Design Automation Conference*, June 1988.

[Bar88b]  Barth, R., B. Serlet, and P. Sindhu, "Parameterized schematics," in *Proc. of the 25th Design Automation Conference*, June 1988.

[Bar88c]  Barth, R., L. Monier, and B. Serlet, "Patchwork: layout from schematic annotations," in *Proc. of the 25th Design Automation Conference*, June 1988.

[Ein77]  Einarson, N. S., *Printed Circuit Technology*. Printed Circuit Technology, Burlington, Massachusetts, 1977.

[Gaj83]  Gajski, D. D., and R. H. Kuhn, "New VLSI tools," *IEEE Computer*, vol. 16, no. 12, pp. 11-14, December 1983.

[Gla85]  Glasser, L. A., and D. W. Dobberpuhl, *The Design and Analysis of VLSI Circuits*. Addison-Wesley, Reading, Massachusetts, 1985.

[Gri61]  Grim, R. K., and D. P. Brouwer, "Wiring terminal panels by machine," *Control Engineering*, pp. 77-81, August 1961.

[Joh86]  Johnson, R. W., J. L. Davidson, R. C. Jaeger, and D. V. Kerns, "Silicon hybrid wafer-scale package technology," *IEEE Journal of Solid-State Circuits*, vol. SC-21, no. 5, pp. 845-851, October 1986.

[Kat85]  Katz, R. H., *Information Management for Engineering Design*. Springer-Verlag, Berlin, 1985.

[Lat81]  Lattin, W. W., J. A. Bayliss, D. L. Budde, J. R. Rattner, and W. S. Richardson, "A methodology for VLSI chip design," *LAMBDA*, vol. II, no. 2, pp. 34-44, Second Quarter 1981.

[Lym80a] Lyman, J., "Techniques of automatic wiring multiply," in *Microelectronics Interconnection and Packaging*, edited by J. Lyman, McGraw-Hill Publications, New York, pp. 148-152, 1980.

[Lym80b] Lyman, J., "Advances in materials, components, processes ensure hybrid prosperity in the LSI age," in *Microelectronics Interconnection and Packaging*, edited by J. Lyman, McGraw-Hill Publications, New York, pp. 62-74, 1980.

[Nai85]  Nair R., A. Bruss, and J. Reif, "Linear time algorithms for optimal CMOS layout," in *VLSI: Algorithms and Architecture*, edited by P. Bertolazzi and F. Luccio, Elsevier Science Publishers, Amsterdam, the Netherlands, pp. 327-338, 1985.

[New81]  Newton, A. R., "Computer-aided design of VLSI circuits," *Proc. of the IEEE*, vol. 69, no. 10, pp. 1189-1199, October 1981.

[Nie83]  Niessen, C., "Hierarchical design methodologies and tools for VLSI chips," *Proc. of the IEEE*, vol. 71, no. 1, pp. 66-75, January 1983.

[Pre87]  Preas, B. T., "Benchmarks for cell-based layout systems," in *Proc. of the 24th Design Automation Conf.*, pp. 319-320, 1987.

[Seq81]  Sequin, C. H., "Generalized IC layout rules and layout representations," in *VLSI 81*, edited by J. P. Gray, Academic Press, New York, pp. 13-23, 1981.

[Seq83]  Sequin, C. H., "Managing VLSI complexity, an outlook," *Proc. of the IEEE*, vol. 71, no. 1, pp. 149-166, January 1983.

[Ser87]  Serlet, B., "Fast, small, and static combinatorial CMOS circuits," in *Proc. of the 24th Design Automation Conference*, pp. 451-458, June 1987.

[Swi86]  Swinehart, D., P. Zellweger, R. Beach, and R. Hagmann, "A structural view of the Cedar programming environment," *ACM Transactions on Programming Languages and Systems*, vol. 8, no. 4, pp. 419-490, October 1986.

[Wes86]  Weste, N., and K. Eshraghian, *Principles of CMOS VLSI Design: A Systems Perspective*. Addison-Wesley, Reading, Massachusetts, 1986.

## Further Reading

[Bre72]  Breuer, M. A., editor, *Design Automation of Digital Systems, Volume 1, Theory and Techniques*. Prentice-Hall, Inc., Englewood Cliffs, New Jersey, 1972.

[Lym80c] Lyman, J., editor, *Microelectronics Interconnection and Packaging*. McGraw-Hill Publications, New York, 1980.

[Mea80]  Mead, C., and L. Conway, *Introduction to VLSI Systems*. Addison-Wesley, Reading, Massachusetts, 1980.

[Oht86]  Ohtsuki, T., editor., *Advances in CAD for VLSI, Volume 4; Layout Design and Verification*. Elsevier Science Publishing Company, Amsterdam, the Netherlands, 1986.

[Sou81]  Soukup, J., "Circuit layout," *Proc. of the IEEE*, vol. 69, no. 10, pp. 1281-1304, October 1981.

# Chapter 2

# Interconnection Analysis

**Donald L. Hanson**

*Integrated CMOS Systems*
*Sunnyvale, California*

## 2.1 Introduction

The analysis of interconnection and physical structures determines how well a logical design can be built in a packaging technology. There are two important practical questions addressed by interconnection analysis:

- Given that the logical design or the technology, or both, are under development, what choices should the designers make in order to increase compatibility and make implementation easier?

- Given the logical design and an implementation technology, what is the probability that each element of the design can be constructed in the technology?

The first question concerns the design of packaging technologies that are well suited to the types of systems that need to be designed. Typically, the implementation technology is given and the system is designed for that technology. Interconnection analysis lets the system designer measure the fit of an implementation to the technology. By measuring the partitionability and wirability of the logical design in the technology, the designer can modify the implementation to make best use of the available technology. Alternatively, the logical design and the technology can be developed together. This allows for a compromise of the trade-offs between the design and the technology in which it will be implemented. Again, the interconnection analysis determines partitionability and wirability of the currently proposed logic design, subject to the constraints of the currently proposed technology. (In practice, fixing the logical design and changing the implementation technology does not work well. Many implementation problems that are easy to fix by modifying the system design are very difficult to fix by changing the packaging technology.)

The second question concerns the implementation of a particular system. Once the logical design and the implementation technology have been specified, the physical design (partitioning, placement and routing) must be done. The system's physical design depends on good predictions of its partitionability and wirability; bad estimates may result in a bad physical design that must be redone at the expense of time and money.

The rest of this chapter develops techniques for modeling and analyzing interconnect. Section 2.2 introduces ideas and terms used in analyzing designs. Section 2.3 discusses the two major analytical methods for interconnection analysis techniques: *stochastic modeling* and *deterministic modeling*. Section 2.4 discusses empirical methods for modeling the properties of interconnect.

## 2.2  Overview and History

Interconnection analysis considers both the logical design specification to be implemented and the physical implementation technology in which the logical design is to be built.

The part of the logical design under analysis is called the complex. Inputs and outputs of the complex are called pins. A complex consists of a collection of logical components called *blocks*, which may be simple components like logic gates or complex components made of a number of subblocks. Blocks are wired together by nets through terminals. The block structure of the complex defines a hierarchy, called the logical hierarchy to distinguish it from the hierarchy of packages in the physical implementation. All blocks at the same depth in the hierarchy are said to be at the same level, with the highest level being the complex itself. An example hierarchy is shown in Figure 2.1. Although each block in the logical hierarchy has all the structure of a complex, the term *complex* is reserved for the structure to be analyzed. Any block may become the complex of interest at some time during analysis.

Each net is represented as a set of terminals and pins which are to be electrically common in the finished design. A net may be divided into two-point connections called interconnections, wires, or subnets. In general, a net has $c$ terminals and pins ($c \geq 2$) and so has $c-1$ wires.

The physical implementation technology is the set of packages used to contain the components of the logical design. A package may contain subpackages and the wiring necessary to implement nets that connect the subpackages. Packages also have I/O pins that connect nets to the outside world. The set of

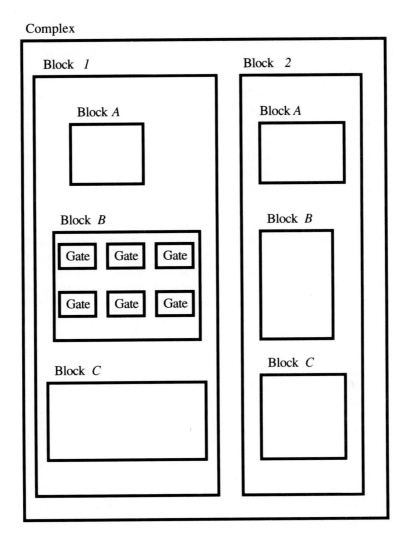

**Figure 2.1**   A Hierarchical Structure

package types is arranged in a physical hierarchy similar to the logical hierarchy. The lowest level packages correspond to the lowest level blocks in the logical hierarchy.

A package has limited capacities for subpackages, I/O pins, and wiring. The logical design must be partitioned into packages so that all the capacity constraints are obeyed. Wiring capacity is measured in terms of area, but since the physical wires are generally of uniform width, a length measure is usually used. The wire lengths or interconnection lengths are measured by the

Manhattan distance ($d = \Delta x + \Delta y$) between the ends of the wires. The wiring capacity of a package can be compared with the total estimated wire length of the complex, which can be estimated as the product of the number of interconnections in the complex and the estimated average interconnection length (AIL) for the complex.

Once the logical design of a system is completed, it is translated into a physical design, including partitioning, placement and routing. Partitioning divides the complex into groups of blocks such that each group can be implemented by a single package in the implementation technology. Placement determines the locations within that package for each block it contains. Routing translates all of the interconnecting nets of the complex into wiring in the packages. Each of these process steps can be accomplished by some combination of automatic and manual methods.

Packages tend to have the same basic characteristics, even though there may be tremendous physical differences in size and structure. Each package is designed as a rectangular array of subpackages. The rows and columns of subpackages are arranged with space in between, called channels, for wiring. A channel is divided into tracks of wires. Some packages also allow wires to pass over or under the subpackages to shorten connections. Wires are almost always implemented as vertical and horizontal segments; often wire segments of different directions are made on different layers of wiring. The conducting layers are separated by insulators; vias tunnel through the insulator to connect wires on different layers.

The most common package type is the integrated circuit (IC), or chip. Chips which have a regular array of subpackage locations are referred to as gate arrays because the subpackages are typically logic gates. The term *masterslice* is sometimes used interchangeably with gate array, although it refers to a particular feature of the implementation of the chip.

A good mathematical model of interconnection structure must preserve the important characteristics of the design while eliminating unnecessary information. The model must characterize three things:

- Logical design — the interconnection structure of the complex.

- Physical design — the results of the partitioning, placement, and routing processes.

- Physical structure — the constraints imposed by the implementation technology.

Depending upon the design approach in use, the physical structure, or the logical design combined with the physical structure, represents the critical factor in design. However, all three areas must be considered during analysis.

The important features of the logical design for interconnection analysis are structural, not functional. Such things as the number of blocks, the number of nets, and the interconnection network need to be represented in any useful abstraction. The function of the blocks and their operating speeds do not matter. Such things as critical performance paths may, however, be used by the physical design processes and, therefore, in the analysis.

The physical design abstraction needs to model the results of the design process. The modeling may try to recreate to some extent the design process itself; however, too much detail in this respect can result in essentially performing the task rather than modeling or predicting the results. The result of the analysis is figures of merit such as block and pin utilization of a package or estimated routed net lengths.

The physical design abstraction takes two basic forms, depending upon the use of the model. The first method attempts to model the results obtained from an actual physical design process. This approach is used in the analysis if the physical design process is known. The second method attempts to model the results that are desired from the physical design process. This approach would be used to help develop physical design algorithms and to analyze the design when the partitioning, placement or routing processes are unknown.

The physical structure model needs to abstract the organization and the limitations of the packaging hierarchy. These constraints are then applied through the physical design model to assess how well the complex can be implemented in the given technology.

Interconnection analysis has been studied in a variety of research efforts, using both theoretical and empirical techniques. The early theoretical work was based upon an assumed randomness in computer logic designs. Gilbert [Gil65] and then Donath [Don68] used a graph model approach and studied the characteristics of all possible graphs. The idea was to approximate a complex as a random example of an interconnection graph. Unfortunately, the results obtained did not agree with the even earlier empirical results of E. F. Rent at IBM. Although his work was not published, Rent observed the relationship between the pin count (*IO*) and the block count (*BK*) of a complex by plotting a series of curves with the empirical data from various IBM computer designs. He was able to obtain a reasonable fit to these curves with a simple relation known as *Rent's Rule*:

$$IO = AS \cdot BK^r \tag{2.1}$$

*AS* represents the *average size* of the blocks in the complex, while $r$ is a mystery quantity and has been labeled the *Rent exponent*. Rent's Rule has been the subject of several studies [Lan71, Rad69 and Rus72a] that have tried to validate the rule and evaluate the parameters *AS* and $r$.

Similarly, early examinations of the characteristics produced by the physical design processes (partitioning, placement and routing) were based upon assumptions of randomness in those processes. Again, empirical data indicated that the randomness assumption was not valid.

These early theoretical and empirical studies prompted a continuing effort to develop accurate models of the structure of computer logic designs and the characteristics of the results of the physical design processes. The remainder of this chapter is devoted to the presentation and discussion of research efforts pertinent to interconnection analysis.

## 2.3  Theoretical Studies

The theory of interconnection tries to produce closed form mathematical descriptions of the characteristics of complexes and physical implementation technologies. A good model must retain essential information while eliminating unnecessary detail, and it must be solvable.

Two types of models have been developed to describe interconnect. One type is *stochastic* − the characteristics of the logical design are approximated by probability distributions of random variables. Circuit, pin, and wiring requirements as well as other relationships are then derived from these stochastic models. The other type is *deterministic* − the interconnect structure is described by exact equations. Once a model's equations are formulated, the derivation of results is similar for both types of models.

In either case the model contains simplifying assumptions about the interconnect structure. These assumptions primarily make the mathematics of the derivations tractable, but they may not apply to actual design environments. Applicability of the assumptions to real examples determines the applicability of the conclusions.

### Stochastic Models

**Characteristics of Computer Logic.**   An early attempt to formalize the apparent characteristics of computer logic designs was published by Donath [Don70]. This paper proposes a model of the logic design process which leads to some of the empirically observed characteristics in computer logic designs. The first premise is that the design process determines the structural characteristics of the implementation, *not* the functional requirements of the design.

Donath defines a top-down, hierarchical design approach in which each step of the expansion of the hierarchy is modeled by the substitution of a pattern of interconnected blocks for one of the current blocks in the complex. The pattern specifies how to expand (unbundle) the current block's inputs and outputs and how to wire the new blocks. The pattern used at each step of the expansion is selected from a pattern library by a stochastic process, without regard to any functional attributes of the design. Figure 2.2 shows a sample

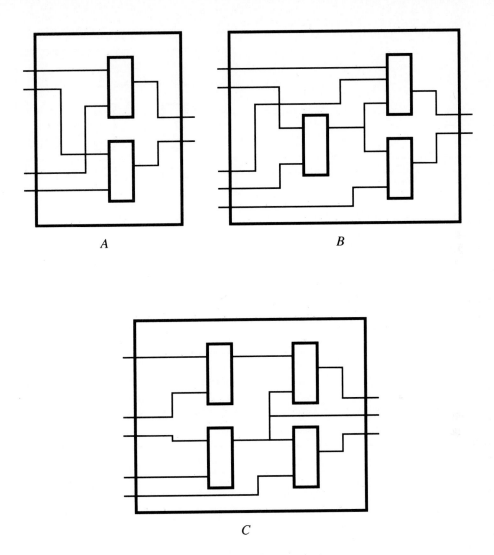

**Figure 2.2**   A Sample Pattern Library

pattern library.  Figure 2.3 is an example of expansion using this library. The initial complex consists of one block with two inputs and one output.  The first level of expansion substitutes Pattern *B* for the initial block, while the three patterns were used once each for the second level of expansion. Based upon this model, and the assumption that all patterns in the library contain the same number of blocks, Donath demonstrates that the expansion process approaches a stationary stochastic state. He notes that the rate of approach is exponential with the number of expansion steps, which implies that the expansion comes

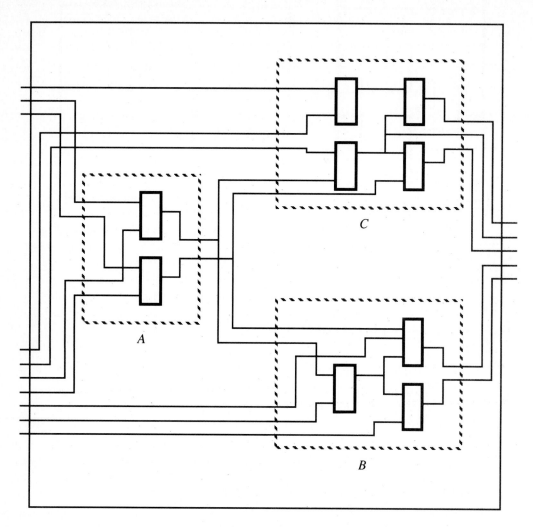

**Figure 2.3** A Two-level Expansion

reasonably close to the stationary state in a few steps. The results of the paper are derived from this stationary state.

There are three main results which are derived from the design process model. The first is a relationship between $IO$, the number of pins, and $BK$, the number of blocks in any portion of the complex. The relation has the form of Rent's Rule:

$$IO = AT \cdot BK^r \qquad (2.2)$$

The factor $AT$ is given as the average number of terminals per block in the complex. The Rent exponent $r$ is claimed to be a characteristic parameter of the design process and its value can be established empirically for a complex given $AT$, $BK$, and $IO$. The model definitions and the derivation imply that $0 \leq r \leq 1$, which, unfortunately, does not provide much insight into the characteristics of a complex. In fact, almost any interconnection of blocks will conform to the above relation with a value of $r$ in the defined range.

The second result of the paper relates $LP$ (the average length in blocks of a path through the complex) to $BK$ (the number of blocks in the complex):

$$LP = C \cdot BK^{\alpha} \qquad (2.3)$$

The exponent $\alpha$ is a second characteristic parameter of the design process. The coefficient $C$ did not appear in the original version of this formula. Path length is defined only for combinational logic because the derivation assumes that paths do not have cycles. Path length is used as a measure of the speed of a path — longer combinational paths imply more delay. As with the block-to-pin ratio exponent, $0 \leq \alpha \leq 1$, which is not very illuminating since it places no real restriction on the value of $LP$.

These two relations have the potential to provide interesting and useful information, but only if the acceptable ranges of $r$ and $\alpha$ are somehow restricted. The third result of the paper does so by deriving an approximate relation between $r$ and $\alpha$:

$$r + \alpha = 1 \qquad (2.4)$$

Since low values of both parameters are desirable, Equation 2.4 implies a trade-off between a large block-to-pin ratio, needed for high density packages, and short logic paths, which give short delays.

The achievement of this paper is to establish a unifying theory for the apparent behavior of the block-to-pin ratio and its relation to performance in computer logic designs. There are, however, three important features of the model that should be carefully considered before using its results:

- The modeling of the design process as a random selection of patterns from a fixed pattern library.

- The assumption of a stationary stochastic state and its relation to the size and structure of the complex.

- The path lengths through the combinational logic and their relation to the characteristics of the complex.

Random selection of patterns with fixed structural characteristics and without regard for their functional attributes is definitely not an accurate model for the computer logic design process. It may, however, be acceptable for the purposes of this analysis because the equations model only the characteristics of the completed implementation and not the design process that created it.

Suppose an actual computer logic design was available and that it was unexpanded by a process that was exactly the reverse of Donath's. That is, for each un-expansion step, a fixed number of interconnected blocks is replaced by one block and the new block provides the same terminals as the group that is removed. It is easy to see that such a process would yield, in reverse order, the expansion path for Donath's design process to follow in order to produce the given complex. Hence, although his process may not accurately model actual computer design practices, the results Donath obtains can be used as a reasonable approximation to the characteristics of actual complexes.

Using a stationary stochastic state to model the interconnection characteristics of a large design is reasonable because the characteristics of the design will not be significantly altered by the addition of a small amount of additional logic. The assumption of a fixed pattern size does not correctly model the actual design process, but, as discussed above, it is not unreasonable to suggest that it will produce results with characteristics similar to the actual situation. The exponential approach of the expansion to a stationary state places some limitations on actual complexes for which the theory is applicable. Since the stationary state is never actually reached, there is some question as to how close the complex must be to this state for the results presented to be meaningful. Very large complexes should conform well to the assumptions, but as the size of the complex decreases, the accuracy of the model decreases.

The requirement of a cycle-free complex has some interesting consequences for the derivation of the path length relation (Equation 2.3). The elimination of storage elements from consideration alters the structure of the complex. One way to satisfy the restriction is to simply remove the storage elements, thus adding I/O connections to the complex where storage elements were connected. Such an alteration significantly changes the block-to-pin ratio and therefore the value of $r$. Another approach is to connect storage element inputs to the corresponding storage element outputs, removing the storage element cycles without affecting the block-to-pin ratio. Unfortunately, this proposal does not remove all of the cycles in the complex, since the contents of a storage element may be used to determine its future contents. Finally, the design can be divided into pieces such that each piece is a cycle-free portion of the combinational logic of the entire complex. This technique also tends to alter the block-to-pin ratio by introducing artificial I/O connections between the pieces. In addition, the pieces may tend to be so small that they cannot be considered to be in a stationary stochastic state. Therefore, the path length relation of Equation 2.3 can only be considered an approximation since the requirement of a cycle-free structure cannot be met without significantly changing the complex to satisfy the requirements of the model.

The third result of the paper, Equation 2.4, is a step toward refining our understanding of the block-to-pin ratio and its relation to performance. However, since it is an approximate derivation based upon the approximate result of

Equation 2.3, the restrictions it places on the values of $r$ and $\alpha$ are weak. Although Equation 2.4 presents an interesting relation, limited confidence can be given to this approach as a means of refining the range of values for the exponents $r$ and $\alpha$.

**Connection Length Estimation.**    The problem of estimating the required connection length for a complex was also addressed by Donath [Don79]. This paper formulates a model for the prediction of the average interconnection length for a digital computer logic design. The model attempts to account for both the non-random interconnection structure of a complex and the non-random subpackage locations obtained from a purposeful placement procedure.

The model of a complex is a hierarchical structure where the block-to-pin ratio for each block in the hierarchy conforms to Rent's Rule. The value of the exponent $r$ is assumed to be constant for all levels in the hierarchy. The number of interconnections within each block is assumed to be proportional to the total terminal count within that block, and the constant of proportionality is assumed to be the same for all levels in the hierarchy. In addition, the number of sub-blocks $q$ within each block is limited to exactly four. A package hierarchy is defined where each package has a capacity of exactly four subpackages, thus providing a parallel to the hierarchical structure of the complex.

A placement procedure is defined which ensures that the subblocks within a given block of the complex are placed on the same package within the package hierarchy. Subblocks are placed randomly within a package. This placement procedure assumes that blocks that are strongly connected logically are placed close together in the implementation.

Although this model is stochastic, the limitation of four subblocks per block allows Donath to use enumeration techniques, rather than probability theory, to derive his results. The predictions he obtains for average interconnection length are summarized in Table 2.1. $AIL$ is a function of the number of blocks $(BK)$ and the Rent exponent $(r)$. Donath's derivations were carried out for placement on both linear and square arrays of physical locations. For comparison, expressions for $AIL$ are also given by Donath, assuming random placement on both types of arrays.

Donath notes that these expressions for $AIL$ are an upper bound. The paper presents empirical data which verifies the dependence of $AIL$ on block count and Rent exponent and also indicates actual values for $AIL$ of approximately one-half of the predicted values. This formula does a good job of predicting interconnection length taking into account the non-randomness of computer logic designs and placement procedures. The discrepancies between the model and the empirical data can be attributed to simplifying assumptions made in the model.

Another effort to model interconnection length was presented by Feuer [Feu82]. He related the block-to-pin ratio of the design to the distribution of the interconnection lengths. The model represents the completely placed complex,

| square array placement: | |
|---|---|
| $AIL \propto BK^{r-1/2}$ | $r > 1/2$ |
| $AIL \propto \log(BK)$ | $r = 1/2$ |
| $AIL \propto f(r)$ | $r < 1/2$, f independent of $BK$ |
| **linear array placement:** | |
| $AIL \propto BK^r$ | |
| **random placement:** | |
| $AIL = \sqrt{BK}/3$ | for square array |
| $AIL = (BK + 1)/3$ | for linear array |

**Table 2.1**   Average Interconnection Length

with blocks positioned uniformly on a two-dimensional infinite grid. Portions of the complex are represented by circles of a given radius measured as a Manhattan distance. All nets are assumed to have been decomposed into interconnections of two terminal wires.

Two functions represent the partitioning properties and the interconnection length distribution. The partitioning function $I(R)$ specifies the number of connections with one end inside a circle of radius $R$ and the other end outside of the circle. The distribution function $Q(d)$ specifies the number of connections between any two blocks which are a distance $d$ apart. A constraint on the number of wires incident upon each block (which are dictated by fanin and fanout limitations) leads to the requirement that $Q(d)$ must decrease faster than $1/d^2$. The form of $I(R)$ is assumed to be that of Rent's Rule, Equation 2.1.

Based on the above model, Feuer derives two relationships:

$$Q(d) \propto d^{2r-4} \tag{2.5}$$

$$AIL \propto BK^{r-1/2} \tag{2.6}$$

These results are subject to the condition that the partitioning function, $I(R)$, is described by Rent's Rule with $1/2 < r < 1$.

The work of Feuer provides a different, although not entirely independent, approach to the interconnection length prediction problem. His result is in agreement with that of Donath in the limit of large $BK$, although the exact relations are slightly different. The added feature of Feuer's derivation is the expression for the distribution of interconnection lengths. These results are based upon the assumption that Rent's Rule is a valid model for the interconnection structure of a complex which has been placed on the grid. Modeling all wires as two-point connections and assuming an infinite grid makes the model most useful on large, random complexes.

**Wiring Area Estimation.**    The estimation of the wiring area required by computer logic designs is a key function of interconnection analysis. A good estimate of wiring space can be useful both to evaluate the fit of a logic design onto a packaging technology and to evaluate the wiring capabilities of new technologies.

Heller *et al.* [Hel77] addressed the problem of estimating wiring space requirements for a regular array of blocks in a package with wiring channels. The paper presents a model and corresponding wiring space predictor for a one-dimensional array of blocks, and proposes a generalization of the results to a two-dimensional array.

The model incorporates the assumptions of the non-random structure of the complex and non-random placement of Donath [Don79]. The model assumes a relationship between the average interconnection length and *BK*, the number of blocks in the complex. The model also assumes that all nets can be decomposed into independent interconnections without introducing any noticeable error in the analysis. Interconnecting wires are modeled as originating stochastically with a Poisson distribution from some block, covering a random distance (with average *AIL*) and terminating at some second block. Based upon this model, an expression is derived for the probability of wiring completion of some number of blocks in a limited number of wiring tracks. This result applies to a one-dimensional array of blocks.

A decomposition technique is applied to the two-dimensional array in order to approximate it with a pair of one-dimensional problems. The one-dimensional result is then applied to this decomposition to obtain wiring space estimates for a two-dimensional array of blocks. A scheme for stochastically generating sample complexes is defined, and a wealth of comparisons are presented between these samples and the results of the two-dimensional model. An encouraging comparison is also made between the model's predictions and the actual wiring results from a sample of real integrated circuit (IC) designs. The model correctly predicts the relative difficulty of wiring completion in these ICs, but not the correct levels of difficulty.

Heller's work attempts to characterize the wiring structure of digital logic designs with a few representative parameters, such as *AIL* and the Poisson parameter for the generation of wires. The results are generally favorable; however, there does not seem to be a strong agreement of the model's predictions with real situations. The applicability of the assumptions for Heller's logic design abstraction, independent interconnections and Poisson distribution, limits the validity of his results to special classes of interconnection structures. The adaptation of the one-dimensional model to the two-dimensional problem introduces errors that could be corrected with an improved two-dimensional model.

A refined adaptation of Heller's model to the two-dimensional case was presented by El Gamal [ElG81]. This model defines a regular two-dimensional array of blocks. The generation and length of interconnecting wires are modeled just as Heller *et al.* [Hel77] modeled them. The path followed by each wire is established randomly, with the restriction that its endpoints be separated by a Manhattan distance which is equal to the path length. One of the four possible directions for these trajectories, illustrated in Figure 2.4, is randomly chosen for each wire. The actual path is established to provide the desired length of the wire in the chosen direction from the originating block. This model results in the formal definition of an array of segments which would result if a typical complex of $BK$ blocks were placed and routed in a square array. Here a segment refers to that portion of a horizontal or vertical wire that runs between two adjacent block locations.

El Gamal derives from this model several interesting relationships concerning the wiring space requirements of the complex. An expression is given for the minimum number of wire segments, and hence the minimum wiring area, required for the square array of $BK$ blocks. The distribution of the segments is also modeled, such that the minimum number of wiring tracks required in any wiring channel can be predicted with respect to a desired probability of completion. This probability of success is achieved if each wiring channel in the array has at least the specified minimum number of tracks.

The essential results of El Gamal's work are that the overall minimum wiring area is of the order of $N^2$ where $BK$ blocks have been placed in an $N$ by $N$ array ($N^2 = BK$). However, the distribution of the wiring requires a larger than minimal channel size in order to accommodate the most congested area on the array. This leads to a wiring space requirement of the order of $N^2\ln^2(N)$ for a regular array of blocks like a gate array. Various compaction approaches are discussed which allow some of the unused wiring space in the regular array to be saved (the difference between $O(N^2\ln^2(N))$ and $O(N^2)$. The work of El Gamal establishes a basis for the estimation of wiring area in masterslice technologies.

A further refinement of the stochastic modeling style as presented by El Gamal has been published by Sastry and Parker [Sas86]. They presented three distinct models of problems in the wirability analysis of gate arrays: a prediction of required channel widths; a means to determine the probability of routing completion given the predicted number of routing tracks per channel; and a new model and derivation for the wire length distribution function used to compute the channel size and routability estimates.

Sastry and Parker use a model of the interconnections on a placed package which is very similar to the model of El Gamal. Two-point wires emerge stochastically with a Poisson distribution from a block on a two-dimensional grid, follow a minimum rectilinear path on the grid, and terminate at another block

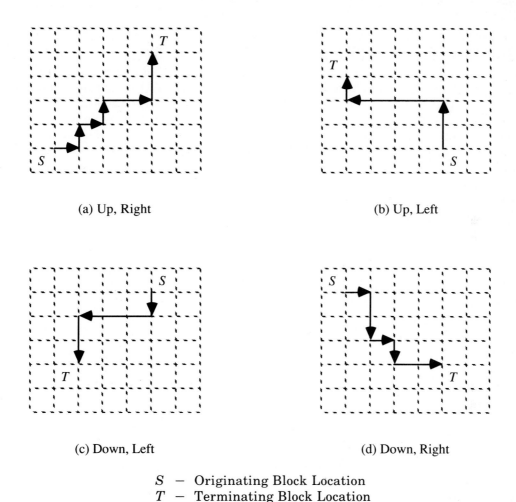

(a) Up, Right

(b) Up, Left

(c) Down, Left

(d) Down, Right

$S$ — Originating Block Location
$T$ — Terminating Block Location

**Figure 2.4**   The Four Directions for Wire Trajectories

an average interconnect length away. The characteristics of the wire length distribution are modeled later. Noting the limited number of wire patterns to be found in minimum rectilinear paths (again similar to the model of El Gamal), a rigorous derivation of the expected width, in routing tracks, of the channels is presented. Sastry and Parker defined the width $W$ and the height $H$ of a channel intersection as random variables. Formulas are developed for these random variables that represent the track requirements at the channel intersection of interest. The estimate of the required channel sizes is then obtained by finding the expected values of the random variables $W$ and $H$. The result is sufficiently formidable to motivate a development of a simplified, asymptotic approximation which is of more practical use. An analysis of sample numerical data indicates that for packages of more than 20 channel intersections on a side, the simplified formula provides an excellent estimate of the actual results.

The second section of the paper addresses the routability of the package. The problem is formulated as the probability that the wiring can be completed given the chip's supply of routing tracks. Sastry and Parker define this probability independently for the vertical and horizontal channels. They point out that the vertical and horizontal track requirements are in fact dependent, but since the dependency is not known, the derivation would become intractable. Rather, they suggest that their results will be somewhat pessimistic due to the simplification.

Sastry and Parker define the routabilities $R_v$ and $R_h$ to be the probability that the wiring track requirements at any channel intersection are less than or equal to the wiring track supply:

$$R_v \propto Pr\{W \leq \mu E(W)\} \text{ for } vertical \ routability$$

$$R_h \propto Pr\{H \leq \mu E(H)\} \text{ for } horizontal \ routability \qquad (2.7)$$

where $E(x)$ denotes the expected value of random variable $x$. The $\mu$ represents the fraction of the estimated track requirement which should be supplied to achieve the $R_v$ or $R_h$ routability level.

A formidable derivation leads to a complex set of formulas for the routability. The authors again provide an asymptotic approximation for the results and demonstrate its excellent behavior for packages with 20 or more channel intersections on a side. In addition, they make the following two observations based on the behavior of the routability estimate with varying package size:

- As the package size increases, the probability that the expected value of $W$ or $H$ will provide enough routing tracks decreases. The implication is that larger packages may be less routable.

- The value of $\mu$ to use to achieve a given level of routability decreases as the package size increases.

Finally, Sastry and Parker present a linear abstraction version of the models used by Nishizawa [Nis71] and later by Feuer [Feu82] to explore the interconnection structure of the logic design. The model represents blocks as points on a continuous line and examines the number of interconnecting wires crossing a coordinate $(x)$ on the line. A random variable $L(x)$ is defined to give the distance from a gate to the $x$ boundary for gates which drive a wire that crosses the boundary. The probability that a wire starts in a given differential interval on the line and then crosses $x$ is developed, and the density and distribution functions of $L(x)$ are derived. Rent's Rule is then imposed on the developing formula and the resulting wire length distribution identified as the Weibull distribution with two parameters. The average value of this distribution is given in terms of the parameters, and two methods are presented to determine the parameters from empirical data.

The Weibull distribution function was tested against the empirical data provided by Donath [Don81]. The actual data fits the Weibull distribution well. When the resulting parameters were used to estimate values for the average interconnection length, the results were much closer to the actual values than those of Donath's Pareto-Levy distribution. All three estimates were still low by 14% to 25%.

The usefulness of this paper is to estimate the channel sizes required for a given logic design on a target package, characterize the placed logic design with the Weibull distribution, and apply this distribution to the routability prediction model. The result is information to guide logic design decisions or track supply decisions. Unfortunately, the only method given for evaluating the Weibull parameters is based on the net lengths obtained from an actual layout. A method which provides for the description of the wire length distribution as a function of the characterization of the logic design before layout is still needed.

The results of Sastry and Parker almost take stochastic modeling far enough to serve as a useful tool in layout feasibility prediction and interconnection analysis. The final step is a small but elusive one: to adequately relate the interconnection structure of the logic design to the wire length distribution obtained from placement and routing. Once attained, the impetus of stochastic modeling will change to the relaxation of assumptions and other refinements in an effort to produce more and more accurate predictions.

## Abstractions in Stochastic Modeling

The abstractions developed, and their applicability to practical design problems, are the important aspects of stochastic modeling. The six papers reviewed here present an evolution of the stochastic abstractions for logic designs, physical design processes, and physical structures.

The interconnection structure of the complex is perhaps the most difficult aspect of the computer design to characterize successfully. The early abstractions were based upon Rent's Rule. Donath's expansion process predicts this relation as a basic characteristic of the logic design. Unfortunately, the unknown nature of the value of the Rent exponent $r$ limits the ability of the abstraction to represent the design. Later work in wiring area estimation developed the model of interconnection structure as a collection of independent two-point wires, with the number of wires incident on each block modeled as a Poisson distribution. The applicability of both the independence assumption and the Poisson distribution is questionable for actual logic designs. Both the Rent's Rule model and the independent Poisson distribution model draw their parameters from empirical data and can therefore be adapted to new designs. It is questionable whether the abstractions retain enough of the information about the actual design to correctly influence the model and its results.

The effects of purposeful placement are modeled in two ways in the work we have examined. The first model [Don79] assumes that blocks are placed in groups according to the hierarchical structure of the complex and then placed randomly within these groups. The second model abstracts the effects of placement into the average interconnection length and an associated interconnection length distribution. The assumption is that the pertinent characteristics are embodied in the placement process and not directly related to the interconnection structure of the complex. An attempt is made in these approaches to relate the average length to the exponent $r$ in Rent's Rule. The corresponding distribution functions are also defined in terms of $r$. Consequently, the results tend to be tied to the Rent's Rule characterization of the logic design as well as to the placement process. None of these attempts is entirely satisfactory.

The routing operation is modeled by assuming that the minimum length path will always be chosen for each interconnection. The typical routing abstraction accounts for no congestion or interference from already routed wires when modeling the path for a given interconnection.

The global physical structure is modeled as a simple abstraction to a one- or two-dimensional regular array of logic blocks. The only purpose in modeling the physical structure in these papers is to establish a topology for the allowable routing channels for the interconnecting wires. The abstraction of the physical structure typically does not account for the various physical constraints imposed on the design by the implementation technology. This lack of detail in the physical technology model is acknowledged by some of the researchers and is dismissed by claiming that the analysis they present is only a first cut at the investigation and that the details of the implementation can safely be ignored.

The goal of stochastic modeling is to arrive at closed-form results. Such results represent the general case by averaging the plausible cases. They can serve as useful guidelines but may not apply to specific designs. One way to improve the quality of these results is to add parameters that can be measured from real designs. The parameters can be used to adapt the closed-form results to individual situations, which is particularly useful for highly atypical designs.

To have a closed-form result, the model must be simple and carefully structured. The simplifications required to allow closed-form results tend to average the details of the real situation.

In conclusion, the generalized abstractions used for stochastic modeling tend to produce generalized theoretical results. These results can be used to make global predictions and to provide insight into the underlying structures of computer designs and the computer design process. The results are not very useful in specific cases, since they lack sufficient detail to accurately represent actual design instances.

The various abstractions and their usage in the papers are summarized in Table 2.2. Three areas of abstraction are presented along with a brief summary of the format and content of the results developed in each of the papers.

## Deterministic Models

**Characteristics of Computer Logic.** An early deterministic model of the structure of computer logic designs was presented by Donath [Don69]. This work devises a plausible structure for the logic design which predicts the Rent's Rule relation between block count and pin count.

Donath develops a model for a hierarchical structure of the complex. Each block includes a fixed number $q$ of subblocks, and there are $\lambda$ levels in the hierarchy. A structure which conforms to this definition is called a $q$-regular $\lambda$-level hierarchical structure. Constraints which are designed to model certain characteristics of the complex are then imposed upon the structure. For example, the number of connections at each level must be sufficient to connect all blocks at that level. However, actual connectedness is not imposed because of mathematical difficulties. A connection is assumed to be a two-point wire and can be taken to be a subnet rather than representing an entire net. The average number of terminals per block is specified as $AT$. Finally, an encoding hypothesis is introduced. This hypothesis states that only a fraction of the terminals contained within one hierarchical level will need to be connected to terminals not contained in that level. That is, information transfer between levels is encoded.

Based on formal mathematical definitions of the above model, Donath derives essentially two results. The first is concerned with the structure of the complex needed to exhibit a block-to-pin ratio which conforms to Rent's Rule. The appendix of the paper demonstrates that a randomly constructed complex does not exhibit such a relationship. A complex which conforms to the hierarchically structured model is shown by Donath's Theorem 1 and its corollary to exhibit a block-to-pin relation in the form of Rent's Rule with the following limits on the Rent exponent ($r$):

$$0 \le r \le 1 + ln(1 - 2(q-1) / (ATq)) / ln(q) \qquad (2.8)$$

Donath presents a table of the maximum values for $r$ given values for the parameters $AT$ and $q$. Some sample values are illustrated in Table 2.3.

The second result is an estimate of the number of additional blocks required to produce a hierarchically structured complex rather than a complex designed for minimum block count. The estimate is based on the enumeration of a constraint limited class of graphs, where the constraints are intended to impose conformity with Rent's Rule. The number of these graphs which conform

| Form of Abstraction | Don 70 | Don 79 | Feu 82 | Hel 77 | ElG 81 | Sas 86 |
|---|:---:|:---:|:---:|:---:|:---:|:---:|
| **Logic Design:** | | | | | | |
| 1. hierarchical expansion | O | | | | | |
| 2. hierarchical, $IO = AS \cdot BK^r$ nets $\propto$ terminals | | O | | | | |
| 3. two-point wires, $IO = AS \cdot BK^r$ | | | O | | | |
| 4. Poisson distribution of independent wires | | | | O | O | O |
| **Physical Design:** | | | | | | |
| 1. hierarchical grouping, random within groups | | O | | | | |
| 2. wire length dist., partitioned by $IO = AS \cdot BK^r$ | | | O | | | |
| 3. average interconnection length ($AIL$) & dist. | | | | O | O | O |
| **Physical Structure:** | | | | | | |
| 1. uniform linear array | | O | | O | | |
| 2. uniform 2-D array | | O | O | | O | O |
| **Results:** | | | | | | |
| 1. $IO = AT \cdot BK^r$, $LP = BK^\alpha$, $r + \alpha = 1$ | O | | | | | |
| 2. $AIL \propto BK^{r-1/2}$, $r > 1/2$ | | O | | | | |
| 3. $Q(d) \propto d^{2r-4}$, $AIL \propto BK^{r-1/2}$, $1/2 < r < 1$ | | | O | | | |
| 4. probability of success, given a wiring distribution | | | | O | | |
| 5. min. area $O(N^2)$, masterslice $O(N^2 \ln^2(N))$ | | | | | O | |
| 6. channel width estimate, routability estimate, Weibull dist. of lengths | | | | | | O |

**Table 2.2**   Abstractions Used in Stochastic Models

| $q$ | $AT = 4.0$ | $AT = 5.0$ | $AT = 6.0$ |
|---|---|---|---|
| 2 | 0.585 | 0.678 | 0.737 |
| 3 | 0.631 | 0.718 | 0.771 |
| 4 | 0.661 | 0.743 | 0.792 |
| 5 | 0.683 | 0.760 | 0.807 |

**Table 2.3**   Sample Maximums for the Rent Exponent

to the hierarchical structure divided by the total number of these graphs is found to be on the order of $ln(BK)$. That is, according to Donath, a given computer design will take approximately $ln(BK)$ times more blocks in a hierarchical implementation than it will take in a minimum block count implementation.

**Block-to-Pin Ratio.**   Radke [Rad69] explored the use of Rent's Rule as a predictor of the pin requirements for a complex and developed some refinements to the relation. In order to model the complex and its partition, Radke uses a simple description of the already partitioned design. The total number of blocks ($BK$) and the total number of terminals on blocks ($TB$) are defined, as well as the total number of pins ($IO$) for the complex. In addition, he defines $XB$ as the average number of blocks per package and $XN$ as the average number of pins per package. The block and pin counts are denoted $XB(n)$ and $XN(n)$, respectively, for a partition onto $n$ packages. (Note: $BK = nXB(n)$.)

As a first step, Radke assumed that the total package pin count (*i.e.*, $nXN(n)$) will increase by $D$ pins if the package count ($n$) is increased by one. This assumption yields the following difference equation:

$$nXN(n) - (n+1)XN(n+1) = -D \tag{2.9}$$

He then claims that the solution to this equation for restricted values of $XB$ is:

$$XN = D + ((TB - D{\cdot}BK)/BK)XB \tag{2.10}$$

Based on this solution, Radke concludes that the parameters in Rent's Rule should have the following values:

$$AS = TB/BK \equiv AT \tag{2.11}$$

$$r = (TB - D{\cdot}BK)/TB = 1 - D/AT \tag{2.12}$$

$AT$ is the average number of terminals per block, and $r$ is a function of $AT$ and the new parameter $D$.

The result is that characterization of the connectivity of the logic design with a value for $D$, makes Radke's modified block-to-pin relation, Equation 2.13, a more useful tool for prediction of pin requirements than is Rent's Rule, because a method is provided for assigning a value to the exponent.

$$XN = AT{\cdot}XB^{1-D/AT} \qquad\qquad (2.13)$$

Radke's second step is the observation that even his modified form of Rent's Rule overestimates pin requirements for small package counts (large packages). Based upon a simple degenerate example, he concludes that a correction factor which somehow represents the natural structure of the design should be subtracted from Equation 2.13. He somewhat arbitrarily chooses a cubic rational polynomial, Equation 2.14, which introduces a new parameter, $a$. This new parameter is taken to represent the efficiency or quality of the partition. The result is that if a measure of the effectiveness of the partition can be determined and assigned to $a$, then Equation 2.14 is an even more useful predictor of pin requirements than is Equation 2.13:

$$XN = AT{\cdot}XB^{1-D/AT} - (XB-1)^3/(1 + aXB)^s \qquad (2.14)$$

where $s$ is a function of $IO$ and $a$ such that $XN=IO$ when $XB=BK$. Radke's overall result is an elaboration of the formula for Rent's Rule which more accurately fits the empirical data.

Hitchcock [Hit70] takes a different approach to the question of pin requirements and the block-to-pin ratio. This paper derives a formula for the block-to-pin ratio that is based upon the distribution of the terminals on the nets and the partition of the complex.

Hitchcock's abstraction of the situation is very similar to Radke's. The model is a partition of $BK$ blocks onto $n$ packages with an average of $XB(n)$ blocks per package ($BK=nXB(n)$). The number of terminals in a net is denoted by $t$. The number of nets with $t$ terminals is $K_t$, and the largest number of terminals in any net is $T$.

The first step in the development of his formula is to establish the maximum possible number of pins per package. This maximum occurs when each terminal also requires a pin and is referred to as the unit logic pin requirement. The total number of pins in the unit logic case ($P_u$) is independent of the package count ($n$) and is given by Equation 2.15. The average unit logic pin count per package for a partition onto $n$ packages is then $P_u/n$.

$$P_u = \sum_{t=1}^{T} tK_t \qquad\qquad (2.15)$$

Hitchcock's next step is to observe that there are two basic mechanisms for reducing the unit logic pin count. The first is for all terminals within one package which are on the same net to share one package pin (shared pin) for that net. The second mechanism occurs when *all* terminals of a net are contained in the same package. This situation cannot occur for those nets which contain pins of the complex. In this case, the net is said to be buried, and the package pin for the net is saved. Hitchcock enumerates all possible ways of partitioning the design onto $n$ packages and counts the number of pins saved by sharing and by burying in each case. The results are formulas for the average number of pins per package saved by sharing ($S(n)$) and the average number of pins per package saved by burying ($B(n)$). The total average pin requirement per package ($XN(n)$) for a partition onto $n$ packages is given by Equation 2.16.

$$XN(n) = P_u/n - S(n) - B(n) \qquad (2.16)$$

Hitchcock continues by analyzing the relative importance of sharing and burying as the means to reduce pin requirements. His conclusion is that sharing is the more dominant phenomenon when $XB(n)$ is small compared to $BK$. However, as the number of packages ($n$) is decreased and $XB(n)$ approaches $BK/2$, the burying phenomenon becomes the more dominant. The result is that Equation 2.16 exhibits a curvature much like the empirical curves for block-to-pin ratio; however, Hitchcock's formulas have a strong tendency to overstate the actual pin requirements. This overstatement is because an actual complex exhibits a specialized interconnection structure which is not represented well by an average over all possible partitions of $BK$ blocks onto $n$ packages.

**Interconnection Count.**    Nishizawa [Nis71] attempts to derive formulas for not only the package pin requirements, but also for the number of interconnections within and external to the packages. His model is similar to that used by Feuer [Feu82]. The logic design is represented by a complex of $BK$ equal-sized blocks with the fanin and fanout of each given as an average. The fanout is a two-part parameter consisting of the average number of outputs from a block and the average fanout of each output. It is assumed that there are no primary I/Os; that is, all nets are contained within the complex.

The physical design is modeled as blocks placed uniformly on a plane, with each block occupying a unit area. A package is represented by a circle of radius $R$, and therefore it contains $\pi R^2$ blocks. Partitioning is modeled by constructing circles on the plane. The routed wires are represented by an average interconnection length, $AIL$, and an associated distribution function.

Nishizawa selects an exponential function for the interconnection length distribution; however, any function could be used. He then proceeds to derive formulas, in the form of integrals, for:

- The number of package input pins, $NT_i(R)$.
- The number of package output pins, $NT_o(R)$.
- The number of wires within a package, $NC_i(R)$.
- The number of wires between packages, $NC_p(R)$.

The average number of pins required per package is the sum of $NT_i(R)$ and $NT_o(R)$. These formulas are all given in terms of the basic parameters of his logic design abstraction: $BK$, number of outputs per block, fanin, fanout, the package radius ($R$), and $AIL$ with its distribution.

Nishizawa extracts the necessary parameters from a total of four partitions of two actual logic designs. Since he uses an exponential function for the interconnection length distribution, he only needs to specify $AIL$, which he does by producing a family of curves for various values of this average. His curves are plots of average package size $(\pi R^2)$ versus the pin and interconnect counts defined by his formulas. The four actual example points are plotted on these families of curves, and Nishizawa concludes that for $AIL$ in the range 1.2 to 1.3 units, his formulas correctly model the characteristics of the designs. Also included with the I/O pin count curves is a curve representing Rent's Rule. The Rent curve is similar to Nishizawa's curves, but has slightly less curvature and corresponds to an $AIL$ of less than 1.0 unit.

## Abstractions in Deterministic Modeling

Deterministic modeling tends to be more empirically oriented than does stochastic modeling. As a result, the abstractions used for deterministic modeling are structured around parameters that can be measured from actual designs. The applicability of the abstractions is still the important factor in determining the validity of the results. The abstractions and parameters of successful models must be able to accurately represent the characteristics of the complex.

The interconnection structure of the complex is characterized in the examined work by a small number of simple parameters. The number of blocks, the number of terminals on a block, and the number of blocks connected to a net are samples. Donath defines an elaborate mathematical hierarchical structure; however, the parameters of this structure are merely the average number of terminals per block $(AT)$ and the number of blocks that are a group $(q)$. Radke and Donath both include a more abstract concept in their models, the encoding concept. The idea is that if blocks are packaged together, some portion of their interconnections will not require package pins. This concept is represented in both papers by a parameter which specifies the amount of encoding taking place. The same concept is embodied in the sharing and burying ideas presented by Hitchcock. Nishizawa parameterizes the complex with a block count $(BK)$ and the average fanin and fanout of the blocks.

The effects of the physical design processes are also characterized predominantly by simple measurable parameters. The partitioning distribution function is specified and then used deterministically in the derivations.

The various abstractions and their usage in deterministic modeling are summarized in Table 2.4. The two areas of abstraction are presented along with a brief summary of the format and content of the results developed in each of the papers.

The deterministic modeling effort of Donath is directed toward a closed-form result which specifies the relation between block count $(BK)$ and pin count $(IO)$, just as his stochastic modeling effort. In order to achieve this end, Donath abstracts a rigid hierarchical structure which does not correspond to the real

| Form of Abstraction | Don69 | Rad69 | Hit70 | Nis71 |
|---|:---:|:---:|:---:|:---:|
| Logic Design: | | | | |
| 1.   $q$-regular $\lambda$-level hierarchical structure; encoding hypothesis | O | | | |
| 2.   $BK$, $IO$, encoding parameter $D$ $TB$, quality parameter $a$ | | O | | |
| 3.   $BK$, pin count on the nets | | | O | |
| 4.   $BK$, pin count on both nets and blocks ($IO=0$) | | | | O |
| Physical Design: Partitioning | | | | |
| 1.   $n$ packages, $XB(n)$, $XN(n)$ | | O | O | |
| 2.   package is circle of radius $R$, contains $\pi R^2$ blocks | | | | O |
| Placement and Routing | | | | |
| 1.   uniform placement on plane $AIL$ and distribution | O | O | O | O |
| Results: | | | | |
| 1.   $IO = AP{\cdot}BK^r$, subject to Equation 2.8 | O | | | |
| 2.   Equations 2.13 and 2.14 | | O | | |
| 3.   Equation 2.16 | | | O | |
| 4.   I/O and interconnect count formulas | | | | O |

**Table 2.4**  Abstractions Used in Deterministic Models

world. However, just as before, the results of the abstraction seem to be an acceptable approximation to the structure of a real complex. His result is essentially the same as in [Don70]. Radke makes no unrealistic assumptions, but his effort is really only some heuristic curve-fitting techniques attempting to match the empirical curves more closely than does the basic form of Rent's Rule. The result is a more elaborate formula with new parameters. Unfortunately, the approach given for evaluating the new parameters is empirical in nature. The overall achievement is not significantly different from the original form of Rent's Rule.

Hitchcock presents a refreshing new approach which is not based upon Rent's Rule. Unfortunately, he makes the assumption that the average behavior over all possible interconnection structures will represent the general case for the interconnection structure of computer logic complexes. This is another instance where the specialized structure found in computer logic designs is not represented well by a random or generalized model. Finally, Nishizawa also presents a new approach which is not related to Rent's Rule. His results give favorable comparisons with some empirical data as well as curves similar to those from Rent's Rule. His assumption of no I/O for the complex is a little weak; however, he assumes an infinite placement plane, implying an extremely large complex with the pins found at the edges. Nishizawa's results are based on an exponential distribution of interconnection lengths which has not been demonstrated to be applicable. On the other hand, his results are quite reasonable, indicating that perhaps an exponential distribution is not unrealistic. This work has the added attraction of providing interconnect count formulas as well as the block-to-pin relation.

In conclusion, it seems that parameterized models can be utilized to abstract the characteristics of computer logic designs. Such models have the inherent ability to adapt to variations in actual designs and are useful as predictors of technology requirements given logic designs. Unfortunately, deterministic modeling has so far addressed essentially the block-to-pin relationship with varying degrees of success. Our ability to predict wiring requirements for packages is still limited to the general stochastic results and other forms of guesswork.

## 2.4  Empirical Studies

The initial empirical work on the structure of computer logic designs was made in an unpublished study by E. F. Rent at IBM in the early 1960s. His formulation of what is now called Rent's Rule was passed along to researchers both inside and outside IBM, and several additional studies of the block-to-pin relationship were performed. More recently, other researchers have made empirical measurements of other characteristics of computer logic.

### Block-to-Pin Ratio

Landman and Russo [Lan71] present the results of an extensive study of the block-to-pin relationship for four actual complexes. The examples range from a 671 block complex with 1 gate per block to a 437 block complex with up to 29 gates per block. An automatic partitioning program is used to obtain an assortment of partitions for these examples.

First a minimum package count partition is found subject to a package pin capacity but with an unlimited package block capacity. Then the minimum block capacity which still permits this partition is found. This process is performed for several pin capacities on each of the sample complexes. The resulting average block count versus average pin count data points are then fit to a curve to extract the values for the *r* and *AS* of Rent's Rule.

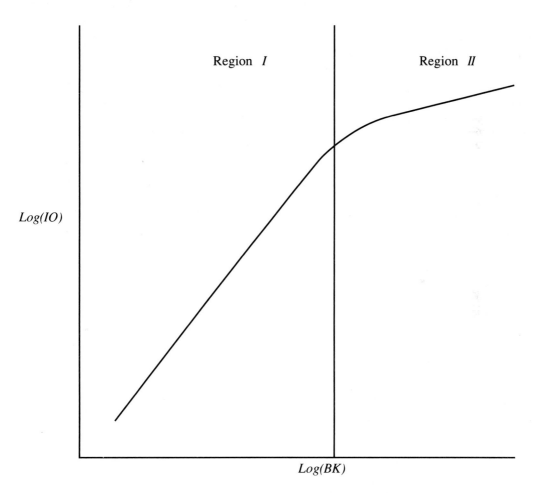

**Figure 2.5**   Empirical Block-to-Pin Ratio Graph

Landman and Russo observed two regions in the block versus pin curves, as illustrated in Figure 2.5. Region I, corresponding to smaller average block counts, fits the Rent equation very well, with values for *r* between 0.57 and 0.75. Region II, corresponding to large average block counts and a small number of packages, exhibits more curvature than the Rent equation would predict. One theory is that this behavior is an artifact of the partitioning program for large packages, but the authors rejected this idea and instead postulated a mechanism similar to that presented by Hitchcock [Hit70] to explain the large curvature. They attempted to fit curves for sharing and for burying pins to the data with some success. Better agreement was found by merely refitting Rent's Rule to the Region II data.

This work shows that the block-to-pin relationship does exhibit a power law behavior for a complex partitioned onto relatively small packages. However, when the packages become large, the pin requirements seem to decrease, which indicates a reduction in the Rent exponent for large packages. This phenomenon needs further study, but it does correspond to the relatively small pin requirements for microprocessor and other VLSI chips. An additional result of Landman and Russo's study is that the block capacity of the packages needs to be approximately 1.5 to 2.0 times the average block count per package for the pin limited partitioning problem in Region I.

## Characteristics of Computer Logic

A companion study to [Don70] was published by Donath and Hitchcock [Don71]. This study attempts to analyze several example logic designs in an effort to measure the average path length dependence on the number of blocks in a design and thereby verify the results derived in [Don70].

The analysis is based upon cones-of-influence. A cone includes all logic that directly or indirectly drives an output of the complex; the leaf nodes of the cone are blocks whose inputs are either inputs to the complex or storage elements. The cone can be used to calculate the average path length ($LP$) in blocks of all paths from a primary input or a storage element to the target block. The number of blocks in the cone-of-influence is used as an estimate of $BK$, the number of blocks in the complex, assuming that the cone-of-influence of the target block can be taken as a sample complex.

A data point ($LP$, $BK$) was obtained for each block of a sample design, and a least squares fit was made to Equation 2.17 with these data points.

$$\log(LP) = \alpha\log(BK) + \log(C) \qquad (2.17)$$

The parameter $\alpha$ is defined in [Don70] to be a characteristic parameter of the complex, while $C$ is claimed to have the value 1.0. In this manner, it was possible to show that Equation 2.3 had the correct form and to obtain values for $\alpha$ and $C$, for each of five examples. The values for $C$ were found to be approximately 1.0, as the theory in [Don70] suggests. Previous measurements of block-to-pin ratios for the same five examples supplied values for the Rent exponent $r$. The $r + \alpha$ value (Equation 2.4) was found to range from 1.09 to 1.28 for the examples, as compared to the suggested theoretical value of 1.0. Since Equation 2.4 is approximate, these results can be taken as a verification of the theory in

[Don70]. In any event, the cone-of-influence approach for the extraction of small sample complexes does seem to select samples which exhibit a correlation between the perceived performance level and the block-to-pin ratio of that complex.

Russo [Rus72a] explored the relationship between performance and the block-to-pin ratio for several example designs. He used the same partitioning system, techniques, and some examples previously used by him and Landman [Lan71]. The results demonstrate a correlation between the Rent exponent and combinational logic delays. Low performance designs exhibit a Rent exponent value in the range 0.47 to 0.63, while high performance designs have exponent values of 0.69 to 0.75. Russo then made the argument that high density LSI technologies will be limited to low-speed designs unless the block-to-pin ratio of logic designs can be reduced.

Two known methods of reducing the block-to-pin ratio were discussed by Russo. The first serializes the transfer of information between packages and tends to decrease the performance of the design while decreasing the pin requirements. The second encodes the information that passes between packages. This encoding reduces the pin requirements without affecting performance. Russo modeled data encoding by allowing selected logic blocks to be mapped redundantly onto more than one package during the partitioning process. This technique and its implementation in an automatic partitioning system are discussed in [Rus71] and [Rus72b].

The remainder of the paper is a study of the effectiveness of using mapping to reduce the block-to-pin ratio. One of the examples is mapped onto packages allowing various levels of redundancy. It is shown that the redundant mapping technique can increase the block-to-pin ratio by as much as a factor of two in some cases. It is also shown that the block-to-pin relationship conforms to Rent's Rule and that the introduction of redundancy causes a decrease in the value of the Rent exponent. Although Russo's initial premise was that there is a correlation between the block-to-pin ratio and the performance level of the design, the conclusion is that it is possible to adjust the design such that the block-to-pin ratio is decreased while the performance level is essentially unchanged.

### Interconnection Length

An empirical study of the interconnection length distribution for computer logic designs was presented by Donath [Don81]. Based upon his earlier work [Don79], and upon the work of Feuer [Feu82], Donath loosely derives the interconnection length distribution given by Equation 2.18. The fraction of the wires of length $w$ is denoted $f(w)$. The $g$ term is a normalization constant, and $M$ is related to the size of the placement array and to the *quality* of the placement.

$$f(w) = gw^{-\gamma}, \ 1 \le w \le M \qquad (2.18)$$
$$= 0, \ w > M$$

The value for $M$ is about $N/2$ where $N \times N$ is the size of the placement array. The exponent ($\gamma$) is related to the Rent exponent $r$ by Equation 2.19.

$$\gamma = 3 - 2r \qquad (2.19)$$

Donath then analyzes the interconnection length distributions of three actual logic designs. Each example is automatically placed using a hierarchical placement program. The wire length distribution is obtained from these placements and fitted to Equation 2.20, such that at least 95% of the wires were included in the fit and the excluded wires were only those longer than some cutoff value. This cutoff value was then taken to be the value of $M$ in Equation 2.18.

$$\log(f(w)) = \log(g) - \gamma \log(w) \qquad (2.20)$$

The resulting values for the sum $\gamma + 2r$ were in the range 2.91 to 3.15 for the three examples. This range is in agreement with the value of 3.0 specified by Equation 2.19. Donath also noted that the actual distributions exhibited a ratio of $f(w+1)/f(w)$ which increased with $w$. This effect, evidenced by exceptionally long tails of the distribution, shows that the distribution was probably not exponential or Poisson.

**Placement Modeling.**    An alternate approach to the modeling of placement results was presented by this author [Han84]. In this paper, the structure of the logic design and of the package are characterized in more detail than in the other efforts. These detailed characterizations are then applied to a stochastic model of the placement process, and a representation of the resulting placement is obtained. This representation is in the form of estimates of the half-perimeter length of each net in the complex.

The design is characterized by growth functions, which identify a sample portion of a complex that reflects the local influence of the complex over some net. A sequence of samples is extracted by starting with the net of interest (the seed net) and the blocks connected by that net. The sample is then grown by iteratively including nets connected to blocks already in the sample. The accuracy of the analysis is controlled by the depth of search that creates the samples. The logic design is characterized by the block and net populations of these growth samples, each net receiving an appropriately different local characterization.

The package is characterized by parameters that represent the minimum and maximum spacing of blocks placed on the package. This model, along with the selection of blocks connected by a net, is used to establish the absolute minimum and maximum placement half-perimeter of the net in question.

The placement model introduces a definition of placement quality (*PQ*) and models the placement process by estimating where between its minimum and its maximum net lengths a particular net will be placed. The *PQ* is used to refine the bounds between the minimum and the maximum. Then a stochastic selection is made within the refined bounds. This model is applied to each net independently, and the results are the estimated net half-perimeter lengths for each net.

The stochastic placement model precludes an accurate net-by-net prediction of the placement results. It provides a useful model of the average net length and the distribution of net lengths — the estimates can distinguish easy and difficult packages to place and route.

The model was used to analyze a number of example designs. Each of the sample chips was characterized and analyzed according to the placement model. Each chip also had actual placement results available. The comparisons of actual to predicted data indicate that the model is quite capable of segregating designs according to layout difficulty with reasonable accuracy.

## 2.5 Summary

Interconnection analysis is important because it lets the system designer measure the fit of an implementation to the technology. It provides the information that allows the designer to modify the implementation to make best use of the available technology.

This chapter has presented both the theoretical work in the area of interconnection analysis and empirical studies which bear them out.

## Exercises

**2.1** Define stochastic interconnection modeling and deterministic interconnection modeling. What are the advantages and disadvantages of each?

**2.2** Why were random graphs abandoned for modeling the interconnections of complexes? Under what conditions would a random graph be a good model?

**2.3** In the standard formulation of Rent's Rule, Equation 2.1, what would be implied by a value greater than one for the exponent ($r > 1.0$)?

**2.4** Under what circumstances will the Manhattan length of a net exceed the half-perimeter length of the bounding box?

**2.5**† In Donath's paper on estimating connection length, [Don79], he derives $AIL$ to be proportional to $BK$ to the $r - 1/2$ power:

$$AIL = kBK^{r-1/2}$$

where $k$ is the constant of proportionality.

This formula expresses $AIL$ in units of block pitches. Given $S$, the length of the side of a square package in millimeters, provide an expression for $AIL$ in millimeters.

**2.6**    Define an experiment to test the validity of Rent's Rule.

**2.7**‡ Develop a method for extracting the effective value for Rent's exponent from an interconnection structure. The curve-fitting techniques of Landman and Russo [Lan71] suggest that a mechanism for finding samples of the pair block count, I/O pin count for a wide range of block counts is needed. One possibility might be to use the growth function concepts in [Han84] to generate such samples.

# References

[Don68]    Donath, W. E., "Statistical properties of the placement of a graph," *SIAM Journal of Applied Mathematics,* vol. 16, no. 2, pp. 439-457, March 1968.

[Don69]    Donath, W. E., "Hierarchical structure of computers," *IBM T. J. Watson Research Center Report RC 2392,* March 1969.

[Don70]    Donath, W. E., "Stochastic model of the computer logic design process," *IBM T. J. Watson Research Center Report RC 3136,* November 1970.

[Don71]    Donath, W. E., and R. B. Hitchcock, "Path lengths in combinational computer logic graphs," *IBM T. J. Watson Research Center Report RC 3383,* June 1971.

[Don79]    Donath, W. E., "Placement and average interconnection lengths of computer logic," *IEEE Trans. on Circuits and Systems,* vol. CAS-26, no. 4, pp. 272-277, April 1979 (IBM Report 4610).

[Don81]    Donath, W. E., "Wire length distribution for placements of computer logic," *IBM Journal of Research and Development,* vol. 25, no. 3, pp. 152-155, May 1981.

[ElG81]    El Gamal, A., "Two-dimensional stochastic model for interconnections in master slice circuits," *IEEE Trans. on Circuits and Systems,* vol. CAS-28, no. 2, pp. 127-138, February 1981.

[Feu82]    Feuer, M., "Connectivity of random logic," *IEEE Trans. on Computers,* vol. C-31, no. 1, pp. 29-33, January 1982.

[Gil65]    Gilbert, E. N., "Random minimal trees," *SIAM Journal of Applied Mathematics,* vol. 13, no. 2, pp. 376-387, March 1965.

[Han84]   Hanson, D., "Placement modeling," in *Proc. of the IEEE Intl. Conf. on Computer Design*, pp. 351-356, October 1984.

[Hel77]   Heller, W. R., W. F. Mikhail, and W. E. Donath, "Prediction of wiring space requirements for LSI," in *Proc. 14th Design Automation Conf.*, pp. 32-42, June 1977.

[Hit70]   Hitchcock, R. B., "Partitioning of logic graphs: a theoretical analysis of pin reduction," in *Proc. 7th Design Automation Workshop*, pp. 54-63, June 1970.

[Lan71]   Landman, B. S., and R. L. Russo, "On a pin versus block relationship for partitions of logic graphs," *IEEE Trans. on Computers*, vol. C-20, no. 12, pp. 1469-1479, December 1971.

[Nis71]   Nishizawa, M., "Partitioning of logic units," *Fujitsu Scientific and Technical Journal*, vol. 7, no. 2, pp. 1-13, June 1971.

[Rad69]   Radke, C. E., "A justification of and an improvement on a useful rule for predicting circuit-to-pin ratios," in *Proc. 6th Design Automation Workshop*, pp. 257-267, June 1969.

[Rus71]   Russo, R. L., and P. K. Wolff, "ALMS: automated logic mapping system," in *Proc. 8th Design Automation Workshop*, pp. 118-127, June 1971.

[Rus72a]  Russo, R. L., "On the tradeoff between logic performance and circuit-to-pin ratio for LSI," *IEEE Trans. on Computers*, vol. C-21, no. 2, pp. 147-153, February 1972.

[Rus72b]  Russo, R. L., and P. K. Wolff, "A computer-based-design approach to partitioning and mapping of computer logic graphs," *Proc. of the IEEE*, vol. 60, no. 1, pp. 28-34, January 1972.

[Sas86]   Sastry, S., and A. C. Parker, "Stochastic models for wirability analysis of gate arrays," *IEEE Trans. on Computer-Aided-Design*, vol. CAD-5, no. 1, pp. 52-65, January 1986.

## Further Reading

[Bre72]   Breuer, M. A., *Design Automation of Digital Systems: Theory and Practice*, Volume 1, Prentice-Hall, Englewood Cliffs, New Jersey, 1972.

[ElG80]   El Gamal, A., and Z. A. Syed, "A stochastic model for interconnections in custom integrated circuits," *Information Systems Laboratory Technical Report*, Stanford University, 1980.

[Hel78]   Heller, W. R., W. F. Mikhail, and W. E. Donath, "Prediction of wiring space requirements for LSI," *Journal of Design Automation and Fault-Tolerant Computing*, vol. 2, no. 2, pp. 117-144, May 1978.

[Men71a] Mennone, A., and R. L. Russo, "Mapping of computer logic: experiments using the encoding concept to reduce module requirements," *IBM T. J. Watson Research Center Report RC 3409*, June 1971.

[Men71b] Mennone, A., and R. L. Russo, "Selectric seed vertices for multiple mappings using the automatic logic mapping system," *IBM Technical Disclosure Bulletin*, vol. 13, no. 11, pp. 3202-3204, April 1971.

[Sut73]   Sutherland, I. E., and D. Oestreicher, "How big should a printed circuit board be?," *IEEE Trans. on Computers*, vol. C-22, no. 5, pp. 537-542, May 1973.

# Chapter 3

# Logic Partitioning

**William E. Donath**

*IBM T. J. Watson Research Center*
*Yorktown Heights, New York*

## 3.1 Introduction

Logic partitioning divides the components of a large logic circuit into a collection of smaller modules (or subcircuits) according to some criteria. For example, a limit is placed on the physical size of a module, on the number of modules, or on the number of external terminals on a module. Logic partitioning helps solve the following important physical design problems:

- *Circuit packaging.* Integrated circuit packages can hold only a limited number of logic components and external terminals (public pins). The components must be partitioned into subcircuits small enough to be implemented in the available packages. This example is the most direct application of partitioning; several programs have been written for exactly that purpose [Cha68, Don72, Ker70, Men71, Rus71a, Rus71b, Sch72b]. However, some designers reduce the need for automatic partitioning programs by separating the logic by function and designing these subcircuits to fit predefined size and area restrictions.

- *Circuit layout.* Partitioning is often used [Gue69, Sch72a, Bre76, Don75, Lau79] during placement to find strongly connected subcircuits of the logic. A placement algorithm can then place those components together, minimizing both delay and routing area.

- *Prediction and estimation.* Partitioning can also be used to estimate some properties of the physical design [Lan71, Rus72, Don79, Hel78]. For example, Rent's Rule relates the expected terminal count, $T$, of a complex or design of size $C$:*

$$T = AC^p \qquad\qquad (3.1)$$

where $A$ (average size of modules in the complex) and $p$ (Rent's exponent) are constants for a particular circuit. The terminal count can in turn be used to estimate how difficult the circuit will be to route. (See Chapter 2 for more information on Rent's Rule.)

Logic partitioning is NP-complete [Gar79]. In practice the size of partitioning problems makes it impossible to perform the exhaustive search required to find an an optimal partition. Work on partitioning has concentrated on heuristics that give good results in a reasonable amount of time. This chapter compares the different heuristics and evaluates both their performance and ease of implementation.

This chapter is a survey of partitioning techniques; the last such survey was written by Kodres [Kod72], and much has happened since. For example, the partitioning method of Kernighan and Lin [Ker70, Sch72b, Fid82, Kri84] has seen several stages of development and has been transformed into a quasi-linear method. Also the use of partitioning to study Rent's Rule and wirability has yielded important insights. Rent's Rule and these associated experiments are mentioned in Section 3.4. Furthermore, the development of simulated annealing [Kir84, Kir83] as a general design methodology has been an event of considerable significance to partitioning theory. Simulated annealing stochastically explores the search space by allowing the solution to climb past local minima. The work of Greene and Supowit [Gre84] is particularly applicable to partitioning, since it can apply more work in the later stages of partitioning.

The most successful partitioning techniques at this time are simulated annealing and group migration; both of these are described later in this chapter. Both methods can be useful in partitioning circuits. Simulated annealing allows a designer to come arbitrarily close to the optimal solution; therefore, the user can trade off quality and computing time. Group migration gives a designer great control over the computation time, particularly as the number of objects to be partitioned increases.

---

*In Chapter 2 Rent's Rule appears as "$IO = AS \cdot BK^r$." The term "$IO$" is used to designate terminal (pin) count, the term "$AS$" is used to designate the average size of modules (blocks) in the complex, the term "$BK$" is used to indicate the module (block) count of a complex, and "$r$" is used to indicate the Rent exponent.

There is a disappointing lack of data comparing partitioning algorithms. Usually, the work involved in implementing one approach is so large that the additional effort required to compare two methods, particularly tuning each method to achieve the best results, is impractical.

The rest of this chapter describes partitioning in detail. Section 3.2 presents some mathematical formulations for partitioning and uses them to develop theorems. Section 3.3 presents the following partitioning algorithms: the direct method, the group migration method of Kernighan and Lin and its extensions, simulated annealing, and a method based upon eigenvalues and eigenvectors. (Appendices A and B of this chapter give mathematical results related to these algorithms.) Section 3.4 compares these partitioning techniques.

## 3.2  The Mathematics of Partitioning

There are several ways to state the partitioning problem in mathematical form. These formalisms are used to prove theorems about partitioning that help us develop and understand partitioning algorithms. One of the simplest definitions of partitioning is as follows:

Problem 3.1:

Let $V$ denote a set of nodes (that correspond to components), with each node $r$ having area $a(r)$, and $X$ denote a set of nodes (that correspond to public pins or terminals) external to $V$. Let

$$S = (S_1, S_2, \ldots, S_N)$$

be a set of subsets of $V \bigcup X$ (the $S_i$ correspond to the nets). Then, partition $V$ into a set of disjoint subsets $V_1, V_2, \ldots, V_k$ such that

$$\sum_{r \,\in\, V_i} a(r) \leq A_i \qquad (3.2)$$

(where $A_i$ is the area constraint for the module corresponding to $V_i$) and the number of sets in $S$, which have nodes both internal to $V_i$ and external to $V_i$, is constrained by $T_i$, the terminal count.

Problem 3.1 very closely models partitioning required for circuit packaging. A physical example of Problem 3.1 might be the following: $V$ is a collection of IC components to be partitioned over several printed circuit boards (PCBs). Each IC is called $r$, and each IC has an area $a(r)$. $A_i$ is the area constraint for each PCB, and $T_j$ is the connector constraint (*i.e.*, public pin count) for each PCB. $S$ is the set of all nets in the design, and each component is connected to a subset of these nets, called $S_j$. $X$ is the set of all public pins associated with the entire design.

Problem 3.1 quite properly ignores the directionality of nets (the sets $S_i$ represent nets, and typically one pin in the net acts as a signal source, while the others are activated by the source) and the circuit's function. Typically, we partition logic into subcircuits with equal areas $(A_i)$ and terminal counts $(T_i)$. However, this formulation ignores several constraints encountered in real problems. For example, certain components may consume an inordinate amount of power, and we often want to limit the power consumption of each subcircuit. Also, some partitions may put components on critical timing paths far apart, creating unacceptable circuit delays. Some work has been done on partitioning for system performance [Law69] and for placement [Wol78]. Because other constraints may be encountered in real problems, it is important to develop algorithms and programs that can be adapted to meet these special requirements.

We can reformulate the partitioning problem definition for circuit layout. (For more information on the use of partitioning in placement, see Chapter 4.) The intent here is to produce a partition that minimizes the number of strongly connected components that are separated by a partition boundary. During the physical placement of components, such a "weighted" partitioning often helps minimize both delay and routing area.

Problem 3.2:

> Problem 3.2 is the same as Problem 3.1, except that the constraint on the number of nets sharing points inside and outside the partitions $V_i$ is dropped. Instead, we assign to each net a weight $m(S_i)$ which is the number of different partitions that the net connects minus one. If any node in $S_i$ belongs to $X$, then 1 is added to $m(S_i)$. $X$ is treated as $V_{k+1}$. (In other words, all the public pins are grouped into a single partition.) Minimize

$$M = \sum m(S_i) \qquad (3.3)$$

> where $M$ is the number of connected pairs of nodes of the partitioned node.

To be strict, we should attempt to find a solution such that $M \le G$ where $G$ is some integer, but a proof exists that this problem is NP-complete. (The solution exists, however, only for the case where $|S_i|$ is 2 and $|S|$ is significantly greater than $|V|$.) In practice we may first partition using the criterion of Problem 3.2 as a precursor to partitioning to satisfy the criterion of Problem 3.1.

We can redefine the partitioning problem again in terms of graph theory. (See the appendix at the end of Chapter 10 for more information on the relationship of graph theory to circuit layout.)

Problem 3.3:

> Let $V, E$ denote a graph, where $V$ is the set of nodes and $E$ is a subset of $V \times V$. Let each node $r$ have area $a(r)$, and give each edge $e$ a weight $w(e)$. Let $P(r)$ be the partition to which node $r$ belongs. Find a partition $V_1, V_2, \ldots, V_k$ of $V$ that minimizes area

$$\sum_{r \in V_i} a(r) \le A_i \qquad (3.4)$$

and such that the sum $W$ of the weights of the edges that have their endpoints in different partitions is as small as possible:

$$\sum w(e: e = (r, s) \text{ and } P(r) \neq P(s)) \leq W \qquad (3.5)$$

Problem 3.3 is more divorced from real partitioning problems than are Problems 3.1 and 3.2. Attempts to solve Problem 3.3 rather than Problem 3.1 or 3.2 have led to poor partitioning results. If one simply replaces each net $S_i$ with its complete graph with weights 1 on all the edges generated, one may obtain highly suboptimal partitions [Sch72b]. The partitions can be improved by using an edge weight less than 1 for this transformation. The question is what weight is best? Charney and Plato used a weight of $2/(n-1)$ [Cha68], and a weight of $2/n$ was used in some placement studies [Han72]; $n$ is the number of pins in the net. However, the weight

$$w = \frac{4}{n^2 - mod(n, 2)} \qquad (3.6)$$

can be shown to be the best weight by the following theorem:

Theorem 3.1:
Given a set $S$ of $n$ nodes and a complete graph on $S$, consider a partition of $S$ into $k$ groups each holding at least one node of $S$. Let $c$ be the number of edges of the complete graph cut by the partition; then

$$k - 1 \geq \frac{4c}{n^2 - mod(n, 2)} \qquad (3.7)$$

for all partitions and the relation is strict equality for at least some partitions. A proof of this theorem and a more general version of it are given in Appendix A at the end of this chapter.

An immediate corollary of Theorem 3.1 is as follows. If we substitute edges with weights computed from Equation 3.6, then

$$M \geq W \qquad (3.8)$$

(where $M$ is the result of Equation 3.3).

We can also recast the partitioning problem as an eigenvector problem. Assign to each node $r$ of the graph of Problem 3.3 a set of coordinates $X(r, 1), X(r, 2), \ldots, X(r, k)$ which have the property that

$$\sum_r X(r, h) \cdot X(r, m) = \begin{cases} 1 & if\ h = m \\ 0 & otherwise \end{cases} \qquad (3.9)$$

and that minimizes the following sum with respect to the $X(r, h)$:

$$\sum_{h=1}^{k} \sum_e w(e) \cdot (X(r, h) - X(s, h))^2 + \sum_{h=1}^{k} \sum_r U(r)\, X(r, h)^2 \qquad (3.10)$$

The $U(r)$ can have any values as long as the sum over all $U(r)$ is zero:

$$\sum_{r} U(r) = 0 \qquad (3.11)$$

When the value of Equation 3.10 is minimized, this problem is equivalent to the following eigenvalue problem of a matrix $D(r, s)$. For $r \neq s$, let $D(r, s)$ be the sum of the weight $w$ of all edges connecting $r$ and $s$. For the diagonal elements let $D(r, r) = - \sum_{s,\, s \neq r} D(r, s) + U(r)$.

We can use the $D$ matrix to evaluate the partitioning. Let $d_1 \geq d_2 \geq \cdots \geq d_k$ be the $k$ largest eigenvalues of the matrix $D$. Let the areas of the nodes $a(r)$ all be constant 1. Finally, let the size of the partitioned units $V_1, V_2, \ldots, V_k$ all be equal. We can then prove the following theorem:

Theorem 3.2:

$$W \geq \frac{-|V| \cdot (d_1 + d_2 + \cdots + d_k)}{(2k)} \qquad (3.12)$$

A proof of this theorem is given in Appendix B at the end of this chapter.

We then want to vary the $U(r)$ to maximize the value of Equation 3.10. This has the effect of increasing the value of the right side of Equation 3.12 without changing the value of the left side. Methods for doing so have been developed [Cul75]. Partitioning methods based on this result have been developed [Don72] and are discussed later in this chapter.

## 3.3  Partitioning Methods

There are four major partitioning algorithms:

- The *direct method* starts with a seed node of (or set of seed nodes for) each cluster and assigns a node at a time to one partition, using preferences to satisfy the constraints. Various embellishments have been made to this method. For example, after the initial partition is complete, the algorithm can be restarted by selecting new seeds for the clusters; this may produce a better partition [Kod72, Men71].

- The *group migration method*, also known as the Kernighan-Lin algorithm, starts with some partition, usually generated randomly, and then moves components between partitions to improve the partitioning. For more details, see [Ker70, Don74a, Wol78, Fid82].

- A family of *metric allocation methods* attempts to find a metric other than the structure of the interconnection graph which in some way reflects the direct and indirect connectedness of the nets. In these approaches nodes are put together on the basis of the metric, not on the basis of their connectedness. Therefore, the number of connections is only indirectly minimized. Several approaches for doing the actual partitioning exist [Cha68, Don72].

- The *simulated annealing* method is a non-convex optimization algorithm. The partitioning problem is cast in two parts: a *cost function*, which classifies any feasible solution, and a set of moves, which allow movement from solution to solution. The algorithm starts at a random solution and makes stochastically chosen moves to modify that solution. Initially the moves which are accepted include a high proportion of moves which increase the solution's cost. As the algorithm progresses, the proportion of such moves is decreased until finally almost no moves that increase the cost are accepted. Simulated annealing is discussed in detail in Chapter 4.

**Direct Partitioning**

The mathematical model for direct partitioning is Problem 3.1 with simplification that the $A_i$ (areas of the partitioned modules) and $T_i$ (terminals of the partitioned modules) are not dependent on the module $i$. In other words, the area of each module must be less than $A$, and the terminal count of each module must be less than $T$.

The direct partitioning method, particularly as developed for the ALMS system [Men71, Rus71a, Rus71b], consists of three phases:

- An initial partitioning method that proceeds bottom-to-top. The initial groups (a node or a set of nodes) can be generated either manually or automatically. These groups should be considerably smaller than the desired size of the final groups. These groups can be, in the simplest case, just the nodes of the graph. We do not require that the groups be disjoint. Different groups could contain some common nodes. In this case the procedure could be called mapping instead of partitioning. Allowing some redundancy can improve the quality of a partition [Men74].

- A seeding step that determines the initial node of each module. In the ALMS program seeding is done differently in the first attempt at partitioning and in subsequent attempts; this allows the program to explore a larger part of the search space.

- The group allocation procedure, where the groups are partitioned into the modules.

The discussion of the direct method in this chapter is brief; for a more extensive description of the program, consult [Men71] and [Rus71b].

**Group Allocation.**    Assume that at least some, if not all, of the modules contain groups or nodes. Initially the groups or nodes represent seeds; in later passes they represent intermediate partitions. Also, there exists a set of unallocated groups or nodes, called the *candidate set*, each of which has at least one net in common with the nodes already allocated to modules. Now, allocate each set of groups in the candidate set to modules.

For each module $i$ and group $j$ in the candidate set, we define $F_{ij}$:

$$F_{ij} = (P_{ij} + kp)(B_{ij} + kb)$$

where $P_{ij}$ is the number of terminals that would be added to module $i$ if group $j$ were included in module $i$ and $B_{ij}$ is the amount of area that would be added to module $i$ if group $j$ were included. If either $P_{zj}$ or $B_{zj}$ is greater than the remaining capacity of module $z$, then $F_{zj}$ is said to be undefined. Note that $P_{ij}$ may be negative (which would be a particularly favorable case) and that $B_{ij}$ may be smaller than the area contained in $j$ because of possible redundancy. For each group $j$ there are three cases:

(1)    $F_{ij}$ may be undefined for all $i$, so that the given group will not fit on any given module.

(2)    $F_{ij}$ is defined for only one $i$, and the group will fit on only one module.

(3)    $F_{ij}$ is defined for at least two values of $i$, and the given group will fit on at least two modules. In this case, we compute $R_j$, the ratio of the two smallest values of $F_{ij}$:

$$R_j = \frac{minimum_i(F_{ij})}{next\ smallest_i(F_{ij})}$$

$R_j$ tells us the value of immediately assigning a group. If $R_j$ is close to 1, then there are two almost equivalent possible assignments, and there is no hurry to assign this particular group. If $R_j$ is close to 0, then one assignment is much preferable to the other and there is some advantage in assigning it now.

We want to allocate the most advantageous group at each step. If any group satisfies case 2 above, that group which satisfies case 2 and has the lowest $F_{ij}$ is allocated to module $i$. If no group satisfies case 2, then that group $j$ which has the smallest value $R_j$ is allocated to that module $i$, for which its $F_{ij}$ is minimum.

The allocation procedure terminates in a successful mapping when every allocatable group has been allocated to a module. The procedure also terminates when all groups satisfy case 1 (that is, no group can be allocated to any given module). In this case there are two possibilities:

(1)    The number of modules in existence is equal to the limit that has been imposed. In this case one might either want to terminate or go to another seeding/allocation phase.

(2)    The number of modules has not yet reached the limit imposed upon the program. In the latter case, which can happen only on the first allocation phase, the algorithm picks out the node with the largest number of terminals on it and uses it to seed another module.

**The Seeding Procedure.** The ALMS program partitions in several passes. Each pass performs a seeding step, which allocates one group to each module to start the partitioning, followed by an allocation procedure.

In the first allocation, seeding is done one module at a time. The group with the largest terminal count not allocated yet is taken as the seed. The allocation program puts whatever it can into the module, and then another module is started. The process continues until the program runs out of either modules or nodes. In succeeding passes the program uses heuristics to start each module with groups that are as close to the center of the module as possible.

The centering heuristic uses data gathered during allocation on the sequence in which the groups are allocated to each module. Each allocated group is assigned an order number $O_j$ for its allocation in its groups, and each module has $J_i$ groups allocated to it. The heuristic uses a module allocated about 1/3 of the way through the allocation pass as the seed for the next pass. For each module we define

$$L_i = \left\lfloor \frac{N\, J_i}{D} \right\rfloor \tag{3.13}$$

$$C_i = \left| L_i + J_i - \left\lfloor \frac{D\, L_i}{N} \right\rfloor \right|$$

(where $\lfloor X \rfloor$ means $X$ truncated to an integer). Typical values are $N \approx 1$ and $D \approx 3$. This formula states that we select the group with order number $C_i$ as the seed group for the next allocation pass; it assures that $C_i$ is greater than zero. When $N = 1$ and $D = 3$ the function shows the following correspondence:

$$J_i = 1 \rightarrow C_i = 1 \tag{3.14}$$
$$J_i = 2 \rightarrow C_i = 2$$
$$J_i = 3 \rightarrow C_i = 1$$

A simpler formula for $C_i$ that shows the same correspondence is:

$$C_i = \left\lfloor \frac{(N\, J_i) + D - 1}{D} \right\rfloor \tag{3.15}$$

Although this procedure is heuristic, some heuristic must be used to efficiently solve the problem, and this heuristic is both intuitively appealing and simple to compute. An alternative is a heuristic based on the center of gravity of the connectivity graph. Exact determination of the center of gravity of a graph is more expensive but feasible. (See the exercises at the end of the chapter.)

**The Grouping Procedure.** The methods used for grouping attempt to recover some of the structure of the logic by backtracking through the logic. Experience suggests that nodes which electrically drive, or *feed*, the same node probably belong in the same group. An *initiate* net is a net which is to be an output of the group. All nodes feeding the initiate net are grouped together with the initiate node, unless they are themselves initiate nodes. Typically, nodes occurring in storage elements and nodes connected to primary outputs of the logic are treated as initiate nodes. From there on, however, the rules are elastic. In reference [Men74] the status of initiate nodes was assigned on the basis of much trial and error. It appears from that report that considerable benefit was obtained from using redundancy; however, it is possible that the mappings (components duplicated in modules) caused some modules to contain too many nodes.

**Implementation.** The partitioning program contains a central loop over the candidate set of groups — those groups which are connected in any way to the nodes already allocated to modules — in which all the $F_{ij}$ are calculated for this candidate set and the group next to be assigned is selected. The assignment of one group to a module, which is the result of this loop, may change the evaluation function for all members of the candidate set, which is also changed. No information is preserved from one iteration to the next.

## Partitioning by Group Migration

**The Algorithm.** This method was first developed by Kernighan and Lin [Ker70]. Their problem formulation was as graph partitioning, like that of Problem 3.3, and was later extended [Sch72b] to a net-graph partitioning formulation like that of Problem 3.2. The more recent work of Fiduccia and Mattheyses [Fid82] assures that the central loop in the algorithm requires only $n \log(n)$ time. Krishnamurthy [Kri84] extended this approach to use more look-ahead and gave a detailed pseudo-code implementation of his algorithm, which can very readily be used as the basis of a program. The method is particularly well suited for *bisection*, dividing the graph into two modules, but can easily be generalized to partitioning into unequal pieces, which could then be used as the basis of a hierarchical partitioning scheme.

The group migration algorithm works as follows:

(1)   Start with some partition of nodes into two groups $A$ and $B$. This partition could very well be a simple, random partition.

(2)   For every node $a$ in $A$ and every node $b$ in $B$, calculate the algebraic changes in terminal counts $D_a$ and $D_b$ that would occur if node $a$ were moved to $B$ or node $b$ moved to $A$. Set the queue to empty and $i$ to 1.

(3)   Select from all the pairs of nodes $(a, b)$ the node pair $(a_i, b_i)$ that gives the most improvement in total terminal count and add that pair to the queue. Save the algebraic improvement in terminal count as $g_i$. (We will discuss below how to find the pair that gives the most improvement.)

(4)    Remove $a_i$ from $A$ and $b_i$ from $B$, and recalculate the values $D_a$ and $D_b$ for the members of $A$ and $B$. If $A$ and $B$ are both not empty, then increment $i$ and repeat step 3.

(5)    Find $k$ such that

$$G = \sum_{j=1}^{k} g_j \qquad (3.16)$$

is a minimum and move $a_1, \ldots, a_k$ to $B$ and $b_1, \ldots, b_k$ to $A$. If $G$ is negative and $k$ greater than zero, repeat the process and go to step 2; otherwise, stop.

Each iteration is relatively expensive (at least proportional to the number of nodes $n$ in the graph and possibly $n^2$). It is reassuring that the process usually converges after relatively few iterations [Ker70]. In this version of the algorithm, if the average number of connections per node is relatively small, the computation is proportional to $n \log(n)$, but for arbitrary connectivity the cost grows proportional to $n^2$.

The Kernighan-Lin algorithm was improved by Fiduccia and Mattheyses [Fid82], who reduced the time per pass to $O(n \log(n))$. Krishnamurthy [Kri84] later added lookahead to the Fiduccia-Mattheyses algorithm.

## Partitioning by Metric Allocation

The earliest algorithm measuring direct and indirect connectivity with a metric is that of Charney and Plato [Cha68], who used an electrical analog of the network to choose partitions. They pointed out that solving the analog network minimizes the distance squared between components. Their actual partitioning model was somewhat different but still depended on electrical analogs. This chapter presents a model related by means of lower bounds to the partitioning problem. The partitioning metric is calculated from eigenvalues of the network. A complete description of the method is given by Cullum, Donath, and Wolfe [Cul75]. Here, we describe partitioning once the values of the metric have been computed. Our partitioning method works for any metric; the metric of Charney and Plato could be used in place of the eigenvalue metric. Our partitioning algorithm invokes a function that computes the metric for a pair of nodes. The metric must preserve partial ordering (as does a distance measure). It should also be quick to calculate since it will be used many times.

Our partitioning algorithm starts with a set $V$ of nodes and a set $S = S_1, S_2, \ldots, S_n$ of nets. We define a metric $m(r, s)$ over any $V \times V$ and a weight $a(r)$ on each of the nodes. The algorithm partitions the nodes $V$ into subsets $V_1, V_2, \ldots, V_k$ such that the sum of the areas in $V_i$ is $\leq A$ for all $i$ and the number of nets with members both internal to $V_i$ and external to $V_i$ is $< T$ for all $i$. The merging algorithm below reports whether it can partition the nodes into $k$ groups to satisfy the requirements:

(1)    Set $L$, a list of edges, to empty.

(2)    Compute for each net $S_i$ in $S$ the metric $m(r, s)$ over $S_i \times S_i$, and find a minimum spanning tree $ST$ over the net $S_i$. Add the edges $(i, j)$ in $ST$ to $L$.

(3)    Sort the edges of $L$ in order of increasing metric.

(4)    Initialize a set of groups of nodes, such that each group has just one member in $V$ and all members of $V$ belong to one group. The group structure is arranged in such a way that it remembers how groups were merged during the following process, so that an unmerging is feasible in the order that the merging was done.

(5)    Consider each edge $(r, s)$ in $L$ in increasing order of the metric and consider the groups $G(r)$ and $G(s)$, to which $r$ and $s$ belong, respectively. If $G(r) \neq G(s)$, then consider merging them. If the size of the merged groups does not exceed $A$, and the terminal count of the merged group does not exceed $T$, then merge the two groups into a new group.

(6)    If there are just $k$ groups in existence, return and report success.

(7)    Take the smallest of the groups, and attempt to merge it with any of the largest $k$ groups. If it is possible to merge it with any or some of the groups, merge it with the one that yields the least increase (or most decrease) in terminal count, and go to step 6. Otherwise, do one of the following:

   (7.1) If the group under consideration contains more than one component, break it into two groups, increase the total group count by 1, and go to step 6.

   (7.2) If the group contains just one node, return and report failure.

## Partitioning by Simulated Annealing

This approach is described in detail in Chapter 4 of this book, so we will confine this discussion to a brief overview. This method was first used for computer-aided design problems by Kirkpatrick *et al.* [Kir83]. A problem formulated for simulated annealing defines an objective function for the quality of a solution and a set of moves that transform one valid solution into another. The objective function is analogous to energy in a physical system, and the moves are analogous to changes in the energy of the system. Moves are selected randomly, and the probability that a move is accepted is proportional to the system's current temperature. At high temperatures moves that make the total cost higher may be accepted, while at low temperatures such moves are unlikely to be accepted. Stochastically accepting moves that increase the solution's cost allows simulated annealing to explore the design space more thoroughly and extricate the solution from a local optimum. The hope is that the number of feasible solutions located near low-cost solutions is larger than those removed a few steps from the low-cost solutions.

Early work on simulated annealing used the Metropolis algorithm [Met53]. This algorithm generates a move at random, evaluates the cost of the move, and then accepts it or rejects it. In the later stages many more moves are rejected than are accepted, greatly increasing the computation time required. Greene and Supowit [Gre84] proposed an algorithm whereby a list of moves is generated and the moves are taken from the list by a random selection process. This technique works extremely well for partitioning into a small number of clusters, provided one considers one's move set to be the move of a single object from its current cluster to another cluster and if the average degree of the nodes is small.

## 3.4  Comparison of Partitioning Algorithms

Partitioning algorithms can be compared on their difficulty of implementation, their performance on common partitioning problems, and their runtimes. The group migration method is the easiest to implement, while the metric allocation method is the hardest since it requires some numerical programming. Group migration probably also has the fastest running time; however, it is less easily generalized to many groups or limitations on pin counts than the direct metric allocation methods. Metric allocation is considerably more costly in computing time than the direct and group migration methods, since it requires either the calculation of eigenvalues or the inversion of matrices. However, because of its ability to consider indirect linkages between nodes, it often gives superior results (about a 10% improvement in placement distance in one case [Han76]).

Table 3.1 compares the direct method (as implemented in ALMS) and the metric allocation method. The same graph was partitioned three times using the direct method and four times using the metric allocation method. A summary of the results is shown. The graph considered here has 1239 nodes with a total area of 1283. The graph has the relatively low Rent exponent of 0.47. The experiments for the direct method were done with ALMS by B. S. Landman, and the experiments for the metric allocation method were done by this author. The metric allocation method took considerably longer than the direct method. Mennone and Russo [Men74] published a graph which may serve as a standard. In that work they also showed the considerable benefit of allowing redundancy. Partitioning results could probably be substantially improved by a method of automatically introducing redundancy. Further improvement could be gained by taking advantage of the input graph's hierarchical structure.

| Method | Number of Modules | Maximum Terminal Count of Modules | Maximum Area of Modules |
|---|---|---|---|
| direct | 4 | 61 | 330 |
| | 6 | 43 | 330 |
| | 10 | 36 | 208 |
| metric allocation | 2 | 55 | 836 |
| | 3 | 50 | 500 |
| | 5 | 39 | 321 |
| | 7 | 36 | 300 |

**Table 3.1**  Comparison of the Direct and Metric Partitioning Methods

## 3.5  Summary

Partitioning methods have become more sophisticated over time as computation time has become more affordable. Simulated annealing requires vastly more CPU time than does the direct method, but simulated annealing works from costs more directly related to the underlying problem and is able to explore a much larger fraction of the design space. Unfortunately, comparison of the available algorithms has not kept pace with their development, so we cannot always judge the cost-effectiveness of the different methods. As partitioning algorithms become more sophisticated, we can expect to make even more use of the information contained in the user's input, particularly the design hierarchy, and to make more effective use of CPU cycles to exploit that information.

## Appendix A:  Weighting the Edges of a Net

We want to prove that the subdivision of a complete graph of $n$ nodes into $k$ non-empty subsets cuts at most

$$\frac{n^2 - mod(n, 2)}{4}$$

edges. We will actually prove a more general theorem, but first we introduce a lemma:

Lemma 3.1

Given a tree $T$ of $n$ nodes and the set of all paths $P(u, v)$ on the tree, where $u$ and $v$ are two distinct nodes of the tree $T$, the number of paths containing a given edge $(s, t)$ of the tree $T$ is, at most,

$$\frac{n^2 - mod(n, 2)}{4}$$

Proof: The edge divides the tree into two parts: one holding $m$ $(0 < m < n)$ nodes, the other $n - m$ nodes. There are exactly $m(n - m)$ paths on the tree containing the edge $(s, t)$. It is easy to see that, for $m$ an integer and $0 < m < n$,

$$m(n - m) \leq \frac{n^2 - mod(n, 2)}{4}$$

where the equality is met when

$$n \; even \; \text{and} \; m = n/2$$

$$n \; odd \; \text{and} \; m = \frac{n - 1}{2} \; or \; \frac{n + 1}{2}$$

We now use the lemma to prove our theorem:

Theorem 3.3

Given a finite set $S$ of $n$ elements and a metric $m(r, s)$ on the space $S \times S$ such that

$$m(r, r) = 0$$

$$m(r, s) = m(s, r)$$

$$m(r, s) \leq m(r, u) + m(u, s)$$

let

$$M(F) = \sum_{s < t} m(s, t)$$

denote the sum of the metrics over all the connections of the full graph, let $ST$ be a spanning tree on $S$, and let

$$M(ST) = \sum_{(s, t) \in ST} m(s, t)$$

Then,

$$M(ST) \geq \frac{4 \cdot M(F)}{n^2 - mod(n, 2)}$$

Proof: Consider all the paths $P(s, r)$ between nodes of the tree $ST$. These are themselves trees, and they have a corresponding metric $M(P(r, s))$. We find

$$\sum_{r < s} M(P(r, s)) \leq M(ST) \cdot \frac{n - mod(n, 2)}{4}$$

using Lemma 3.1. However, from the triangle inequality we can see that

$$\sum_{r < s} M(P(r, s)) \geq M(F)$$

Therefore

$$M(ST) \geq \frac{4 \cdot M(F)}{(n^2 - mod(n, 2))}$$

Corollary:

For each $n$ there exists a metric and tree $ST$ such that the equality of the theorem holds.

Proof: For $n$ nodes, divide the set $S$ into two parts, each containing either exactly half the nodes (in case $n$ is even), or one part containing $(n+1)/2$ nodes and the other $(n-1)/2$ nodes. Let the metric have the value 0 for all pairs of nodes belonging to the same part, and 1 for all pairs of nodes not belonging to the same part. Consider the spanning tree $ST$ which has only one of its edges connecting the two parts with the other edges internal to one of the parts. Then,

$$M(ST) = 1$$

$$M(F) = \frac{n^2 - mod(n, 2)}{4}$$

We leave as an exercise for the reader the proof of the initial statement (Equation 3.7) of Theorem 3.1 in Section 3.2.

## Appendix B: The Relationship Between Partitioning and Eigenvalues

We can recast Problem 3.3 as an eigenvalue problem. We present here a simple proof of Theorem 3.2; a more general version of the theorem can be found in [Don73].

Consider the nodes $r = 1, 2, \ldots, |V|$ (where $|V|$ is the size of the set $V$) to be imbedded into a $k$-dimensional space, such that for each node $r$ the coordinate $x(r, i)$ ($i = 1, 2, \ldots, k$) is given. Denote by $D(r, s)$ (as in Section 3.2) the sum of the weights of the edges on $(r, s)$, and assign to the $x(r, i)$ coordinates the following orthonormalization property:

$$\sum_r x(r, i) \, x(r, j) = 1 \ iff \ i = j \tag{3.17}$$

$$= 0 \ otherwise$$

Let the $x(r, i)$ coordinates also have the property that they minimize the following expression $E$:

$$E(x) = \sum_{r<s} \sum_i D(r, s) \, (x(r, i) - x(s, i)) + \sum_r \sum_i U(r) \, x(r, i)^2 \tag{3.18}$$

where the term in the $U(r)$ is optional, but is required to have the property that

$$\sum_r U(r) = 0 \tag{3.19}$$

(Typically, one makes the $U(r)$ initially 0 and then adjusts its value to maximize $E(x)$). Minimizing Equation 3.18 for $x$ requires finding the highest $k$ eigenvalues $d_1, d_2, \ldots, d_k$ of the matrix equation

$$\sum_s D(r, s)\, x(s, i) + U(r)\, x(r, i) = d_i\, x(r, i) \tag{3.20}$$

(Equation 3.20 is obtained by inserting the constraints of Equation 3.17 into 3.18 using a Lagrangian multiplier and differentiating with respect to $x(r, i)$. The derivation is left to the reader.) We find that if $d_1 \geq d_2 \geq d_3 \geq \cdots$ then

$$E(x) = -(d_1 + d_2 + \cdots + d_k) \tag{3.21}$$

Now, consider any other set of spatial coordinates $y(r, i)$ that satisfy the constraints of Equation 3.17:

$$\sum_r y(r, i)\, y(r, j) = 1 \ iff \ i = j$$

$$= 0 \ otherwise$$

Since $E(x)$ is a minimum, we must have

$$E(y) \geq E(x) \tag{3.22}$$

Construct an orthonormal set of $y$ by considering some partition of $V$ into $k$ equal subsets. ($|V|$ is an integral multiple of $k$.) We define $y(r, i)$ as:

$$if \ r \ in \ Vi \ then \ y(r, i) = \frac{k}{|V|} \tag{3.23}$$

$$else \ y(r, i) = 0$$

We leave as an exercise for the reader the proof that

$$E(y) = \frac{2\, k\, W}{|V|} \tag{3.24}$$

where $W$ is just the sum of the weights of the edges that are cut by the partition (see Exercise 3.2).

Using Equations 3.20, 3.21, and 3.24, we find that

$$W \geq -(d_1 + d_2 + \cdots + d_k)\, \frac{|V|}{(2k)} \tag{3.25}$$

Eigenvectors have been used for both placement and partitioning [Hal70, Don72, Don75].

## Exercises

**3.1**  Complete the proof of Appendix A by proving Equation 3.7.

**3.2**  Prove Equation 3.24. Notice that $\sum_i y(r, i)^2$ is a constant for all $r$; therefore, the term in $U(r)$ drops out due to the condition in Equation 3.19. Notice also that $y(r, i) - y(s, i)$ is non-zero only if $r$ and $s$ belong to different partitions.

**3.3**  Derive Equation 3.20 by inserting the constraints of Equation 3.17 into 3.18 using a Lagrangian multiplier and differentiating with respect to $x(r, i)$.

**3.4†**  Invent a seeding procedure for the direct partitioning method presented in Section 3.3 using a heuristic based on the center of gravity of the connectivity graph. Exact determination of the center of gravity of a graph is expensive but feasible. (A possible procedure is to label all groups in the module which have connections externally as having distance 0 to the outside; then label with distance 1 all the groups connected to the distance 0, and so on.)

**3.5†**  Demonstrate a circuit partitioning instance where Kernighan and Lin's partitioning algorithm will outperform Fiduccia and Mattheyses' partitioning algorithm. Similarly, demonstrate a circuit partitioning instance where the reverse happens.

**3.6†**  Describe how a group migration algorithm can be used as the basis for a hierarchical partitioning scheme.

**3.7†**  Describe how a group migration algorithm can be used as the basis for a placement algorithm.

**3.8†**  Define a functional interface for a partitioning program. The interface should be general enough to be implemented by any of the algorithms described in this chapter.

**3.9‡**  Implement the Kernighan and Lin group migration partitioning algorithm. Obtain the placement benchmarks [Pre87] from the Microelectronics Center of North Carolina, P.O. Box 12889, 3021 Cornwallis Road, Research Triangle Park, NC 27709. Test your implementation on these benchmarks.

## References

[Bre76]  Breuer, M. A., "Min-cut placement," *J. of Design Automation and Fault Tolerant Computing*, vol. 1, pp. 343-362, 1976.

[Cha68]    Charney, H. R., and D. L. Plato, "Efficient partitioning of components," in *Proc. of the 5th Annual Design Automation Workshop*, pp. 16-0 to 16-21, 1968.

[Cul75]    Cullum, J., W. E. Donath, and P. Wolfe, "The minimization of certain nondifferentiable sums of eigenvalues of symmetric matrices," in *Mathematical Programming Study*, vol. 3, pp. 35-55, 1975. See also J. Cullum and W. E. Donath, "A block lanczos algorithm for computing the $q$ algebraically largest eigenvalues and a corresponding eigenspace of large, sparse, real symmetric matrices," in *Proc. IEEE Conf. on Decision and Control*, Phoenix, Arizona, pp. 505-509, 1974.

[Don72]    Donath, W. E., and A. J. Hoffman, "Algorithms for partitioning of graphs and computer logic based on eigenvectors of connection matrices," *IBM Technical Disclosure Bulletin 15*, pp. 938-944, 1972.

[Don73]    Donath, W. E., and A. J. Hoffman, "Lower bounds for the partitioning of graphs," *IBM Journal of Research and Development*, vol. 17, pp. 420-425, 1973.

[Don74a]   Donath, W. E., "Algorithm for partitioning improvement," *IBM Technical Disclosure Bulletin*, pp. 2654-2655, 1974.

[Don75]    Donath, W. E., "Hierarchical placement method," *IBM Technical Disclosure Bulletin*, vol. 17, pp. 3121-3125, 1975.

[Don79]    Donath, W. E., "Placement and average interconnection lengths of computer logic," in *IEEE Trans. on Circuits and Systems*, vol. CAS-26, pp. 272-277, 1979. See also "Equivalence of memory to random logic," *IBM J. of Research and Development*, vol. 18, pp. 401-407; "Stochastic model of the computer logic design process," IBM T. J. Watson Research Ctr., Yorktown Hts., New York, Report RC 3136, November 1970.

[Fid82]    Fiduccia, C. M., and R. M. Mattheyses, "A linear time heuristic for improving network partitions," in *Proc. 19th Design Automation Conf.*, pp. 175-181, 1982.

[Gar79]    Garey, M. R., and D. S. Johnson, *Computers and Intractability, a Guide to the Theory of NP Completeness*. W. H. Freeman and Co., San Francisco, California, pp. 209-210, 1979.

[Gre84]    Greene, J., and K. Supowit, "Simulated annealing without rejected moves," in *Digest Intl. Conf. on Computer Design*, pp. 658-663, October 1984.

[Gue69]    Guenther, T., "Die raeumliche anordnung von einheiten mit wechselbeziehungen," *Elektronische Datenverarbeitung*, vol. 11, pp. 209-211, 1969.

[Hal70]    Hall, K. M., "*r*-dimension quadratic placement algorithm," *Management Science*, vol. 17, pp. 219-229, 1970.

[Han72]    Hanan, M., and J. M. Kurtzberg, "A review of the placement and quadratic assignment problems," *SIAM Review*, vol. 14, pp. 324-342, 1972.

[Han76]    Hanan, M., P. K. Wolff, Sr., and B. J. Agule, "A study of placement techniques," *J. of Design Automation and Fault Tolerant Computing*, vol. 1, pp. 28-61, 1976.

[Hel78]    Heller, W. R., W. F. Mikhail, and W. E. Donath, "Prediction of wiring space requirements for LSI," *J. of Design Automation and Fault Tolerant Computing*, vol. 2, pp. 117-144, 1978.

[Ker70]    Kernighan, B. W., and S. Lin, "An efficient heuristic procedure for partitioning graphs," in *The Bell System Technical Journal*, vol. 49, pp. 291-307, 1970. See also "Partitioning graphs," in *Bell Laboratories Record*, pp. 8-12, January 1970.

[Kir83]    Kirkpatrick, S., C. D. Gelatt, Jr., and M. P. Vecchi, "Optimization by simulated annealing," *Science*, vol. 220, no. 4598, pp. 671-680, 1983.

[Kir84]    Kirkpatrick, S., "Optimization by simulated annealing: quantitative studies," *Journal of Statistical Physics*, vol. 34, pp. 975-986, 1984.

[Kod72]    Kodres, U. R., *Partitioning and Card Selection.* Design Automation of Digital Systems (edited by M. A. Breuer), Prentice-Hall, Inc., Englewood Cliffs, New Jersey, pp. 173-212, 1972.

[Kri84]    Krishnamurthy, B., "An improved min-cut algorithm for partitioning VLSI networks," *IEEE Trans. on Computers*, vol. C-33, pp. 438-446, May 1984.

[Lan71]    Landman, B. S., and R. L. Russo, "On a pin versus block relationship for partitions of logic graphs," *IEEE Trans. on Computers*, vol. C-20, December, pp. 1469-1479, 1971.

[Lau79]    Lauther, U., "A min-cut placement algorithm for general cell assemblies based on a graph representation," *J. of Digital Systems*, vol. 4, pp. 21-34, 1979.

[Law69]    Lawler, E. L., K. N. Levitt, and J. Turner, "Module clustering to minimize delay in digital networks," *IEEE Trans. on Computers*, vol. C-18, January, pp. 47-57, 1969.

[Men71]    Mennone, A., and R. L. Russo, "Selecting seed vertices for multiple mappings using the automated logic mapping system," in *IBM Technical Disclosure Bulletin 13*, pp. 3202-3204, April 1971. See also "Partitioning and mapping of logic graphs: The ALMS /360 algorithm for seeding of modules," IBM T. J. Watson Research Ctr. Research Report RC-2785, February, 1970.

[Men74]   Mennone, A., and R. L. Russo, "An example computer logic graph and its partitions and mappings," *IEEE Trans. on Computers*, vol. C-23, pp. 1198-1204, November 1974.

[Met53]   Metropolis, N., A. Rosenbluth, M. Rosenbluth, A. Teller, and E. Teller, "Equation of state calculations by fast computing machines," *J. Chem. Phys.*, vol. 21, pp. 1087-1092, 1953.

[Pre87]   Preas, B. T., "Benchmarks for cell-based layout systems," in *Proc. of the 24th Design Automation Conf.*, pp. 319-320, 1987.

[Rus71a]  Russo, R. L., P. H. Oden, and P. K. Wolff, Sr., "A heuristic procedure for the partitioning and mapping of computer logic graphs," *IEEE Trans. on Computers*, vol. C-20, pp. 1455-1462, December 1971.

[Rus71b]  Russo, R. L., and P. K. Wolff, Sr., "ALMS: automated logic mapping system," in *Proc. of 8th Annual Design Automation Workshop*, pp. 118-127, 1971.

[Rus72]   Russo, R. L., "On the tradeoff between logic performance and circuit-to-pin ratio for LSI," *IEEE Trans. on Computers*, vol. C-21, pp. 147-153, 1972.

[Sch72a]  Schuler, D. M., and E. G. Ulrich, "Clustering and linear placement," in *Proc. of 9th Annual Design Automation Workshop*, pp. 50-56, 1972.

[Sch72b]  Schweikert, D. G., and B. W. Kernighan, "A proper model for the partitioning of electrical circuits," in *Proc. of 9th Annual Design Automation Workshop*, pp. 57-62, 1972.

[Wol78]   Wolff, P. K., Sr., A. E. Ruehli, B. J. Agule, J. D. Lesser, and G. Goertzel, "Power/timing: optimization and layout techniques for LSI chips," *J. of Design Automation and Fault Tolerant Comp.*, vol. 2, pp. 145-164, 1978.

## Further Reading

[Aho74]   Aho, A. V., J. E. Hopcroft, and J. D. Ullman, *The Design and Analysis of Computer Algorithms.* Addison-Wesley, Reading, Massachusetts, 1974.

[Don74b]  Donath, W. E., "Equivalence of memory to 'random logic'," in *IBM J. of Research and Development*, vol. 18, pp. 401-407, 1974. See also "Stochastic model of the computer logic design process," IBM T. J. Watson Research Ctr., Yorktown Heights, New York, Report RC-3136, 1970.

[Law62]   Lawler, E. L., "Electrical assemblies with a minimum number of interconnections," *IEEE Trans. on Elect. Computers*, vol. EC-18, pp. 86-88, 1962.

[Lei80]    Leiserson, C. E., "Area efficient graph layouts," *21st Annual Symposium on Foundations of Computer Science*, IEEE, pp. 270-281, 1980.

[Tho79]    Thompson, C. D., "Area-time complexity for VLSI," in *Proc. 11th ACM Symposium on Theory of Comp.*, pp. 81-88, 1979.

[Val81]    Valiant, L. G., "Universality considerations in VLSI circuits," *IEEE Trans. on Computers*, vol. C-30, pp. 135-140, February 1981.

[Wil64]    Williams, J. W. J., "Alg. 232," *Communications of ACM*, vol. 7, pp. 347-348, 1964.

# Chapter 4

# Placement, Assignment and Floorplanning

**Bryan T. Preas**

*Xerox Palo Alto Research Center*
*Palo Alto, California*

**Patrick G. Karger**

*Tektronix/CAE Systems Division*
*Beaverton, Oregon*

## 4.1 Introduction

This chapter describes the positioning functions, collectively called *placement*, within cell-based layout systems. These functions include component placement, pin assignment, gate assignment, and floorplanning. Automatic placement is defined and the data abstractions, or *model*s, are discussed. Important placement algorithms, and applications of the algorithms within layout systems, are described. This is a broad topic and is difficult to cover in a single chapter. Therefore, this chapter should be viewed as an introduction; the reader is encouraged to read the references for more detail. The references were chosen to illustrate concepts and to serve as an introduction to the literature rather than constitute a catalog of the placement literature. Very few comparative studies have been published, and those that are available are now old. As a result, comparisons must be subjective, but we believe them to be accurate.

Physical design consists of transforming a circuit design specification into a physical representation that can be used to manufacture the specified electronic circuit. The speed with which this transformation takes place is greatly enhanced by the use of automatic layout techniques. Automatic layout is a subset of the physical design process that automatically maps a structural representation of the circuit, or the equivalent information, into a physical representation. This *physical representation* consists of geometric coordinates for all of the circuit elements and the wiring that interconnects the elements. The *structural representation* that is used as input to the layout process consists of the following items: a list of circuit elements, or *components*, that are to be included in the layout, and a list of signal sets indicating the terminals, or *pins* on the components, that are to be made electrically common by the layout process. The *interconnection nets*, or simply *nets*, are the connections among the signal sets. Breuer, Soukup and Ohtsuki provide good overviews of the physical design process [Bre72, Sou81, Oht86].

Automatic layout consists of two primary functions: determining the positions of components on a layout surface, called *placement*, and interconnecting the components with wiring, called *routing*. Although placement and routing are intimately related and interdependent, historically they have been separated because of computational complexity. (There is at least one attempt at performing automatic placement and routing in parallel [Loo79], but this approach does not have a large following. Iteration between placement and routing has also been tried [Ada82].)

Automatic placement, the focus of this chapter, determines the locations of the components within the circuit being designed, subject to the *constraints* imposed by the designer and the design rules imposed by the fabrication process and by physical principles. Design rules are discussed in Chapter 8.

Good placement is a key aspect of automatic layout, but it sometimes receives insufficient attention. A poor placement can leave the router with a difficult or impossible task; a good placement can make the router's job easy. Also, since placement directly determines the minimum length of the interconnection wiring for the circuit, and since wiring delay can be the dominant part of the response time of the electrical signals carried by the wiring, placement often determines the performance of the resulting circuit.

Typically, the design of a digital system and the associated placement subproblem is defined hierarchically [Nie83]. This approach greatly simplifies the placement problem since placement algorithms can operate on one hierarchical cell at a time [Pre78]. The placement of (sub) components within a single, higher level component is normally considered as a separate problem. Because of the confusion of components at different levels of the design hierarchy, the term *placeable objects* is used to refer to the components being placed.

The term *placement*, in normal usage, can have at least three different meanings depending on the context. One context is the outcome of the placement process. In this case placement refers to the positions of components in the circuit design, for example, "The *resulting placement* is better than the previous." Another context is that of finding positions and orientations of components on a layout surface; this context is termed *component placement*. The third context, called *comprehensive placement*, is that of finding locations for components as well as the assignments of nets to pins and logical gates to physical gate sites. This concept includes component placement, pin assignment, gate assignment, and floorplanning. In this chapter the context is usually sufficient to differentiate among these concepts. When there is possible confusion, the term *resulting placement, component placement*, or *comprehensive placement* is used.

This chapter first defines the placement problem and then categorizes and reviews the placement techniques that are available. The chapter concentrates on placement of components in the upper and intermediate levels of the design hierarchy. Specifically excluded from this discussion is the generation of the lowest level or *leaf cell*s of a hierarchical structure; the leaf cells are considered to be atomic objects. In typical standard cell or gate array designs, the leaf cells are small- to medium-scale-integration gates found in a cell library or generated using the techniques described in Chapters 6 and 7. In general cell designs, the placeable objects may be very large and complex. The focus of this chapter is concepts and algorithms for automatic placement; the issues associated with interactive placement are not addressed.

Section 4.2 describes the placement problem, while Section 4.3 describes abstractions of the components being placed, the interconnection nets, and the layout surface on which the components reside. As a result of a technology-independent presentation, the concepts discussed in this chapter are applicable to electronic circuit design for a wide range of circuit fabrication technologies and design styles. This range includes printed circuit boards (PCBs) and hybrid chip carriers and silicon integrated circuits (ICs) such as gate array, standard cell, and general cell designs. Section 4.8 tailors the basic algorithms, which are the focus of Sections 4.4 through 4.7, to the specific applications, design styles, and fabrication technologies.

Component placement methods fall into two groups: constructive and iterative. *Constructive placement* methods, described in Section 4.4, produce a *complete placement* (all components have assigned positions) based on a *partial placement* (some or all components do not have assigned positions) as input. The *iterative placement* methods, discussed in Section 4.5, improve a complete placement by modifying it to produce a better, complete placement. Comprehensive placement systems must also address *pin assignment* (the assignment of nets to pins) and *gate assignment* (the assignment of logical gates to functionally equivalent physical gates which share a common physical component). Pin and gate assignment methods are described in Section 4.6. The size and shape of the circuit elements are determined as part of the process of general IC design; floor-planning, described in Section 4.7, aids this process.

Because a large number of techniques for placement, assignment and floor-planning have been developed, this chapter imposes a taxonomy and explains the base algorithms.

## 4.2  Description of the Placement Problem

**Component Placement**

The *placement problem* consists of mapping the components in a structural description onto positions on a layout surface. A position consists of a location and an orientation (rotation and reflection). Pins on the components define locations where the circuitry within the component connects to interconnection routing among the circuitry. *Public pin*s designate the interface area between the circuitry inside and the environment outside of the component in question; they are the pins of the component at the next higher level of hierarchy. The subsets of pins, termed *signal set*s, which are to be connected by wiring to form electrically common interconnection nets, are part of the circuit's structural description.

The actual goal of placement is to determine positions for the components that permit completely automatic routing of the interconnection wiring in a small area. During placement it may be necessary to honor other (possibly conflicting) goals, such as minimizing layout area, minimizing cross talk among the signals carried by the wiring, equalizing heat dissipation across the layout surface, or maximizing circuit performance (speed, throughput, or noise margins). Since such goals are difficult to cast into objective functions that can be evaluated by a computer, more restricted objective functions must be substituted. When a placement operation derives a good placement as measured by a restricted objective, it is hoped that the placement is also good with respect to the actual goal. The restricted objective function, then, is defined on the interconnection nets, the components, and the layout surface and is used as a metric to compare alternative placements. Ideally, the restricted function should accurately reflect, or at least be correlated with, the actual goal and should be fast to compute. Conflicting goals have lead to a plethora of objective functions; some are discussed in the next section.

It is possible to define an objective function for specialized placement problems in such a way that an optimum solution can be determined. For example, methods are available for determining the optimum placement of transistors within a restricted domain [Ueh81, Nai85]. Chandrasekhar [Cha82] illustrates the complexity of finding the optimum placement for only two rectangular components. However, in general, placement problems are much more complicated; practical gate array and standard cell designs can contain a few hundred to many thousand components. This complexity results in huge placement state spaces. For example, a design with 1000 components (modest by today's standards) has approximately $10^{2562}$ different placements. To put this number in perspective, if a computer that could evaluate a trial placement in one microsecond had begun enumerating those placements at the beginning of the universe, today it would have evaluated $10^{24}$ states. To further complicate matters, it has been shown that the general placement problem is non-deterministic polynomial (NP)-complete. An introduction to NP-completeness may be found in Garey and Johnson [Gar79]. Chapter 10 of this book presents good discussions of the complexity of placement as well as other design automation problems. A more general discussion of combinatorial complexity is provided by Tarjan [Tar83]. Because the general placement problem is NP-complete and large numbers of components are involved, optimum solutions cannot be guaranteed. Hence, methods based on heuristics must be used for all but the smallest problems.

### Pin and Gate Assignment

In addition to determining positions for components, comprehensive automatic placement systems must address pin assignment and gate assignment issues. These problems are closely related to component placement and, conceptually, share many attributes.

Pin assignment determines an assignment of nets to functionally equivalent pins (groups of pins that have a functionally symmetric effect such as the input pins of a NAND gate) or equipotential pins (pins which are physically connected within the component and are therefore electrically common). Here, logical pins in the structural description can be considered the placeable objects and the components can be considered as restricted layout surfaces which contain positions for these functionally equivalent or equipotential physical pins. A good pin assignment improves routability and can have a significant impact on difficult problems.

A gate is an element of a cell. Some fabrication technologies (for example, the TTL 7400 family of SSI and MSI parts) make it economical to package multiple gates within a single cell. Gate assignment consists of mapping the logical gates in the structural description onto the *functionally equivalent gates* of the components. Gate assignment is analogous to component placement where the logic gates are the placeable objects and the gate sites on the components specify positions where the gates are to be placed.

### Floorplanning

Early in the IC design process, designers must make far reaching decisions based on incomplete or tentative information. The external interface of the cells that correspond to the constituent components (size and shape of components and the positions of the pins) must sometimes be determined before the cells have been designed. Floorplanning consists of finding positions for the constituent components and determining constraints on external interfaces for the corresponding cells that have flexibility in their interface. An objective function is necessary to measure the quality of the floorplan. In floorplanning, constraints are usually imposed on the component interfaces as well as on the attributes of the cell being planned. In some floorplanning applications, hierarchical partitioning (assigning functionality among the components) is considered part of the problem to be solved.

Floorplanning is related to placement, and many of the same techniques are used. However, the extra degree of freedom (the flexibility of the interfaces of the cells that comprise the design) significantly expands the size of the design space that must be exposed to component placement.

## 4.3  Abstract Models of Placement

In order to apply an algorithm to a real problem, the problem must be transformed into an abstract model. Abstraction replaces an object with a simplified model that defines the interaction of the object with its environment; details of internal organization are deleted. Abstraction reduces the amount of data for the problem, adds structure to the problem, and makes the problem easier to solve. After the algorithm is applied in the abstract domain, the abstract results must be transformed back into the real domain. This section describes the abstract models that are important to placement algorithms. The important aspects of electronic circuits that must be modeled are the circuit elements, the interconnections, and the carrier (*i.e.*, physical surface). The models of these aspects, cells and components, pins, interconnection nets, objective functions, and layout surfaces, as well as their relation to the physical circuit are discussed.

### Cells and Components

Automatic placement systems typically use an object/instance paradigm to describe the circuit structure. The definition of a circuit element (the object) is called a *cell* and describes how the circuit element is constructed from more primitive circuit elements and wiring. Cells are placeable objects which may be obtained from a library or constructed by the methods described in this book. An example cell, consisting of a D latch, is shown in Plate 4a and Figure 4.1. These show two views of the cell. Plate 4a shows the mask layers required to fabricate the latch in a standard cell design, while Figure 4.1 illustrates the properties of the cell that are important for automatic placement. Many of the details of the circuit element are abstracted away for placement functions; only those properties that are important to the electrical and physical interface with the rest of the circuit and the layout surface are retained. These properties include the physical size and shape of the object as well as the location in three dimensions (geometric coordinates and the physical interconnection layer) of the signal and power supply pins. The cell definition may also include (mandatory) constraints or (optional) hints on placement of the components that are associated with the cell. For example, bonding pads might be restricted to lie only on the periphery of an integrated circuit, and some gate arrays have input/output drivers that must reside in special positions close to the bonding pads. At higher levels of hierarchy, cells may be more complex, composite objects, but the aspects that must be modeled remain the same.

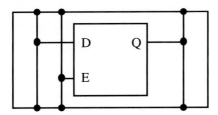

**Figure 4.1**    This cell definition of the D latch corresponds to the mask layout of Plate 4a. Only the information important to placement is retained by the cell abstraction.

Depending on the accuracy required by the layout system, other details of the cell may be included in its definition. If a placer is used in a layout system that can exploit the connections among equipotential pins within a cell, then those interconnections should be included in the cell definition. The locations of material within the cell that can influence routing outside (or over) the cell may be necessary. Electrical characteristics such as power dissipation can be important.

An important aspect of the cell-based design style is that cells exhibit local autonomy over their internal area and function. Thus, a means for controlling the interaction between the inside and outside of a cell is needed. Pins provide the appropriate abstraction. Pins define the area at which the internal circuitry of a cell connects to the external circuitry. Electrical or functional characteristics, such as electrical strength or drive of output signals or capacitance of input signals, may be necessary for the pin abstraction. If a placement algorithm can optimize the performance of the design, then setup, hold and delay parameters are required as well.

While cells carry the information pertaining to the definition of circuit elements, the components carry the information pertaining to their application; thus a component is an instance of a cell. The pins of the cell are bound to the pins specified in the structural representation. Since the components are cells that are placed, they must carry placement information such as their current positions. A component should also carry application-oriented placement constraints. For example, performance constraints may dictate that two components sharing signal sets must be placed near each other. Akers provides a discussion of positional constraints and methods to determine if these constraints can be satisfied [Ake81].

## Interconnection Nets

The circuit elements, as represented by their abstractions (cells and components), must be interconnected according to the signal set specifications. It is the responsibility of the placement function to place the components so that the interconnection wiring can be effectively routed. Thus, an important factor within a placer is the modeling of the interconnection.

It is important to model the topology of the interconnection nets since the wiring paths will be determined by a router and are unknown during placement. If we assume the pins of a signal set are the vertices of an undirected graph, then the connections among them form the edges of the graph called an interconnection tree. Automatic layout systems, the circuit elements, routers, and fabrication technologies combine to impose restrictions on the form of a given interconnection tree. Figure 4.2 shows examples of the interconnection forms. Topologies for interconnection nets with only two pins are simple; only one edge is meaningful. Topologies for nets with three or more pins are more complex. The most general topological form, called a *Steiner tree* [Cha72] (Figure 4.2a), permits vertices of the connection graph to be at pins as well as at locations other than the pins and places no restrictions on the *degree* (the number of incident edges) of the vertices. This Steiner tree model is typical of connections within integrated circuits. A slightly more restrictive form (Figure 4.3b) is the Steiner tree with a trunk which is constrained to have a single central trunk with branches to connect the vertices. This form is most appropriate for gate array or standard cell design styles. A more restrictive interconnection method is the *spanning tree*, shown in Figure 4.2c, in which the vertices are restricted to be at the pin locations [Pri57]. Other restrictions may apply. For example, wire-wrap fabrication imposes constraints on the degree of the vertices because the posts that implement the pins have a fixed height and thus can have a limited number (typically three) of connected wires. An even more restricted method of interconnection is shown in Figure 4.2d; the *chain* allows no branching of the interconnection wiring (*i.e.,* degree of vertices $\leq 2$). This topology corresponds to a traveling salesman path [Lin65]. Some placement algorithms model the interconnection tree as a complete graph (Figure 4.2e) for computational simplicity or mathematical necessity. For example, equations describing the interconnection cannot be formed unless the interconnection topology is constant. Placement systems may directly model the interconnection topologies introduced here, or they may use the approximations described in the next section.

## Objective Functions

The quality of a placement is based on many factors; some examples are completion rate after routing, electrical performance characteristics, and layout area. Metrics based on these factors can compare alternative placements. These metrics are used by placement algorithms to compute a *score* which measures

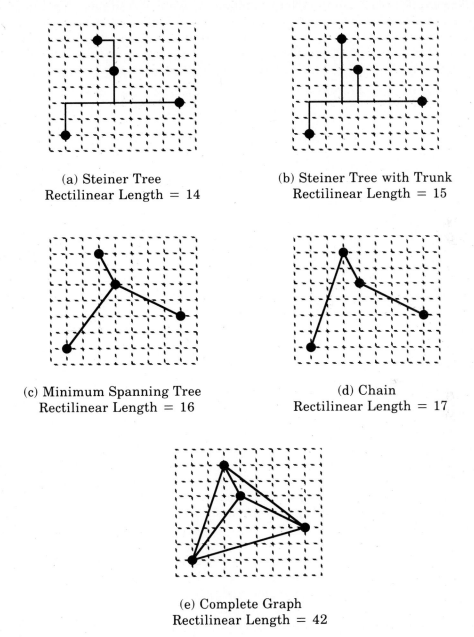

(a) Steiner Tree
Rectilinear Length = 14

(b) Steiner Tree with Trunk
Rectilinear Length = 15

(c) Minimum Spanning Tree
Rectilinear Length = 16

(d) Chain
Rectilinear Length = 17

(e) Complete Graph
Rectilinear Length = 42

**Figure 4.2** These interconnection topologies are important for placement.

the quality of a placement. A large number of metrics, ranging from simple to complex, are divided into two classes: those that assume that the net routings do not interact with other nets and those that account for this interaction, or *congestion*. Some congestion metrics also consider the interaction of nets with the components or predefined features on the layout surface.

**Net Metrics.** The simplest net metrics assume that the nets can be routed without interfering with each other or with the components. One widely used metric is *wire length*, which is the sum of the lengths of the interconnection trees over all nets. The length of an interconnection tree is the sum of the individual edges within the tree. This metric is fast to compute, and the algorithm is easy to implement. Wire length accounts for interaction among the nets only indirectly, since less wire means less interaction. An approximation to wire length of an individual net is the half-perimeter of the smallest rectangle enclosing the pins of the signal set [Sch76]. This approximation is the same as the rectilinear, minimal Steiner tree length for two- or three-pin nets. The approximation for four- and five-pin nets is within twice the width of the enclosing rectangle of the Steiner tree length [Han66]. A more accurate, but more time-consuming, approximation to the Steiner tree length is the minimal spanning tree length [Hwa76].

Another technique to measure placement quality models the connections as springs that exert forces on components [Fis67, Qui75]. This model leads to a force metric where a good placement is one that minimizes the sum of forces on the components. Related methods exploit the convexity of this metric to perform quadratic minimization [Bla84, Bla85b].

**Congestion Metrics.** The net metrics quantify only the amount of wiring; they do not measure where the wiring is located. This can lead to heavy wiring buildup or congestion as demonstrated by Figure 4.3. In this example, a smaller wire length leads to incomplete routing because wiring is in the "wrong" place. The second major class of interconnection metrics incorporates the interaction among the nets, the components, and the layout surface in the measure of placement quality. An intuitive measure of congestion is illustrated in Figure 4.4. This *rat's nest* drawing showing the minimum spanning trees of the signal sets provides a visual indication of placement quality; a good placement looks less congested. Congestion may be a global measure such as the number of nets that cross a *cutline*, as described in [Bre77, Lau79] or a local measure such as *track density* within a routing channel [Per76]. A technique to minimize the maximum wiring density by accounting for wiring using a two-dimensional array of cells is presented in [Jun81, Got86]. Area is discussed as a congestion metric in [Pre77]. An even more complex congestion measure is *routability* [Per76]. In this case, the placement metric measures cyclic routing constraints because extra routing area is required to resolve constraint cycles. (See Chapter 5 for a discussion of routing constraints.) Karger and Malek [Kar84] propose an approach that uses wire length as a metric and models routing congestion as constraints. Congestion metrics often correlate better with routability than do wire length metrics because the areas where routing resources are needed are included [Wan80, Pat84].

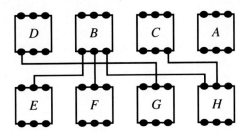

(a) Two Tracks Required. All Connections Routed.

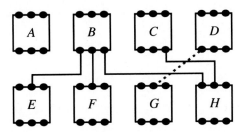

(b) Shorter Wire Length. Three Tracks Required.
A failure occurs if only two tracks are available.

**Figure 4.3**  Wire length reduction can cause routing failures. This illustrates that placement that minimizes wire length can require more area than a placement with longer wire length.

### Layout Surface

The characteristics of the physical surface (or *carrier*) on which the circuit elements are placed must be modeled. The abstraction of the carrier is called the *layout surface*. Carrier models can be divided into two categories: geometric and topological. Automatic layout systems based on the geometric model tend to accept the layout surface geometry as a fixed constraint and optimize the completion of the interconnection. Conversely, systems built on the topological model assume completion of all of the interconnections as a constraint and optimize the layout area.

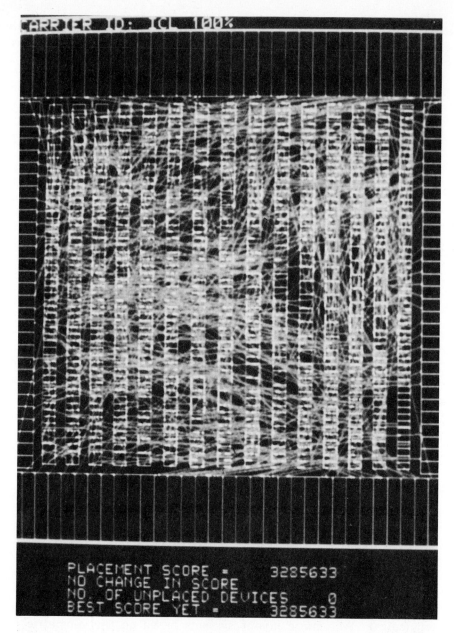

**Figure 4.4**    A "rat's nest" diagram shows the minimum spanning tree connections of all the nets of a gate array placement. Congested areas are apparent. (Photograph courtesy of Tektronix/CAE Systems Division, Beaverton, Oregon.)

**The Geometric Model.** Geometric models are appropriate for gate array and PCB design styles where placement aspects such as size, shape, and public pin positions do not change during the layout process. Some geometric models place components on a continuous plane and use geometrical interference checking; other models restrict components to fixed positions called *slot*s. This gives rise to a placement approach which assigns components to slots that carry the geometric coordinates. This approach is simple when all components uniformly fit in all slots. However, placement becomes much more complicated when components vary in size or can be assigned to only a subset of the slots. Some placers arbitrarily restrict otherwise general slots to hold certain types of components [Smi80]. Such arbitrary restrictions tend to reduce the quality of the resulting placement.

**The Topological Model.** Topological models are appropriate for more flexible design styles such as standard cell and general cell IC designs. With these design styles, the size, shape, and public pin positions of the cell being laid out, as well as the components' positions, are determined by the layout system. Furthermore, these positions are interdependent in complex ways and vary during the layout process. For example, routing areas can be made the exact size necessary to accommodate the interconnections. This flexibility suggests a *topological model* composed of directed and undirected graphs. Such a model provides an efficient representation of placement, is easy to modify as the placement changes, and allows rapid computation of the geometric functions of topology.

A placement composed of rectangular components can be explicitly represented as a *rectangular dissection*. This dissection can be represented as an undirected plane graph (called a *channel intersection graph* or *floorplan graph*) where the vertices represent the intersections of the dissection and the edges represent the adjacencies of the intersection. Figures 4.5a and 4.5b illustrate an example of a placement and floorplan graph, respectively. The necessary graph theory background for the topological model can be found in [Deo74].

A placement composed of rectangular components can also be represented by a pair of directed acyclic graphs (one for the horizontal direction and one for the vertical direction), as shown in Figures 4.5c and 4.5d. The vertices of the horizontal graph represent the vertical lines of the dissection, and the edges indicate if a line is to the right of or above another line. A similar description holds for the vertical graph. These graphs, called *channel position graphs* or *placement graphs*, allow the positions of the components to be easily computed.

Brooks *et al.* [Bro40] introduced these concepts and Ohtsuki *et al.* [Oht70] first applied them to IC layout. The topological model is widely used for placement and floorplanning [Zib74, Kan76, Pre78, Pre79a, Lau79, and Che83]. As a variation on the basic model, Otten introduced a restricted rectangular dissection, called a slicing structure [Ott80]. This restriction limits the flexibility of floorplanning and placement but simplifies the model. An extension of the basic,

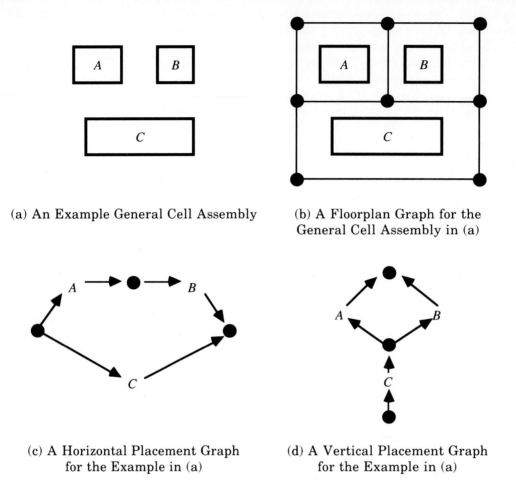

(a) An Example General Cell Assembly

(b) A Floorplan Graph for the
General Cell Assembly in (a)

(c) A Horizontal Placement Graph
for the Example in (a)

(d) A Vertical Placement Graph
for the Example in (a)

**Figure 4.5**  A Topological Layout Surface Model for a General Cell Assembly

rectangular model to represent the placement of a subset of rectilinear shaped components (arbitrary rectangles with arbitrary rectangles removed from 0 to 4 of the corners) is presented by Preas and Chow [Pre85, Cho85].

### Summary of Placement Models

This section describes the abstract models that placement algorithms must manipulate in order to solve a placement problem. These models provide the foundation for the component placement, assignment and floorplanning algorithms discussed in Sections 4.4 through 4.7.

## 4.4  Constructive Placement Algorithms

An overview of component placement techniques is provided in this and the following section. The techniques reviewed do not form an exhaustive list, but rather demonstrate the major approaches. Constructive placement techniques are the topic of this section, and iterative placement techniques are the topic of the following section.

Constructive placement algorithms share the characteristic that their input is a partial placement and their output is a complete placement. Although some constructive placement algorithms permit a *seed* placement as an initial condition, the ability to operate on unplaced components differentiates these algorithms from the iterative algorithms that are described in the next section. Constructive placement algorithms are used for initial placement and are normally followed by one or more (iterative) placement improvement algorithms.

Constructive placement techniques as a group are discussed in [Han76a, Han78]. In these references constructive placement algorithms, as represented by cluster growth (called *constructive initial placement*), are compared with various placement improvement algorithms. Richard [Ric84] and Palczewski [Pal84] also discuss constructive placement algorithms as a group, although both use a different taxonomy from that used in this book. [Pal84] also estimates computational complexity for constructive placement algorithms.

The constructive placement algorithms can be divided into the following classes: cluster growth, partitioning-based placement, global techniques, and branch-and-bound techniques. The basic algorithms have been combined and extended in an attempt to improve placement quality. These classes, as well as the combinations of the classes, are discussed in this section.

### Cluster Growth

Cluster growth constructive placement is a "bottom-up" method that consistently considers the most detailed level of abstraction. Cluster growth operates by selecting unplaced components and adding them to a partial, or incomplete, placement. Cluster growth methods are differentiated from other placement methods in that cluster growth selects and places the components independently.

The generic cluster growth algorithm is shown in Figure 4.6, and its operation is illustrated in Figure 4.7. In Figure 4.7, the dots represent the slots in which the rectangular components may be placed. The first step is to determine a seed placement. The components in the seed placement and their positions may be chosen by the user in order to guide the placement process or may be determined algorithmically [Sch76]. Next, unplaced components are sequentially selected and placed in relation to those components already placed. This process continues until all components are placed.

> $seedComponents :=$ Determine component(s) for seed
> $currentPlacement := PLACE\ [seedComponents, currentPlacement]$
> **while** all components are **not** placed **do**
>     $selectedComponent := SELECT\ [currentPlacement]$
>     $currentPlacement := PLACE\ [selectedComponent, currentPlacement]$
> **endloop** { **while** all components are **not** placed }

**Figure 4.6**  Generic Cluster Growth Algorithm

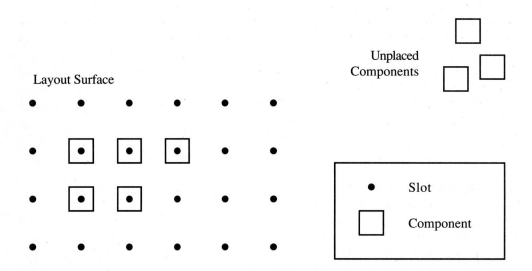

**Figure 4.7**    The cluster growth placement algorithm selects unplaced components and places them in slots.

The *SELECT* function of the cluster growth algorithms determines the order in which unplaced components are included in the placement. The order is determined by how "strongly" the unplaced components are connected to the placed components. The PLACE function determines the best positions for the selected component. The positions adjacent to previously placed components, the *candidate positions*, are investigated by calculating the score that results from placing the selected component at the candidate positions. The component is placed at the candidate position that results in the best score. By necessity, the score is based on incomplete information since unplaced components cannot contribute to the scoring.  It is often necessary to choose among candidate positions that produce the same score.  Such "ties" can be resolved by a secondary metric such as finding the candidate position closest to the "center of gravity" of the component directly connected to the candidate component.  Hanan [Han72a] provides an extensive discussion of selection and placement functions for cluster growth.

The definitions of the selection and placement functions differentiate the various cluster growth techniques.  Two such techniques, cluster development and random placement, are described next.

**Cluster Development.**  The *SELECT* function of the cluster development algorithm [Kur65, Han72a] chooses the unplaced component that is most strongly connected to *all* of the placed components.  The PLACE function chooses a place for the selected component with respect to the other components that are connected to the selected component.  Schweikert describes a closely related cluster development method [Sch76].  This method uses an "inside-outside" selection function, and the position function uses an exhaustive search of the available positions that account for placement restrictions.  Another variation of cluster development selects the unplaced component that is the most strongly connected to any single placed component.  This component is then positioned as closely to its strongly connected mate as possible [Han72a].  Descriptions of other cluster development algorithms appear in [Sch72a, Cox80, Kan83]; these algorithms share the common feature that they position the components in a line and then "fold" the line to form a two-dimensional layout.  A knowledge-based approach to cluster development is reported in [Oda85, Oda87].

**Random Placement.**  Random constructive placement is a degenerate form of cluster growth.  In this method the selection function reduces to randomly choosing an unplaced component and placing it in a randomly chosen position.  A closely allied method is arbitrary constructive placement, in which unplaced components are selected arbitrarily (for example, in sequential order) and are placed in arbitrary (for example, the next available) positions.  While this method is fast and easy to program, it utilizes no circuit-specific information.  Consequently, poor results may be expected [Mag77, Han76a, Han76b].  This method deserves mention only because it is sometimes used to generate complete placements that will be improved by iterative placement.

**Cluster Growth Algorithms.**  Certainly, the complexity of the cluster growth algorithms is dependent on the number of interconnections and the number of pins per interconnection; however, the dominant factor that contributes to complexity is the number of components to be placed.  [Kur65] shows that the cluster development and pair linking algorithms have a computational complexity of $n^2$ where $n$ is the number of components.  Their performance is discussed in [Han72a, Pal84].  These algorithms are easy to implement but produce poor results because placement decisions must be made with incomplete information. Modern systems favor the partitioning-based or global methods for initial placement. These methods are described in the following two sections.

## Partitioning-Based Placement

Placement algorithms based on partitioning divide components into two or more partitions while reserving space (called a *block*) for the components. This process of dividing blocks into smaller blocks and partitioning components into the reduced blocks continues until the number of components per block is small. These algorithms are widely used in modern layout systems. This "top-down" design approach considers higher levels of abstraction before it considers more detailed levels and tends to avoid heavy wiring congestion typically found in the center of the layout surface. These algorithms differ from the cluster growth algorithms in the following way: partitioning-based algorithms consider all interconnections in parallel and then move the components in steps by partitioning the components into specific areas of the layout surface. While determining optimal partitioning is NP-complete, several good heuristics have been developed and are summarized below. (See Chapter 3 for more detail.)

**Partitioning Foundations.** Kernighan and Lin [Ker70] developed a two-way partitioning scheme based on iterative improvement of an initial, possibly random, partitioning. The procedure, based on pairwise exchange, judiciously selects the pairs of components to exchange and allows multiple exchanges to occur before deciding to accept the sequence of exchanges. While the score for the sequence must decrease in order to be accepted, individual exchanges are allowed to increase the score. This allows clusters of tightly coupled vertices to be moved from one partition to the other. Kernighan and Lin also discuss multiple-way partitioning based on reduction to a series of two-way partitionings.

While this heuristic works well for general graph partitioning, it does not take into account a special property of electrical circuits: a group of vertices connected by a single net do not have to be pairwise interconnected but need only be connected by a spanning tree. Schweikert and Kernighan [Sch72b] recognized that any number of pins in one partition can be connected to any number of pins in the other partition by a single connection. Thus, there are only two partitioning possibilities to consider:

- All of the pins in a net are in the same partition, and therefore no nets are cut.
- There are one or more pins in each partition, and the net is cut once.

The modified algorithm keeps track of the decrease in the number of nets cut if a component is moved from one partition to the other. The improvement is judged over a sequence of exchanges so that groups of components which are highly interconnected can move from one partition to the other.

Fiduccia and Mattheyses [Fid82] developed a further modification to the technique. The modification allows only a single component at a time to be moved from one partition to the other. The component is chosen based on its effect on both the net-cut score and the balance of the size of the partitions. This modification results in a linear-time heuristic for a single pass through the components. When Dunlop and Kernighan [Dun85] compared the Kernighan-Lin and Fiduccia-Mattheyses algorithms, they found that the results of Fiduccia-Mattheyses are not quite as good as those of Kernighan-Lin but that the execution time is substantially reduced.

Krishnamurthy and Mellema [Kri83] recognized that it is difficult to judge the quality of a partitioning algorithm without knowing the best partitioning. They developed algorithms which generate partitioning problems with provably optimal partitions for two- and $n$-way partitioning.

**Partitioning-Based Placement Algorithms.** Using the various partitioning algorithms, many min-cut placement techniques have been developed. *Min-cut* techniques address the placement problem by dividing components into two or more partitions so that the number of nets that cross the partition boundaries is minimized. The main differences among techniques concern how finely the blocks are partitioned (*i.e.*, how many components are in a block) before the components are placed, how and where cutlines are generated, and how connections to external components (not in the block currently being partitioned) are handled.

The bipartitioning algorithms, illustrated in Figure 4.8, divide the components into two sets such that the weighted number of connections between the sets is minimized, and the total component area in each set is approximately equal. This process is repeated until each block contains only one component. Breuer [Bre77] describes the concepts of this approach. His technique partitions the components into blocks on the layout surface; each block has an exact physical location. When each block contains only one component, the circuit is completely placed. He defines two algorithms based on this premise: quadrature and slice/bisection (see Figure 4.9). Quadrature forms two blocks by dividing the carrier in half vertically. Each of these blocks is then divided horizontally, forming four blocks. This division continues until each block contains only one component. Slice/bisection uses a horizontal cutline to partition the components into blocks of $k$ and $n-k$ components where $k$ components are placed in the bottom row of components. The block of $n-k$ components is then partitioned in the same manner creating blocks of $k$ and $n-2k$ components. This process continues until the last block contains $k$ components or less. The carrier is then bisected in the vertical direction to form columns. Corrigan [Cor79] describes a bipartitioning placement system with more sophisticated partitioning sequences and gives experimental results from several different sequences.

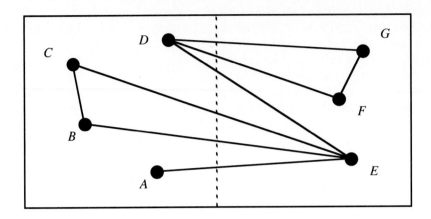

(a)  Original Placement

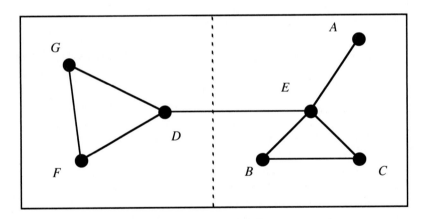

(b)  Improved Placement

**Figure 4.8**    Min-cut placement places components on either side of a cut in such a manner as to minimize the number of interconnections crossing the cut.

Stabler *et al.* [Sta79] developed a partitioning technique that assumes an underlying row structure. The components are partitioned into row sets with an equal number of components in each row set. A branch-and-bound algorithm is used to assign the components to positions in each row set. The row sets are then assigned to physical rows while attempting to minimize the wire length

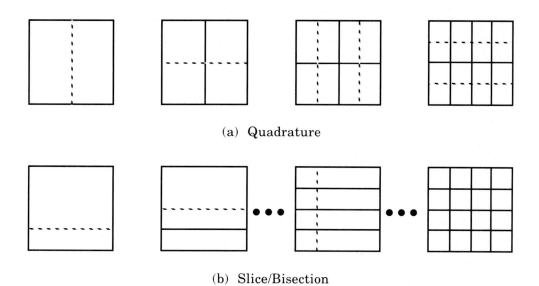

(a) Quadrature

(b) Slice/Bisection

**Figure 4.9** Two cutline strategies are shown. Quadrature alternates the cut direction for each block, while Slice/Bisection changes the cut direction after all cuts have been completed for one direction.

between rows. [Koz83, Kam82] discuss other partitioning-based placement approaches which assign components to blocks, blocks to row sets, row sets to rows, and components to positions within the rows.

Lauther [Lau79] developed a partitioning-based placement technique where the blocks generated during the partitioning process are represented by a polar graph which retains information about the relative positions of the blocks. The partitioning process terminates when each block contains one component.

It is not sufficient to partition components across a cutline in isolation; the signals that enter the blocks must be considered. As a group, min-cut placement algorithms share the problem that components must be partitioned without knowing, in detail, the locations of components external to the block currently being considered. Consider the situation shown in Figure 4.10, where components $A$, $B$, $C$, and $D$ have been partitioned into blocks $L$ and $R$ as shown. Now components $A$ and $B$ must be partitioned into $L_T$ and $L_B$ without knowing the location of components $C$ and $D$ except that they are in block $R$. *Terminal propagation* [Dun85] and *in-place partitioning* [LaP86] predict the positions of external components and reflect those positions into the blocks under consideration.

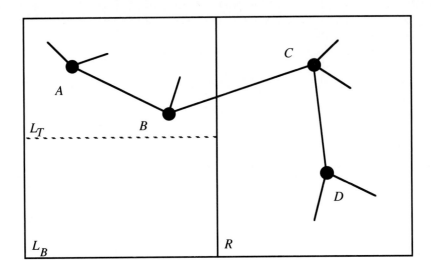

**Figure 4.10**   When partitioning components $A$ and $B$ between blocks $L_T$ and $L_B$ the positions of components $C$ and $D$ may not be known. "Terminal propagation" or "in-place partitioning" must be used to approximate their positions.

**Summary of Partitioning-Based Placement.** The partitioning-based algorithms make use of the interconnection information at a global level and defer local considerations until late in the placement process. These procedures produce good results, but they are computationally expensive [Mur80]. These methods tend to spread the wiring across the layout surface and thus produce very routable placement. However, they have difficulty when components are constrained to fixed positions and when partitions of grossly unequal areas are produced.

**Global Placement**

All placement techniques keep global notions of better or worse placements, but global placement techniques are distinguished by how they move components. Global methods move all of the components simultaneously along an *n*-dimensional gradient. This type of movement contrasts to clustering methods that consider each component (or at most a small number of components) sequentially and to partitioning methods that first divide the components by partitioning and then work with the components individually. It seems that superior results would be obtained by considering all of the interconnections simultaneously and moving all of the components in parallel to find a global solution. However, there is still insufficient data to draw this conclusion. Both global methods described in this book, quadratic assignment and convex function optimization, use a quadratic objective function. Since the quadratic function heavily penalizes long nets, connection lengths tend to cluster tightly around the mean net length. This clustering produces a small standard deviation of net lengths compared to linear metrics, but the effect on routability is yet to be determined. Both global methods also use the complete graph to model the interconnection net topology as described in Section 4.3. Thus, if all nets have the same weight, nets with more than two pins have a disproportionately large impact on the placement [Sch72b]. This disproportionate impact may be compensated for by weighting nets according to the number of pins, $p$, according to $1/(p-1)$, $2/p$, or $(2/p)^{3/2}$ [Cha68, Che84b, Fra86].

**Quadratic Assignment.** The quadratic assignment problem is usually formulated as follows: given a cost matrix $C = [c_{ij}]$ (where $c_{ij}$ is the sum of the weights of the connections between two components $i$ and $j$) and a distance matrix $D = [d_{kl}]$ (where $d_{kl}$ is the distance between positions $k$ and $l$), minimize

$$\sum_{i,j} c_{ij} d_{P_i P_j}$$

over all permutations, $P$, of the components' positions. Hanan and Kurtzberg [Han72a] describe the quadratic assignment method and how a placement problem can be approximated by an associated quadratic assignment problem. However, an optimum solution to the associated quadratic assignment problem does not guarantee an optimum solution to the original placement problem, since only interactions between pairs of components may be specified and since components are modeled as "points." (The linear assignment problem may also be applied in placement algorithms. Akers [Ake81] provides a good description of the linear assignment problem and its application to placement.) A more detailed description of assignment is available in Section 4.6. Although quadratic assignment uses a quadratic metric, it does not exploit the convexity of the metric. Techniques which do exploit convexity are described in the next section.

**Convex Function Optimization.** It is possible to formulate the placement problem as an objective to be minimized and to use mathematical optimization techniques to solve the problem. (A good discussion of mathematical optimization techniques is given in [Wal75].) Optimization techniques work best when both the domain and the objective function are convex; in these cases only a single minimum of the objective function exists. Through appropriate abstractions that produce a convex domain (primarily the absence of slots and the modeling of components as if they had zero size) and the use of a quadratic metric, it is possible to transform a placement problem into one that may be solved using gradient techniques. One of the earliest applications of this technique to the placement problem appears in [Hal70]. Although these methods suffer when transforming the abstract problem formulation back into the physical domain (actual, finite-size components must be mapped into slots on the layout surface),

good results have been reported by Blanks [Bla85b], Frankle and Karp [Fra86], and others.

Quinn uses a force analogy to derive a system of equations in quadratic form [Qui75]. A complete graph with weighted connections is used to model the interconnection nets. Adding repulsive forces serves to incorporate the realism of the carrier into the layout surface, but this destroys the convexity of the resulting metric. Another example of this approach is described in [Joh83]. Cheng and Kuh [Che84b] model the components and their interconnections as a resistive electrical circuit and then use relaxation to solve the resulting equations. This approach requires some fixed-position components; otherwise, all of the components collapse to a single point. A justification for quadratic metrics is provided by Fukunaga [Fuk83], who uses a state space approach to derive a global equation matrix. In this model the eigenvectors corresponding to the two smallest eigenvalues determine the placement. Assignment by relaxation [Hun80] is used to map the global placement solution onto the carrier. A dynamic programming approach to assignment of components to gate array slots has also been proposed [Hil86].

In general these quadratic optimization techniques suffer from an inability to handle practical constraints such as finite component size and constraints imposed on component positions. These drawbacks are addressed in several ways. Blanks [Bla85a, Bla85b] uses quadratic constraints and a two-step procedure to map the ideal, global placement onto the layout surface without violating any physical constraints. Sha and Dutton [Sha85] attempt to alleviate the need for the second step by encoding all geometric information into constraints placed on the scoring function. Frankle and Karp [Fra86] use a multidimensional geometric approach which couples eigenvector probes and linear assignment, providing a more systematic mapping of the ideal placement onto the carrier. Since few comparisons have been performed, it is difficult to evaluate the relative merits of these global methods versus other placement methods.

**Branch-and-Bound Placement**

The branch-and-bound method can be used to find an optimum solution to small placement problems [Han72a, Gil62]. Both Lawler and Wood [Law66], and Hillier and Lieberman [Hil80] provide introductions to general branch-and-bound techniques. The general method is illustrated in Figure 4.11. This algorithm assumes that a feasible placement, *knownPlacement*, and its resulting score are known. Normally this feasible placement is obtained from a constructive placement algorithm. By way of initialization, the *remainingSolutionSet* gets the entire set of solutions under consideration, and the incumbent solution is the *knownPlacement*. Next, a loop is entered which continues until the remaining set of solutions, *remainingSolutionSet*, is empty, indicating that there are no remaining, unfathomed sets.

```
bestScore := ScoreOf [knownPlacement]
remainingSolutionSet := universe
incumbentSolutionSet := knownPlacement
while remainingSolutionSet is not empty do
  activeSet := ChooseASetFrom [remainingSolutionSet]
  remainingSolutionSet := remainingSolutionSet − activeSet
  activeSubSets := PartitionIntoSubsets [activeSet] { branch step }
  for each element in activeSubSets do
    lowerBound := LowerBoundOf [element]
    if lowerBound < bestScore then begin { bound step }
      if element is fathomed then begin
        bestScore := lowerBound
        incumbentSolution := element
      end
      else
        remainingSolutionSet := remainingSolutionSet + element
    end { if lowerBound < bestScore }
  endloop { for each element in activeSubSets }
endloop { while remainingSolutionSet is not empty }
```

**Figure 4.11** A branch-and-bound algorithm is able to guarantee optimum results, but runtime is excessive for reasonably sized problems.

Within the loop, the following operations are performed. One of the member sets of *remainingSolutionSet*, the *activeSet*, is chosen, removed from the solution set, and partitioned into *activeSubSets*. This step is called the branch step. The *activeSet* chosen for partitioning is normally the one with the lowest bound for the objective function for all placements represented by the set. For each of the elements that results from the partitioning, the following steps are performed. If the *lowerBound* of the placements represented by the *element* is less than the currently known *bestScore*, then this *element* is considered further. If this *element* represents a complete placement (in other words, it is fathomed), then this *element* represents an improvement on the currently known best placement. Otherwise, this *element* must be returned to the *remainingSolutionSet*.

The branch-and-bound process prunes the decision tree; this potentially results in reduced computation. The more accurately the lower bound is calculated, the sharper the pruning. However, there is a trade-off. It takes longer to calculate a more accurate lower bound. Because of the need to either calculate accurate lower bounds or explore a large number of branches, this technique is quite time-consuming. Many heuristics have been suggested to prune the tree [Han72b] and thus reduce the number of branches explored. Most of these heuristics have a computational complexity of $n^3$ or $n^4$. Although branch-and-bound algorithms do not appear in modern layout systems (practical problem sizes have grown so large that runtimes are excessive even with the heuristic bounds computation), the technique is still interesting for theoretical studies where optimum solutions are required for small problems.

### Combinations of Constructive Algorithms

While each of the constructive placement algorithms discussed in this section possesses desirable characteristics, none provides the "best" solution. As a result, enhancements or combinations of approaches have been proposed. Wipfler *et al.* [Wip82] reports a combined force and min-cut approach in which the min-cut function is used to account for actual component geometries that the basic force method is unable to handle. Richard [Ric84] uses a combination of partitioning and cluster growth, while Akers reports the use of linear assignment combined with cluster growth [Ake81]. Kozawa *et al.* [Koz83] have developed a system based on a combination of clustering and folding as well as using several different metrics.

## 4.5  Iterative Placement Algorithms

The goal of iterative placement is to transform a complete placement into an improved, complete placement. This process is iterated until some stopping criterion is met. For example, the stopping criterion might be relative or absolute improvement in the placement metric, or perhaps the time expended in the iterative process. Within one iteration, components are selected and moved to alternate locations. If the resulting configuration is better than the previous one, the new configuration is retained. Otherwise, the previous configuration is restored. The improvement process and the most notable iterative improvement algorithms are discussed in this section.

### Three Phases of Iterative Placement

Many different iterative placement techniques exist. Even though they differ substantially, they all share the same underlying structure and have three main phases: selection, movement and scoring. The generic form of iterative improvement placement algorithms is shown in Figure 4.12. These three phases are discussed next.

**Selection.** The *SELECT* function chooses the components to participate in movement. This mechanism reduces the set of all possible combinations of components to move simultaneously to a computationally feasible subset. The selection process may simply select components to be interchanged in a predefined sequence (such as trying all possible pairs) or may involve intelligence to select those components which are placed poorly. Incorporating intelligence into the selection process typically allows placement to converge more quickly but may not improve the quality of the solution. An overly restrictive selection function may miss productive moves and cause the process to stop too quickly or degrade the quality of the solution.

```
        currentScore := SCORE [currentPlacement]
    until stopping criterion is satisfied do
        selectedComponents := SELECT [currentPlacement]
        trialPlacement := MOVE [selectedComponents, currentPlacement]
        trialScore := SCORE [trialPlacement]
        if trialScore < currentScore then
            currentScore := trialScore
            currentPlacement := trialPlacement
        else
            currentPlacement := MOVE [selectedComponents, trialPlacement]
    endloop { until stopping criterion is satisfied }
```

**Figure 4.12**   The generic iterative improvement placement algorithm is defined
in terms of *SELECT*, *MOVE*, and *SCORE* functions.

One possible selection technique is to sequence through all possible $s$-tuples of component combinations. It is questionable whether using values of $s > 2$ achieves significant improvement over simply trying all pairs ($s = 2$). Gilmore [Gil62] and Reiter and Sherma [Rei65] report improvement by using $s > 2$, while [Gar68] reports little improvement at the expense of a significant increase in computation time. (Current wisdom sides with [Gar68], but no recent studies have been published.) For placement algorithms which incorporate computationally intensive scoring metrics, it is impractical to score all of the combinations which are generated by an $s$-tuple selection scheme. It is possible to extend the selection scheme by including a predictor that is computationally simpler than the scoring metric but has a sufficient correlation with the scoring metric to filter a large number of the combinations which are "obviously" going to fail. Such predictor mechanisms can substantially reduce the number of combinations which must be scored but may overlook many potentially successful combinations.

Random pair selection chooses a pair of components at random from all of the available components. At first this selection technique may seem to have little merit; however, one class of placement algorithms, simulated annealing (discussed later in this section), requires random selection to explore the placement space. Another selection technique uses force vectors to determine which components should be moved [Fis67, Han76a]. The interconnections are modeled as springs that transmit forces to the components that are connected to the nets. The components selected are the ones exhibiting strong resultant forces, indicating they are far from their equilibrium positions.

**Movement.**   Once the components are selected for trial interchange, the *MOVE* function determines the new locations for the selected components. If a pair of components is selected, then the components' positions are interchanged. This is called pairwise exchange (see "Iterative Improvement Techniques" below for more information). Of course, it must be physically possible for each component to fit at the location of the other.

Multiple-way exchange involves moving more than two components [Cot80]. In this case each component has multiple positions to which it can be relocated. One movement scheme orders the components from worst placed to best placed and then chooses new positions for the components in this order. Multiple-way exchange allows combinations to be found that pairwise exchange might not find. For example, in Figure 4.13, if pairwise exchange is attempted among components $F$, $G$ and $K$, none of the possible exchanges leads to improvement. However, if $F$, $G$ and $K$ are interchanged as a 3-tuple, a better configuration can be found. Even though examples exist, experience suggests multiple-way exchange leads to only slightly better results at a large increase in computation time. Even if multiple-way exchanges are not efficient, they are important because they allow us to define the concept of *local optima*. A placement is said to be *l-optimum* if it is as good as any other placement that can be obtained by exchanging not more than $l$ components.

The force methods from placement modeling also lead to movement heuristics since the position where the resultant force on a component is minimum can be computed. Heuristics can then be both two-way and multiple-way.

Some iterative placers are limited to moving components among positions on the layout surface which are slots for certain types of components [Smi80]. These positions are based on the initial positions of the components being considered. This restriction eliminates the need to dynamically check for component overlap and results in much faster placement techniques at the sacrifice of placement quality.

**Scoring.** After the selected components have been moved to new positions, the objective function is invoked to measure the quality of the new arrangement. At this stage, the representation of interconnections should be as accurate as possible and be consistent with the time available for iterative improvement. This requirement means that a connection graph should correspond to the routes that will be generated. The scoring metric may be the same as that used by the initial placement or may incorporate more detail concerning the layout surface. The scoring metric for iterative placement should work in concert with any metric used to construct the initial placement; otherwise, any initial placement quality is wasted.

### Iterative Improvement Techniques

Now that the three functions within the iterative placement loop have been described, the important iterative algorithms are reviewed in terms of these functions.

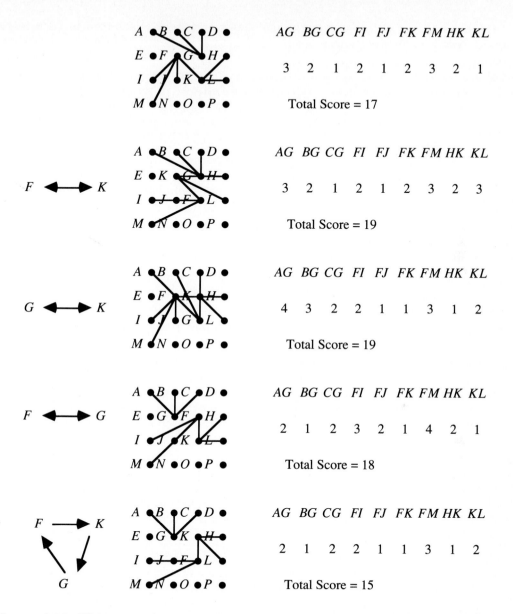

**Figure 4.13**  This example shows a placement reached by a 3-tuple move which pairwise moves cannot reach. The original placement and four modified placements (three 2-way moves and one 3-way move) are shown in the five rows. The placements are shown in the left column, and the rectilinear wire lengths are shown in the right column.

**Interchange.** In *pairwise interchange* each component is selected in turn to be the *primary* component and is exchanged on a trial basis with every other component. If a trial exchange results in an improved placement, it is accepted. Otherwise, the components are returned to their previous positions. Any objective function can be used as the basis for acceptance of an exchange. This technique results in $n(n-1)/2$ trial exchanges, making the computational complexity $O(n^2)$, where $n$ is the number of components. Placement systems using pairwise interchange are discussed in [Han76a, Sch76, Ios83].

*Neighborhood interchange* is similar to pairwise interchange; however, the primary component is interchanged only with components in its vicinity. The vicinity (distance or the number of components included) is a parameter which can be set by the user. If the number of components included in the vicinity is $m$ and there are $n$ components to be placed, the runtime becomes proportional to $(nm)/2$. Placement systems using neighborhood interchange are described in [Han76a, Kho77, Ios83].

A more complicated form of interchange is reported by Cohoon and Paris [Coh86]. Cohoon and Paris use a genetic paradigm to derive a selection mechanism based on random mutation in biological systems.

**Force-Directed Methods.** *Force-directed interchange* uses a force analog to select components to move as well as to determine the location where the components should be moved. However, the new location for the selected component may already be occupied by another component. Furthermore, the location of the selected component may not be a good position for the component currently occupying the selected component's favored positions. In simple force-directed interchange, this problem is avoided by trial interchanging the selected component with the three adjacent components in the direction of the "desired" location. That is, the primary component is interchanged with its nearest horizontal, vertical, and diagonal neighbors in the direction of the desired location. If the desired location is colinear with the current location, then only one interchange takes place; the selected component is interchanged with the component immediately adjacent in the desired direction. Placement techniques using force-directed interchange are described in [Han76a, Ios83]. Experimental results using force-directed interchange are given in [Han76a].

*Force-directed relaxation* is similar to force-directed interchange in the calculation of force vectors for each component. In this technique, however, the primary component is moved to the compatible slot nearest to the desired zero-force point [Han76a]. The component that was occupying that slot is chosen as the next primary component. This results in a series of components to be relaxed. The series terminates when the primary component is moved to an empty slot (that is, a slot which did not have an occupant component). When a series terminates, the new score of the placement is calculated and compared to the previous score. If the placement improved, the series is accepted and all components remain in their current positions. Otherwise, the series is rejected and all components are returned to their previous positions. The process terminates when all components have either initiated or are contained in an accepted series. The primary components used to initiate series are selected in descending order based on the number of signal sets to which each component belongs. Fisk [Fis67] describes a similar technique but without slot definitions. Goto [Got81] describes a generalized, force-directed relaxation technique in which each primary component results in a tree of components to be relaxed instead of a single series. The linear series described above is a special case of Goto's method.

The *force-directed pairwise relaxation* method also uses force vectors to find the zero-force target locations for each component. In this method, however, the primary component is not allowed to initiate a series as in force-directed relaxation. Instead, the primary component $A$ is trial exchanged with a component $B$ in the vicinity of the target location only if the target location of $B$ is in the vicinity of component $A$. The exchange is accepted only if the placement score improves. The components are chosen to be primary components in descending order based on the number of signal sets to which they belong. Once a component has participated in an accepted interchange, it can no longer be a primary component. The placement process ceases when all components have been chosen for relaxation [Han76a, Ios83].

All force-directed techniques share the characteristic that the force is a restricted selection method. Iteration converges quickly, but the process tends to stop even when many productive exchanges still exist.

**Unconnected Sets.** Steinberg describes a technique [Ste61] that selects components by dividing them into sets which have no nets in common; these are called *unconnected set*s of components. Each component in such a set can be trial placed without considering the other components of the set since they are unconnected. Each component in the set can be moved to each of the locations occupied by other components in the set and the score computed. The placement of these components can also be formed by solving the resultant linear assignment problem since the score does not depend on the interaction of the components' positions. After each unconnected set has been processed in this manner, the cycle is complete. The procedure can either be terminated or repeated in an iterative process. The size of the unconnected sets, the number of unconnected sets generated, and the number of cycles performed are parameters. An incremental processing technique (producing a new assignment problem given the results of the previous one) was applied to the unconnected set algorithm in [Car79]. Experience with the unconnected sets technique [Cia75, Han76a, Han76b, Ake81] indicates limited success because components in highly interconnected groupings (a common occurrence in modern circuits) are not moved with respect to each other and nets with large numbers of pins tend to limit component movement.

**Simulated Annealing.** The previous techniques share the characteristic that trials are accepted only if the score does not increase. This limitation usually causes the placement to reach a local minimum but miss the global optimum. Kirkpatrick *et al.* [Kir83] describe an optimization technique, *simulated annealing*, that addresses this problem. This technique uses an algorithm [Met53] developed to find the lowest-energy configuration of a collection of confined molecules.

Simulated annealing is based on the following premise: the optimization of a circuit placement with a very large number of components is analogous to the process of annealing, in which a material is melted and cooled slowly so that it will crystallize into a highly ordered state. The energy within the material corresponds to the placement score. An *annealing schedule* specifies a beginning temperature, a temperature decrement function, an equilibrium condition at each temperature, and a convergence (or frozen) condition. The simulated annealing method begins with a random initial placement. An altered placement is generated, and the resulting change in score, $\Delta s$, is calculated. If $\Delta s < 0$ (the system went to a lower energy level), then the move is accepted. If $\Delta s \geq 0$, then the move is accepted with probability $e^{-\Delta s/t}$. As the simulated temperature $t$ decreases, the probability of accepting an increased score decreases.

The objective function used for scoring may be any of the functions described in Section 4.3; net half-perimeter is often used because it is fast to compute. Simulated annealing has been used extensively as an iterative improvement technique. Figure 4.14 shows the generic iterative placement algorithm rearranged to illustrate the simulated annealing concepts.

Simulated annealing placement is robust over a wide range of design styles and problem types. It usually produces very good results at the expense of *very long* runtime. However, the algorithm does not incorporate constraints or hints well. A seed placement cannot be used, because of the necessity of the random initial placement, and fewer good placements are achieved when component placement constraints are incorporated.

Theoretical studies [Rom85] have shown that simulated annealing can climb out of local minima to find a global optimum if the proper conditions on the annealing schedule are satisfied. In practical problems this search for the global optimum implies an infinite number of iterations at each temperature. Since performing an infinite number of iterations is clearly impossible, heuristics have been developed to reduce runtime. These heuristics range from simple [Kir83, Rom84] to quite sophisticated. The speed-up approaches fall into three categories: move set design, cost function manipulation, and annealing schedule improvements.

Sechen describes a move set design based on range limiting. This discards moves involving components which are more than some specified distance apart; the distance decreases as temperature decreases [Sec84]. This reference also describes cost function manipulations that involve altering the probability of

```
t := t₀
currentPlacement := randomInitialPlacement
currentScore := SCORE [currentPlacement]
until freezing point is reached do
    until equilibrium at current t is reached do
        selectedComponents := SELECT [atRandom]
        trialPlacement := MOVE [selectedComponents, atRandom]
        trialScore := SCORE [trialPlacement]
        Δs := trialScore − currentScore
        if Δs < 0 then currentScore := trialScore
        else begin { Δs ≥ 0 }
            r := uniformly random number { 0 ≤ r ≤ 1 }
            if r < e^(−Δs/t) then currentScore := trialScore
            else
                currentPlacement := MOVE [selectedComponents, trialPlacement]
        end
    endloop { until equilibrium at current t is reached }
    t := αt { 0 < α < 1 }
endloop { until freezing point is reached }
```

**Figure 4.14**  The simulated annealing placement algorithm uses random selection, movement, and acceptance to explore the placement state space.

acceptance or approximating the objective function at high temperatures. Greene and Supowit [Gre84] devised a *rejectionless method* in which the probability of selecting a move is based on its probability of being accepted. This technique requires keeping information about all possible moves, which is both space- and time-consuming. They found that an acceptance rate crossover point exists which is problem dependent. If the acceptance rate is above the crossover point, then standard simulated annealing is faster. Once the acceptance rate falls below the crossover point, the rejectionless method is faster. These results suggest a two-step method where standard simulated annealing is used above the crossover point and the rejectionless method is used below.

For standard cell problems Grover [Gro86] describes a method to bound the error in objective function computation. Execution time is reduced by allowing a larger error at higher temperatures.

Annealing schedule improvements are the third approach to improving simulated annealing performance. White proposes the use of an initial temperature that is greater than the standard deviation of the cost distribution that occurs when all moves are accepted [Whi84]. A widely used temperature decrement is a geometric progression ($t := \alpha t$) where typical $\alpha$ is 0.95. To improve performance, Huang [Hua86] derives the temperature decrement $t := te^{-0.7t}$

based on the condition required to maintain quasi-equilibrium. The equilibrium condition at each temperature may be a fixed Markov chain length [Aar85], a minimum acceptance [Kir83], or a dynamic Markov chain length [Hua86]. A typical stopping criterion is that the average score is unchanged for a few consecutive temperatures.

Even with these speed-up techniques, simulated annealing remains computationally intensive compared to other techniques, but it is robust over a wide range of problem types. Unfortunately, few comparative results exist between simulated annealing and other methods. Experiments reported in [Nah85] compared simulated annealing to other heuristics. The conclusion was that several other heuristics performed as well as or better than simulated annealing for a fixed amount of computation time. It would be more meaningful to allow simulated annealing heuristics to run to completion and give the final results as well as the amount of computation time required to achieve those results. This exhaustive comparison has not yet been reported.

### Summary of Component Placement Techniques

While finding the optimum solution to the general placement problem is NP-complete, the placement methods discussed provide heuristic techniques which attempt to provide a "good" placement. These techniques are divided into two categories: constructive placement and iterative placement. The computational complexity of these heuristics normally ranges from $n^2$ to $n^4$. Many other heuristics are possible with varying complexities, but the methods discussed above represent the most popular placement techniques.

## 4.6  Pin Assignment and Gate Assignment

Techniques for pin assignment and gate assignment are described in this section. Pin assignment performs the following two functions: it assigns nets among the functionally equivalent pins of a component and it assigns portions of nets, called *subnet*s, to equipotential pins. Pin assignment has no effect on the function of the overall circuit because input signals to these pins have a symmetrical effect on the output of the component. Correspondingly, gate assignment maps logical gates in the structural representation onto functionally equivalent physical gates within a component. Of course, pin assignment and gate assignment alter the structural representation; therefore, some form of *back annotation* must be performed in order to update the source representation. Pin assignment and gate assignment typically have small effects on wire length metrics, but can have large effects on congestion and completion rates, especially for difficult placement problems. As an introduction to the main topics of pin assignment and gate assignment, this section first discusses the assignment problem from a mathematical optimization perspective. Then specific pin and gate assignment techniques are discussed.

## The Assignment Problem

The assignment problem from optimization theory provides an introduction to pin assignment and gate assignment. This section discusses the differences between the various formulations of mathematical assignment compared to pin assignment and gate assignment. Mathematical assignment problems are usually formulated in terms of assignees (for example, resources or workers) to assignments (for example, jobs). Assignment involves a one-to-one mapping of assignees to assignments [Hil80].

Linear assignment occurs when a cost $C_{ij}$ can be associated with an assignee $i$ performing an assignment $j$; the objective is to determine the assignments that minimize total costs. Linear assignment is well understood and can be readily solved by linear programming methods [Mun57, Hil80, Hun80]. Restricted forms of placement can be formulated as linear assignment problems. For example, if the components of a set have no nets in common, the cost of placing a member of the set is independent of locations of the other members of the set [Ste61], as described in Section 4.5. Akers [Ake81] describes other linear assignment applications in the placement domain. The quadratic assignment problem is a generalization of linear assignment [Han72a]. In addition to the cost of assigning an assignee to an assignment, there is a cost of interactions among pairs of assignees. Pin assignment and gate assignment problems are a further generalization of quadratic assignment because, in general, pin assignment and gate assignment must handle the interactions of more than pairs of assignees.

## Pin Assignment Techniques

Placements can often be improved by exploiting either symmetries or internal connections that exist within a cell. In some cases, signals which are applied to the input pins of a circuit element have a symmetrical effect on the signals produced at the output pins. Such pins are called *functionally equivalent pins*; an example using an AND gate is shown in Figure 4.15. The address pins of a random access memory (RAM) cell illustrate a more complex example of functionally equivalent pins. Often a placement can be improved by modifying the assignment of nets among functionally equivalent pins. In other cases an electrical connection may exist within a cell; such existing wiring can be exploited to reduce the amount of interconnection routing required by the circuit. Pins which have common internal connections are said to form an equipotential group and are called *equipotential pins*. An example is shown in Figure 4.15.

The process of optimizing the assignment of nets within a functionally equivalent pin group or subnets within an equipotential pin group is called pin assignment. Intelligent assignment of nets or subnets to functionally equivalent and equipotential pins can significantly increase the completion rate of an automatic router. The objectives of pin assignment are to reduce crossovers or reduce congestion. The importance of pin assignment in improving routability is

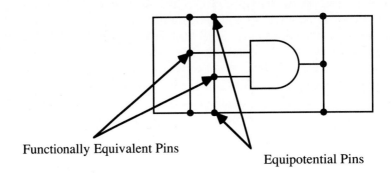

Functionally Equivalent Pins

Equipotential Pins

**Figure 4.15**   Pin assignment nomenclature is defined. Signals on the functionally equivalent pins have a symmetrical effect on the output signals. The equipotential pins are connected within the cell.

shown in Figures 4.16a and 4.16b; such changes in pin assignment can have a significant impact in difficult problems. Despite its importance, pin assignment has received much less attention than component placement. The three major pin assignment techniques which have been published are discussed in the following sections.

**Concentric Circle Mapping.** Koren [Kor72] developed a pin assignment technique which attempts to "planarize" the interconnections; it models pin assignment problems through two concentric circles. Points on the inner circle represent pins on the component currently being considered, while points on the outer circle represent pins on other components to which connections must be made. The problem is divided into two parts: the assignment of pins to points on the two circles and the mapping between points on the inner and outer circles.

Consider the component and pins shown in Figure 4.17a. The inner circle is drawn just inside the component pins, while the outside circle is drawn just outside these pins (Figure 4.17b). Lines radiate from the component center to the component pins and external connections (Figure 4.17c). The points on the inner and outer circles are defined by the intersection of the lines with their respective circles.

To complete the pin assignment, the points on the outer circle are mapped to points on the inner circle, in a cyclic fashion. The best and worst cyclic assignments are shown in Figures 4.18b and 4.18c. Concentric circle mapping then transforms the problem into a set of vectors. The angle of the vector indicates the angle of approach of the incoming connection net to the component being considered, while the magnitude of the vector represents the weight or

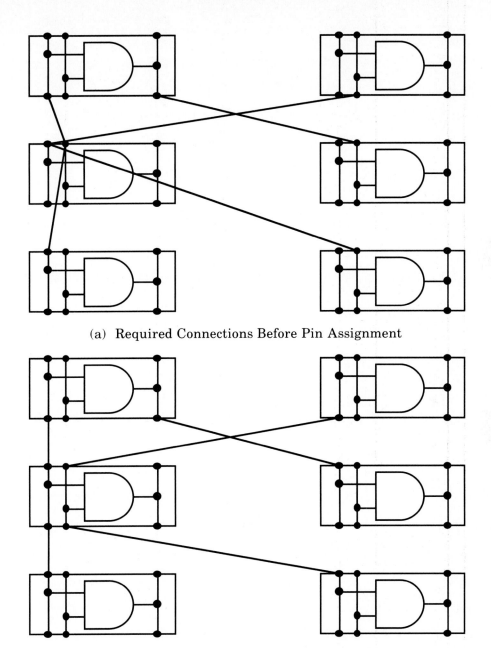

(a) Required Connections Before Pin Assignment

(b) Required Connections After Pin Assignment

**Figure 4.16** This example illustrates the effectiveness of pin assignment.

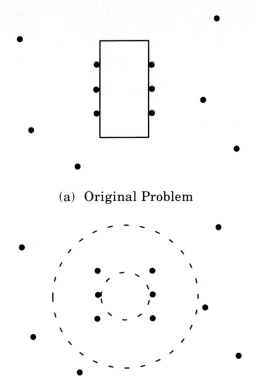

(a)  Original Problem

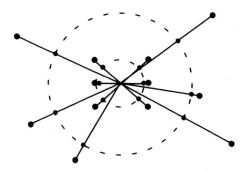

(b)  Insertion of Inner and Outer Concentric Circles

(c)  Radiating Lines Which Map Points onto Circles

**Figure 4.17**   This example illustrates the concentric circle, pin assignment method.  Pins are mapped onto circles.

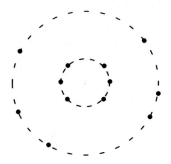

(a)  Points on Circle Mapping Given in Figure 4.17c

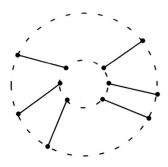

(b)  Best Possible Cyclic Assignment

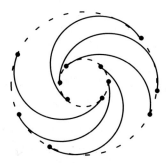

(c)  Worst Possible Cyclic Assignment

**Figure 4.18**  The best and worst possible cyclic assignments are shown for concentric circle mapping.

importance of that connection based on some criterion such as distance. Vector analysis finds the rotational pin assignment.

**Topological Pin Assignment.** Brady [Bra84] developed a technique that is similar to concentric circle mapping. However, multiple pin nets and interference from other components and barriers are more straightforward. If a net being assigned has more than two pins, the pin closest to the center of the primary (*i.e.*, selected) component is chosen. Thus, only a single external pin is considered for each net. The pins of the primary component are mapped onto a circle as shown in Figure 4.19. The pins are assigned nets beginning at the bottom of the circle and sweeping clockwise as indicated by the sweep arm.

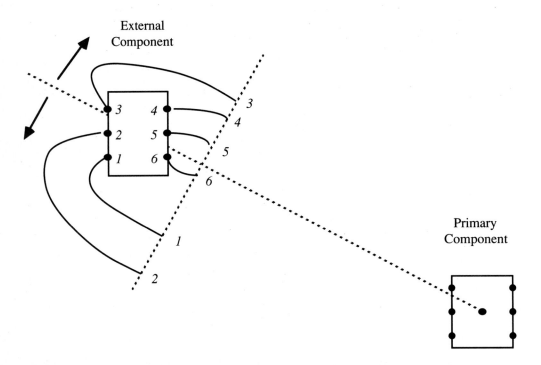

**Figure 4.19**  A sweeping line determines the ordering of multiple external pins on a single component for topological pin assignment.

The nets are assigned based on the order in which the external pins are encountered in the sweep. This results in the same assignment as concentric circle mapping, except when multiple external pins exist on the same component or when barriers and other components are in the path between the primary component and an external pin.

**Nine-Zone Method.** Mory-Rauch [Mor78] developed a pin assignment technique based on zones in a Cartesian coordinate system. The coordinate system is centered inside a group of interchangeable pins on a component called a *pin class.* Each net connected to the pin class defines a net rectangle which is positioned in one of nine zones. (See Figure 4.20.) The positions of these net rectangles are defined relative to the coordinate system defined by the current pin class. Figure 4.20 shows how the net rectangle zones are numbered.

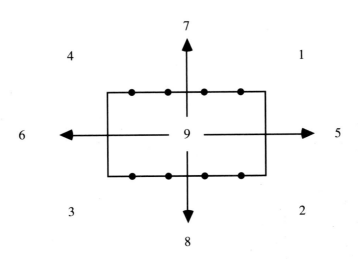

**Figure 4.20**  Nine Possible Positions of Net Rectangle Relative to Current Pin Class

Using specified rules, the nets are ordered for assignment and mapped onto pins, based on distance and relative orientation of the component and net rectangles. The zone of the specified net rectangle determines the topology of the connection in much the same way as the position of the pin on a circle or line determined the topology in the two previous techniques. The goal of this technique is to reduce wire crossings.

**Summary of Pin Assignment Techniques.** Pin assignment is the assignment of nets to pins from functionally equivalent or equipotential pin classes. Judicious assignment of these pins can improve routing by reducing wire length, crossovers, and congestion. Three pin assignment techniques are discussed in this section. The nine-zone method determines the pin assignment topology by dividing the coordinate system into nine possible zones. The concentric circle mapping technique determines the topology based on the relative positions of the pins on inner and outer circles. Topological pin assignment determines the topology based on the relative positions of the pins on a circle and a line.

## Gate Assignment Techniques

A design will typically contain equivalent (interchangeable) logic gates. Some fabrication technologies make it economical to package multiple, equivalent physical gates within a single cell. The prototypical occurrence is in PCB design where, for example, a TTL 7400 cell contains four 2-input NAND gates. In this case each one of these cells possesses four equivalent gates as shown in Figure 4.21.

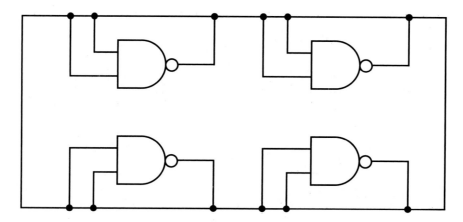

**Figure 4.21**  This cell contains four functionally equivalent NAND gates. The four gates may be assigned to improve the placement.

Gate assignment determines the assignment of logic gates in the structural representation to the physical gates on the cells. Obviously, the best assignment is determined by the interconnection of gates in the structural representation. Even though the logically equivalent circuits are not constrained to be gates, the problem originated in PCB design where the circuits were gates. Hence, the problem is referred to as gate assignment. Here this terminology is retained.

Gate assignment interacts strongly with component placement. It is difficult to determine a good gate assignment until the components have been placed. However, the gates must be assigned to components before placement can be performed. This problem is usually addressed by performing gate assignment in two phases: an initial assignment of gates to components before placement is performed, and an iterative improvement phase after component placement is complete.

A standard approach to initial gate assignment is outlined by Soukup [Sou81]. Assign all logical gates to components that contain only one physical gate. If any component containing more than one physical gate has an available gate, use it for the logical gate that has the most connections in common with gates already assigned to this component. If there is no partially filled component, take the next unused component and assign the first logic gate randomly. This approach is driven by connectivity and hence ignores unused gates. In PCB designs, it is better to have the unused gates distributed around the layout surface for spares. This problem can be addressed by pre-assigning unused gates using any desired distribution.

Once component placement is completed, the initial gate assignment may be improved using a number of techniques. If the equivalent gates are viewed as components and the various equivalent gate definitions are viewed as slots, gate assignment improvement becomes placement improvement and any of the placement improvement techniques can be applied. If the gate assignment has changed, it may be possible to further improve the placement. Some systems [Lue80] iterate between placement improvement and gate assignment improvement. This iteration can be continued until some stopping criteria, such as no more improvement, is achieved. It is also possible for placement and gate assignment improvement to be performed concurrently; every time a trial interchange of components is performed, gate assignment improvement is attempted. However, the authors know of no implementation of such a technique. Nishioka *et al.* [Nis78] describe a technique of gate assignment, component placement, and gate reassignment. They also give experimental results on the effects of placement improvement and gate reassignment.

## 4.7 Floorplanning

During the early stages in the design of electronic systems, decisions are made which have a dramatic effect on the quality (performance, density or area) of the resulting design. Choices must be made in partitioning functions into physical cells and in choosing interface characteristics such as size, shape and pin positions. These choices are difficult because they must be made with relatively little information. The effects are hard to predict, and may not become apparent until much later in the design process. Floorplanning helps solve these problems.

Floorplanning is closely related to general cell placement, which is discussed in the next section. Both problems concern the placement of cells of arbitrary size and (usually rectangular) shape such that the total area occupied by the cells and their interconnections is minimized. Thus, many techniques of general cell placement have been adapted to floorplanning. However, floorplanning has an extra degree of freedom: at least some of the cells' interface characteristics are flexible and must be determined as part of the design process. The flexibility in the interface of the cells is constrained by the function and layout of the cells and must be modeled by the floorplanner.

**Models of Floorplanning**

Because of the similarity to the placement of general cells, many floorplanning techniques use the models described in Section 4.3. The most important exception is in the modeling of the cells. Floorplanning algorithms must model the cells' interface flexibility and any constraints on that flexibility. Three classes of cells are used in floorplanning [Hor81]:

- Some cells are already laid out and are stored in a library. These fixed cells comprise the class used by placement algorithms; all of their interface characteristics are known and fixed.

- The designs of some cells are known, but their layouts are flexible and can be influenced by the results of floorplanning. For example, standard or general cell layout methods can produce a wide range of shapes for a given design. PLAs and memory cells can be distorted through folding or layout design. Several versions of a cell with different characteristics may be stored in a library.

- Cells of a third class are flexible because their designs (and perhaps even the design methods) are not known or are uncertain. In this case it is difficult for designers or algorithms to specify even nominal interface characteristics or constraints thereon.

An important aspect of cell modeling is estimation of area and shape. Four approaches have been reported: experimental, analytical, procedural, and knowledge-based. Experimental approaches develop area estimates through empirical formulas [Ued85, Kur86] and are usually tuned to a design style or even to a particular system. Analytical approaches concentrate on wirability analysis and routing area analysis as described in Chapter 2 and in [Hel77]. Procedural cell models have the ability to sense their context (for example, the neighboring cells and their relation to the subject cell) and optimize their interface characteristics for the current position [Pre87]. A shape model based on rectangular tiles is proposed in [Slu80]. Knowledge-based approaches can operate on more uncertain cell designs [How86].

A more precise method of area estimation is available if a slicing structure is used and constraints on shapes of all of the constituent components of a cell are known. Otten describes a method of computing the bounds on shapes that a cell can assume based on the bounds on shapes of its constituent components [Ott83, Ott86].

**Approaches to Floorplanning**

Many approaches have been proposed for solving floorplanning problems. For this discussion, the methods are grouped as constructive, iterative, and knowledge-based. Constructive methods include cluster growth, connectivity clustering, graph duals, and partitioning/slicing. Iterative methods include interchange and relaxation. Another review of floorplanning with a different taxonomy is available in [Woo86].

**Constructive Floorplanning.** Constructive floorplanning methods share the characteristic that they algorithmically construct a complete placement. Depending on the algorithm, a seed placement may be required as an initial condition.

*Cluster Growth.* Cluster growth floorplanning methods closely follow the cluster growth placement algorithms described in Section 4.4. These methods are inherently sequential and operate in a bottom-up fashion. The placement methods described in [Pre79a, Pre79b] have been adapted to floorplanning by using procedural methods to estimate or define cell shapes. A recursive, "look-ahead" technique is described that improves layout quality at the cost of increased runtime. Another technique grows the floorplan by starting in the lower left corner and adding components on the right and top [Hor81]. This restriction on the growth of the cluster permits a much simplified model for the layout surface but probably degrades layout quality.

*Connectivity Clustering.* Dai and Kuh [Dai86] describe a method based on circuit connectivity. Clusters are mapped onto floorplan templates that have simplified topologies. Complex circuits are planned by recursively applying the same algorithm. A novel feature of this work is that it maintains global routing information as part of the floorplanning process.

*Dual Graph Method.* This method begins with a *block diagram* that represents the top-level components and the busses that connect the components [Hel82, Mal82]. An original graph represents blocks as nodes and the busses as edges. If the primary graph is *planar* (at least one arrangement of nodes exists where the edges do not overlap), then it has a dual where a mesh corresponds to a node on the original graph or a node on the dual corresponds to a mesh on the original. A rectangular dualization is of particular interest because the components (dual meshes) will have a rectangular layout [Koz84]. Nonplanar block diagrams can be approximated by planar ones at the expense of area in the final layout. The solution to a system of layout equations determines cell areas.

*Partitioning and Slicing.* Lauther [Lau79] first applied the min-cut approach to placement of general cells as described in Section 4.4. This approach has been applied to floorplanning by La Potin and Director [LaP86]. They use in-place partitioning to improve the partitioning globally. Cell dimensions are determined in conjunction with routing areas.

**Iterative Floorplanning.** In contrast to the constructive floorplanning methods discussed in the previous section, iterative techniques operate on complete floorplans to improve the layout quality. These iterative methods fall into the categories of interchange and relaxation.

*Interchange Floorplan Improvement.* The iterative component placement algorithms described in Section 4.5 have been adapted to floorplanning. These algorithms have been generalized to handle flexible cell shapes and to define new floorplanning states. Typically, cells are optimized at the time of interchange. In some cases the channel position and channel intersection graphs are manipulated directly [Pre79a, Pre79b] to perform pairwise interchange. In other cases the regularity of slicing structures is exploited to define new floorplanning states; a Polish notation description of the slicing structure may be manipulated to produce new states [Won86]. Simulated annealing [Ott84, Ott86, Won86] and force-directed methods [Ued85] have also been applied as iterative techniques.

*Floorplanning by Relaxation.* Relaxation is differentiated from interchange techniques in that relaxation implies an obvious or preferred next state. Dimensional relaxation has been used to improve a floorplan [Ber85]. This relaxation method consists of modifying the channel graphs of the layout surface as well as the cell shapes.

**Knowledge-Based Floorplanning.** Floorplanning is an extremely complex task; this has led some researchers to apply knowledge-based approaches to the problem. (See Chapter 9 and [Dic86, Wat86] for more information.)

### Discussion of Floorplanning

There are three major thrusts in floorplanning. One thrust accepts that floorplanning is so difficult that a circuit designer must be an integral part of the process; therefore, interactive graphics must be included in addition to algorithmic approaches [Hor81, Woo86, Mac87]. The second major thrust uses floorplanning in the initial stages of design synthesis to develop constraints that can be passed to succeeding stages in the design process [LaP86]. The third major thrust relies on the existence of powerful module generators that can generate modules to specifications. This thrust allows the floorplanner to be integrated with the generator and allows truly automatic layout [Ott86].

It is extremely difficult to compare floorplanning approaches because results are highly dependent on the circuit design, the design styles, and the design methodology. Furthermore, since floorplanning occurs early in the design process, the subsequent layout steps can partially mask the quality of floorplanning. As a result, comparisons of approaches are non-existent.

## 4.8  Applications of Placement Algorithms

So far, discussion has been limited to general models and technology-independent algorithms for component placement, assignment and floorplanning. However, the ways in which the algorithms are applied can vary significantly depending on the problem and the design environment. Some of these characteristics are explored in this section: large number of components, variation in component shapes, and design styles. Specialized techniques must be used if circuit performance is to be optimized.

## Large Number of Components

Most placement algorithms operate on a flat representation of a hierarchical description; this means the placement algorithms do not make use of information at higher or lower levels of the design hierarchy. However, since placement algorithms may be $O(n^2)$ or higher, it may be impossible to consider all placeable objects in a large circuit. At this point the following choices are possible: partitioning, hierarchical placement, hardware accelerators, or multiple processors.

**Partitioning.** In this approach, flat placement problems are divided into separate but interdependent subproblems that are solved sequentially. The subproblems are not independent, since the decisions made in one subproblem affect the others; decisions which improve the placement within one partition might degrade the overall placement. However, dividing the problem reduces the complexity of the problem. For an algorithm which has complexity of $O(n^\alpha)$, $(\alpha > 1)$ a problem which is divided into $p$ parts has a complexity of $O(p(n/p)^\alpha)$ or $O(n^\alpha/p^{\alpha-1})$. Khokhani and Patel [Kho81] describe a method that partitions a circuit into "super nodes" and uses constructive initial placement and iterative placement to improve the assignment of the super nodes to "super locations." Next, the super nodes are decomposed into primitives and placed. Another partitioning approach "slices" a layout recursively to perform the partitioning [Sze80].

**Hierarchy.** In hierarchical placement, problem complexity is reduced by dividing the circuit into multiple levels. The top level divides the circuit into first-level components, which are themselves partitioned into lower-level components. This division continues recursively until (for purposes of placement) primitive placement components are defined. The hierarchical division may be the same as that defined in the structural description, or the layout system may provide automatic hierarchical partitioning [Pay82].

**Hardware Accelerators and Multiple Processors.** It is possible to reduce the execution times through the use of special purpose hardware optimized for placement tasks and/or multiple processors that allow parallel processing. This section concentrates on the problems of mapping placement algorithms onto specialized hardware.

Although an in-depth discussion of the architectures of hardware accelerators is not attempted here, a review is provided. General discussions of special purpose hardware for design automation are given by Adshead [Ads82] and Hancock and DasGupta [Han86]. Blank discusses hardware accelerators for general, computer aided design [Blk84]. A discussion of a placement hardware accelerator implementation with comparative results is given in [Spi85]. An implementation for another placement accelerator with applications to force-directed interchange, force-directed pairwise relaxation, and neighborhood interchange is described in [Ios83]. Other placement accelerators involve extensive concurrency on an array of processors [Chy83]. In these cases, care must be taken to avoid the oscillation which may occur when related problems are being optimized independently.

A great deal of work has been done recently concerning the use of multiprocessors. Hancock and DasGupta [Han86] discuss multiprocessors applied to design automation problems. They also discuss the hypercube multiprocessor configuration [Sei85].

Casotto *et al.* [Cas86] give a good explanation of a parallel algorithm for placement of general cells implemented on an eight-processor Sequent computer using shared memory. Each processor is assigned ownership of some of the components; the processors interact with blocking to avoid collisions. Their experimental results show a factor of six speed-up over a single processor machine.

Kravitz and Rutenbar [Kra86, Rut86] give an excellent taxonomy of techniques for parallel algorithms. They present three different implementations of parallel algorithms for simulated annealing on one to four processors. They present the concepts of object and function decomposition. Object decomposition is an assignment of entities (such as components or nets) to processors, while function decomposition is an assignment of operations (such as wire length calculation) to processors. They also present the idea of static versus dynamic decomposition; this describes whether the assignment of objects or functions to processors should be done once or changes over time. Since the probability of accepting a simulated annealing move changes dynamically with the annealing schedule, they observe that it might be better to use different techniques in different temperature regimes. Experimental results show a speed-up of 2.5x on four processors by using static function decomposition in the hot regime where the probability of acceptance is high and dynamic object decomposition in the cold regime where the probability of an acceptance is low.

Banerjee and Jones [Ban86] discuss implementation of a parallel algorithm for simulated annealing applied to standard cell placement on a six-dimensional (64 processor) hypercube. In their algorithm each processor is assigned 1/64 of the component area and has all of the information pertaining to these components in its own memory. Their scheme allows exchange or displacement (movement of a component to another processor without a reciprocal movement) of components between nearest-neighbor processors using four of the hypercube dimensions. The two remaining dimensions are used for exchanges or displacements across larger distances. When a component move is accepted, the new location is broadcast to all other processors over a Hamiltonian circuit. A later version of their paper [Jon87] gives several improvements, including a more sophisticated broadcast mechanism allowing broadcast to $p$ processors in $\log(p)$ time. Banerjee and Jones do not give experimental data but claim an expected speed-up of between 6x and 20x.

Rose *et al.* [Ros86] also suggest a scheme which uses different techniques in the hot and cold temperature regimes of simulated annealing placement. In the hot regime they use *heuristic spanning* in which each processor forms initial placements independently. While many initial placement techniques could be used, they chose a min-cut-based scheme in which the processors are assigned different seed cells. This scheme essentially speeds up simulated annealing in the hot regimes by replacing simulated annealing with min-cut placement. The best resulting placement is selected for low temperature annealing. The temperature at which to begin low temperature annealing is determined experimentally. In the low temperature regime they use *section annealing* in which each processor is assigned a set of components; interchanges are allowed only between components assigned to that processor. When a move is accepted, the information is broadcast to other processors. In the cold temperature regime, in which true simulated annealing is used, they report a 4x speed-up on five processors.

Devadas and Newton [Dev86] present a parallel simulated annealing algorithm for topological compaction of Weinberger arrays and gate matrices by exchanging the "fundamental units." Their techniques involve two basic concepts. The first is dynamic windowing, in which each processor is assigned a physical window of the layout surface and "owns" moves or "portions of moves" involving that window. The windows move across the layout surface over time so that the moves available to each processor change. The second concept, dynamic partitioning, assigns a set of components to each processor; the processor is responsible for moves involving its associated components. The components assigned to each processor change over time so that different sets of components interact. Devadas and Newton implemented their technique on an eight-processor Sequent computer and give performance results using one to eight processors. Their results indicate that their technique converged on the same solution as on a single processor with a 6x speed-up.

### Variations in Component Shapes

The basic placement algorithms discussed in Sections 4.4 through 4.7 do not consider the shapes of the components or the interaction between components' sizes and wiring requirements. Classical placement algorithms assume that placement quality is a function only of the interconnections and is not affected by the shapes of components. Such assumptions model the components as "point objects." However, algorithms that take component shapes into consideration provide more accurate placement models.

Standard cell and gate array layout systems arrange the components in rows (columns can be transformed to rows by rotation); in effect, the components have only one important dimension. The cells are designed to have a uniform height; widths vary with the complexity of the cell. In these cases, an accurate placement metric must include the width of the rows, as well as the heights of

the routing channels. Simple wire length no longer provides a sufficient placement metric [Per76]. A more accurate placement metric is chip area [Pre77] for
standard cell designs or completion rate for gate array designs.

General cell layout systems must model the components with two dimensions since both their height and width can affect the layout [Lau79, Pre79a,
Pre79b]. In this case, layout area is an accurate metric, but it is very difficult to
compute because of the complex interactions among the components and the
routing areas.

## Layout Design Styles

The demands placed on and the design of placement algorithms vary greatly
depending on the design style that is addressed. The styles considered here are
gate array, standard cell, general cell, and PCB.

**Gate Arrays.** Since gate arrays have fixed, legal gate sites, the placement systems must have knowledge of these sites as well as location restrictions for various cells. For instance, I/O pad sites should be located around the periphery and
I/O drivers near the periphery. Adherence to these restrictions can be accomplished either by having the placement algorithm explicitly take the constraints
into account or by having a second step which maps the components from the
ideal positions generated by the placement algorithms to legal slots based on the
restrictions. The two-step method can degrade the placement since a legal
placement may differ significantly from the ideal placement. Furthermore, the
routing area is fixed before the layout process begins, so it is important to use
the resources wisely. Thus, it may be necessary to model congestion in the
placement objective function. An area is overly congested when the connections
assigned to that area require more routing resources than exist. In gate array
design it is sometimes possible to find an alternate placement with more wire
length but less congestion. Also, care must be taken to avoid fragmenting the
layout surface. Some algorithms do not "look ahead" far enough to reserve
enough contiguous space for large devices.

Newer gate array styles, called *sea-of-gates*, do not have pre-defined routing areas; transistor sites cover the entire carrier. There are two types of sea-of-
gates arrays. In the first type, the cell sites are "porous" so that inter-cell connections can pass through and inter-mix with the intra-cell routing. For this
type, routing resources are precious, and it is extremely important for the placer
to prevent local congestion. In the second type the cell sites are designed tightly
so that no space for inter-cell wiring is left within the cell. Components are
placed only on a fraction of the sites, leaving dynamically created routing areas.
Except for the fact that cells must "snap" to fixed cell sites and the layout size is
fixed, the placement problem resembles that of standard cells discussed in the
next section.

**Standard Cells.** In the standard cell design style all layers are fabricated for each design; there is no concept of legal gate sites or slots as is the case for gate arrays. The cells may have varying heights and widths and are retrieved from a library or generated using the module generation techniques described in Chapter 7. Power is usually distributed through the cells and only minimally intrudes on the routing area.

Positions of components and routing areas are not fixed. The layout size is defined dynamically by the layout program based on the requirements of the specific circuit. While the final topology is arranged with rows of standard cells and routing channels between the rows, the number, lengths, and positions of the component rows and the routing channels can be adjusted dynamically. In order for a net to traverse a cell row, it is necessary to use a feedthrough. Feedthroughs may be either a part of an existing cell or may be a separate cell. It is desirable to minimize the number of feedthrough cells which are inserted in component rows since each feedthrough cell will lengthen the row.

The extra freedom allowed in standard cell design makes the placer's job more complex since now the goals have been extended to include minimizing chip area. For instance, it is necessary to trade off the number and size of the channels and the number of rows. Row lengths must be kept uniform or space is wasted at the end of short rows [Mur86]. Additionally, it is beneficial to consider routing density in the channels. The maximum routing density in each channel directly impacts the area of the layout.

These issues necessitate significantly different capabilities in the placement algorithms and underlying data structures. Metrics must be augmented to account for the dynamic nature of the design style. Due to the complexity of dynamically adjusting cell rows and routing channels, the underlying data structure is typically a topological model.

**General Cells.** General cell placement is a further extension of the topological model. General cells are not restricted to having similar sizes or being arranged in rows; the design may contain a mixture of arbitrarily sized cells as well as rows of standard cells. This lack of restrictions adds considerable complexity. Careful consideration must be given to the placement of the large cells since their positions will have a large impact on the resulting topology. Moving large cells may cause the nature of the problem to change. Also, the placement model must account for routing resources needed in the vicinity of each component. The way in which the various components fit together and the way the routing areas align are extremely important.

The basic placement algorithms discussed in Sections 4.4 and 4.5 have been adapted to general cell placement. The reader is also directed to Section 4.7 on floorplanning because of the many similarities. However, the general nature of the two-dimensional cell placement problem (the lack of a predefined array of slots for gate arrays or rows for standard cell design styles) has led to placement techniques that are applicable to general cell placement. These techniques include cluster growth [Pre79a, Pre79b], clustering, merging and rearranging [Che84a], force-directed [Qui79], global placement [Bla85b], and partitioning [Lau79].

**Printed Circuit Boards.** Printed circuit boards share many attributes with gate arrays: the components may be constrained to fixed sites, and careful consideration must be given to congestion of routing resources. PCB design has several additional problems.

Typically, PCB designs are small compared to gate array and standard cell designs. The number of components usually ranges from 10 to 1000. However, the sizes (except for arrays of elements such as memory devices) and shapes of the components are heterogeneous. Human designers do well in this environment because the designs are small enough to comprehend the underlying structure. Additionally, due to the heterogeneous nature of the components, iterative improvement by interchanging components of the same size does not work as well as in a gate array or standard cell design, where components are more homogeneous in size and shape.

Fabrication technology also differentiates PCB placement from IC methods. For example, the size of a PCB via is usually large compared to that of a wire. This difference means that layer changes require relatively more routing area. Therefore, generating placements that permit routing path topologies requiring few layer transitions is important. Furthermore, some device technologies require pull-up resistors (for example, emitter-coupled logic) or decoupling capacitors. These resistors and capacitors must be treated as special cases by the placement system. New technologies using surface mount devices allow components on both sides of a board; this turns placement into a three-dimensional problem.

These factors must be taken into consideration in order for a layout to be successful. Automatic placement techniques with simplified metrics cannot account for all of the factors which must be considered simultaneously. Attempts at addressing some of these problems are discussed by Odawara *et al.* [Oda85, Oda87]. Their placement strategy is driven by a knowledge base which tries to take the multiple objectives into account.

### Performance-Based Placement

The goal of performance-based placement is to increase the speed of the circuit being designed by manipulation of the placement. There are two primary factors which influence the propagation delay along a path in the circuit: switching delays of the components along the path and interconnection delays caused primarily by resistance and capacitance of the wiring. For an in-depth discussion of propagation delay, see [Pen81]; also see [Kam81, Ng81, Tam83]. Until recently, propagation delays were not important during placement, because the switching delays dominated the interconnection delays (interconnection delay is the only factor that placement can directly influence). However, fabrication processes are improving, and switching delays are dropping substantially. Thus, it is important to consider performance during the placement process since placement determines the minimum interconnection delays of a circuit.

There are many ways to include performance considerations in the placement process. However, since placement can only impact performance characteristics through the interconnections, all of the techniques modify the objective function by influencing the length and topology of the paths which are generated.

The simplest technique allows the designer manually to specify net priorities or weights based on his or her knowledge of the required circuit topology. To control the impact on the placement score, a weighting factor is incorporated as a multiplicative factor of the net length. This method is rather crude because the weight is applied by the placement algorithm regardless of the actual length or delay of a given net.

Alternatively, the designer can specify minimum and maximum net lengths or delays. The placement system must check the designed net length against the length predicted by the interconnection models of the placer. The system then dynamically adjusts the objective function to meet the specified criteria. If delays are used, the placer must model delays as a function of predicted lengths. This model can be simple (delay per unit length) or sophisticated (resistance and capacitance values based on interconnection material length, layer, and loading factors, as well as capacitance delay caused by wires crossing on different layers). A system of this type is described in [Oga86].

A more sophisticated, performance-based placer is able to derive the timing requirements on the nets as a function of the structural representation, the timing parameters of the cells, and the input-to-output delay requirements of the circuit [Bur85]. This system takes the timing information along with the latch and clock information and generates all paths through the circuit. The criticality of each path is determined automatically from the *timing margin* (the difference between the required and predicted delays). Based on the timing margin, the net priorities are automatically set and a layout is attempted. In spite of the complexities, good results have been reported. Camposano and Kunzmann discuss another highly automated, yet substantially different approach [Cam86].

In summary, considering circuit performance during layout is becoming increasingly important as geometrics shrink to the point where interconnection delay is a significant part of the overall propagation delay through the circuit. Many ways exist for handling performance depending on the performance improvements required.

## 4.9  Conclusion

Automated placement is important because it greatly influences the amount and the location of wiring required to interconnect the components. The placement phase impacts the router's ability to complete the required interconnections; it also affects the performance of the circuit design. A good placement must use a metric which closely models the important factors in routing. The placement models and algorithms chosen for a particular layout system depend on many factors. When considering area versus performance trade-offs, it is difficult (if not impossible) to find a single model and algorithm which works best for all circuits encountered.

This chapter describes the major placement algorithms that are available. A large number of choices are available because electronic circuits have a wide variety of attributes that lead to different placement algorithms. Good placement is critical in order to generate high quality layouts. Inclusion of superior placement techniques is therefore essential to the success of a design automation system.

## Exercises

4.1   Explain the differences between the geometric model and the topological model of the layout surface. To what design style does each apply? What are the advantages and disadvantages of each?

4.2   Explain the advantages and disadvantages of simulated annealing for placement.

4.3   Outline an algorithm for gate assignment improvement. Discuss the advantages and disadvantages of your algorithm.

4.4   A complete graph (edges between all pairs of vertices) is sometimes used to model the interconnection topology of nets. When this is done, nets with large numbers of pins have a disproportionate effect on wire length metrics compared to nets with smaller numbers of pins. Derive a factor to normalize the wire length of nets relative to the minimum spanning tree topology.

**4.5**    Assume a design has $n$ components.  A carrier has $m$ positions in which each of the components may be placed.

**a)**    How many arrangements of the component exist?

**b)**    Starting from the same initial placement, how many different pair-wise interchanges can be made?

**c)**    Starting from the same initial placement, how many single component moves can be made?

**d)**    Given an initial placement and a *known* globally optimum placement, how many pairwise exchanges are necessary to transform the initial placement into the optimum placement?

Compute these values for $n = 5000$ and $m = 5500$.  (Stirling's formula gives an approximation you will find helpful: $n! \approx \left\lfloor \dfrac{n}{e} \right\rfloor^{n} \sqrt{2\pi n}$ .)

**4.6†**    The three primary functions of iterative interchange placement are selection, movement and scoring.  Expand the iterative interchange placement algorithm shown in Figure 4.12 by defining the functions described below:

Selection —   Choose all pairs of components.

Movement —   Exchange the selected components.

Scoring —    Approximate the wire length of a net by one-half of the perimeter of the smallest rectangle enclosing the pins of the net.

Incremental scoring is important for reasonable runtimes.  Does your scoring function do work proportional to the changes that were made?

**4.7†**    Consider a routing area with pins (terminals) along the top and bottom sides as shown in Figure 5.6.  Modern channel routers can often route channels in a width that is close to

$$trackDensity \cdot (lineWidth + lineSpacing)$$

where *trackDensity* is the maximum number of nets which cross any vertical cutline and *lineWidth* and *lineSpacing* are design rules.  Thus, track density is often used as a scoring metric for standard cell and gate array design styles.  Develop an algorithm that computes track density as a function of the pin positions and connectivity.

**4.8‡**    Some gate array and standard cell placement algorithms partition components to rows and then place the components on the rows by neighborhood interchange.  Extend the algorithm specified in Exercise 4.7 by adding incremental scoring.  (The scoring algorithm should not recompute the track density at all of the pins on the channel.)  Determine the increase in speed provided by incremental scoring.

**4.9†** Placement quality can be represented visually in a number of ways. The "rat's nest" shown in Figure 4.4 is one way. Another is through histograms of the number of nets that must cross cutlines as a function of cutline position. Develop a placement analysis tool that plots two such histograms (one for each coordinate). What other placement analysis tools would be helpful?

**4.10‡** A fundamental problem with partitioning-based placement is that components must be partitioned across a cutline before components external to the block being partitioned have been placed. Hence, the positions of the external components must be approximated or inferred. Develop an algorithm to reflect the positions of the external components onto the boundary of the block being partitioned.

**4.11‡** Global routing is normally a routing process and is described in Chapter 5. However, for row-based floorplans (gate arrays and standard cells), global routing can be performed as part of the placement process. In theory, better placements will be produced if global routing can influence the placement. Develop a placement algorithm that, in addition to its normal functions, chooses the (equipotential) pins to connect and places feedthrough cells.

**4.12‡** As fabrication processes improve, transistor switching delays are decreasing compared to interconnection delays. Thus, performance-based placement is becoming increasingly important. In order to address this problem, consider the combinatorial logic between two registers. Develop a placer that will minimize the maximum delay through the combinatorial network. How do the results produced by this placer compare to minimum area placements?

**4.13‡** Placement by simulated annealing does well when few restrictions are imposed. It does less well when more restrictions are imposed (some components must be placed at specified positions in specified relations to each other). Experimentally investigate any degradation in placement quality caused by varying numbers of constraints using a simulated annealing placer. Propose modifications that will improve the simulated annealing placer's ability in constrained placement problems.

**4.14†** Consider a placement program based on the concept of expert agents as described in Chapter 9. What expert agents should be included in such a program?

**4.15†** Partitioning-based placement algorithms are effective but are computationally intensive. How can multiple processors be used to speed the partitioning-based placement process?

**4.16†** Investigate the acceptance function for simulated annealing. Traditionally, a negative exponential function has been used. Suggest two others and compare their performance with the traditional one. Similarly investigate the next temperature function. What is the relationship between the number of moves at a given temperature and the difference in two successive temperatures?

# References

[Aar85]    Aarts, E. H. L., and P. J. M. Lararhoven, "Statistical cooling: a general approach to combinatorial optimization problems," in *Philips Journal of Research*, vol. 40, pp. 193-226, 1985.

[Ada82]    Adachi, T., H. Kitazawa, M. Nagatani, and T. Sudo, "Hierarchical top-down layout design method for VLSI chip," in *Proc. of the 19th Design Automation Conf.*, pp. 785-791, 1982.

[Ads82]    Adshead, H. G., "Towards VLSI complexity: the DA algorithm scaling problem: can special DA hardware help?," in *Proc. of the 19th Design Automation Conf.*, pp. 339-344, 1982.

[Ake81]    Akers, S. B., "On the use of the linear assignment algorithm in module placement," in *Proc. of the 18th Design Automation Conf.*, pp. 137-144, 1981.

[Ban86]    Banerjee, F., and M. Jones, "A parallel simulated annealing algorithm for standard cell placement on a hypercube computer," in *Digest of Intl. Conf. on Computer-Aided Design*, pp. 34-37, November 1986.

[Ber85]    Berkcan, E., and E. Kinnen, "IC layout planning and placement by dimensional relaxation," in *Proc. of Intl. Conf. on Computer Design*, pp. 449-452, 1985.

[Blk84]    Blank, T., "A survey of hardware accelerators used in computer aided design," *IEEE Design and Test of Computers*, vol. 1, no. 3, pp. 21-39, August 1984.

[Bla84]    Blanks, J. P., "Initial placement of gate arrays using least-squares methods," in *Proc. of the 21st Design Automation Conf.*, pp. 670-671, 1984.

[Bla85a]   Blanks, J. P., "Near-optimal placement using a quadratic objective function," in *Proc. of the 22nd Design Automation Conf.*, pp. 609-615, 1985.

[Bla85b]   Blanks, J. P., *Use of a Quadratic Objective Function for the Placement Problem in VLSI Design*. Doctoral Dissertation, Department of Electrical Engineering, University of Texas at Austin, 1985.

[Bra84]    Brady, H. N., "An approach to topological pin assignment," *IEEE Trans. on Computer-Aided Design of Integrated Circuits and Systems,* vol. CAD-3, pp. 250-255, July 1984.

[Bre72]    Breuer, M. A., ed, *Design Automation of Digital Systems. Volume 1: Theory and Techniques.* Englewood Cliffs, New Jersey: Prentice-Hall, Inc, 1972.

[Bre77]    Breuer, M. A., "A class of min-cut placement algorithms," in *Proc. of the 14th Design Automation Conf.,* pp. 284-290, 1977.

[Bro40]    Brooks, R. L., C. A. B. Smith, A. H. Stone, and W. T. Tutte, "The dissection of rectangles into squares," *Duke Mathematical Journal,* vol. 7, pp. 312-340, 1940.

[Bur85]    Burstein, M., and M. N. Youssef, "Timing influenced layout design," in *Proc. of the 22nd Design Automation Conf.,* pp. 124-130, 1985.

[Cam86]    Camposano, R., and A. Kunzmann, "Considering timing constraints in synthesis from a behavioural description," in *Proc. of the Intl. Conf. on Computer Design,* pp. 6-9, 1986.

[Car79]    Carter, H. W., M. A. Breuer, and Z. A. Syed, "Incremental processing applied to Steinberg's placement procedure," in *Proc. of the 16th Design Automation Conf.,* pp. 26-31, 1979.

[Cas86]    Casotto, A., F. Romeo, and A. Sangiovanni-Vincentelli, "A parallel simulated annealing algorithm for the placement of macro-cells," in *Digest of Intl. Conf. on Computer-Aided Design,* pp. 30-33, November 1986.

[Cha68]    Charney, H. L., and D. L. Plato, "Efficient partitioning of components," in *Proc. of the 5th Design Automation Workshop,* pp. 16-1 - 16-21, July 1968.

[Cha82]    Chandrasekhar, M. S., and M. A. Breuer, "Optimum placement of two rectangular blocks," in *Proc. of the 19th Design Automation Conf.,* pp. 879-886, 1982.

[Cha72]    Chang, S., "The generation of minimal trees with a Steiner topology," *Journal of the Association for Computing Machinery,* vol. 19, no. 4, pp. 699-711, October 1972.

[Che83]    Chen, N. P., C. P. Hsu, C. C. Chen, M. Tekahashi, and E. S. Kuh, "BBL: A building-block layout system for custom chip design," in *Digest of Intl. Conf. on Computer-Aided Design,* pp. 40-41, 1983.

[Che84a]   Chen, C. C., and E. S. Kuh, "Automatic placement for building block layout," in *Digest of Intl. Conf. on Computer-Aided Design,* pp. 90-92, 1984.

[Che84b] Cheng, C. K., and E. S. Kuh, "Module placement based on resistive network optimization," *IEEE Trans. on Computer-Aided Design of Integrated Circuits and Systems*, vol. CAD-3, no. 3, pp. 218-225, July 1984.

[Cho85] Chow, C. S., *Phoenix: Interactive Hierarchical Topological Floorplanning Placer.* Masters Thesis, Department of Electrical Engineering and Computer Science, Massachusetts Institute of Technology, 1985.

[Chy83] Chyan, D., and M. A. Breuer, "A placement algorithm for array processors," in *Proc. of the 20th Design Automation Conf.*, pp. 182-188, 1983.

[Cia75] Ciampi, P. L., "A system for solution of the placement problem," in *Proc. of the 12th Design Automation Conf.*, pp. 317-323, 1975.

[Coh86] Cohoon, J. P., and W. D. Paris, "Genetic placement," in *Digest of the Intl. Conf. on Computer-Aided Design*, pp. 422-425, 1986.

[Cor79] Corrigan, L. I., "A placement capability based on partitioning," in *Proc. of the 16th Design Automation Conf.*, pp. 406-413, 1979.

[Cot80] Cote, L. C., and A. M. Patel, "The interchange algorithms for circuit placement problems," in *Proc. of the 17th Design Automation Conf.*, pp. 528-534, 1980.

[Cox80] Cox, G. W., and B. D. Carroll, "The standard transistor array (star), part II: automatic cell placement techniques," in *Proc. of the 17th Design Automation Conf.*, pp. 451-457, 1980.

[Dai86] Dai, W., and E. S. Kuh, "Hierarchical floorplanning for building block layout," in *Digest of Intl. Conf. on Computer-Aided Design*, pp. 454-457, 1986.

[Deo74] Deo, N., *Graph Theory with Applications to Engineering and Computer Science.* Englewood Cliffs, New Jersey: Prentice-Hall, Inc., 1974.

[Dev86] Devadas, S., and A. R. Newton, "Topological optimization of multiple level array logic on uni- and multi-processors," in *Digest of Intl. Conf. on Computer-Aided Design*, pp. 38-41, November 1986.

[Dic86] Dickinson, A., "Floyd: a knowledge-based floorplan designer," in *Proc. of the Intl. Conf. on Computer Design*, pp. 176-179, 1986.

[Dun85] Dunlop, A. E., and B. W. Kernighan, "A procedure for placement of standard-cell VLSI circuits," *IEEE Trans. on Computer-Aided-Design of Circuits and Systems*, vol. CAD-4, no. 1, pp. 92-98, January 1985.

[Fid82] Fiduccia, C. M., and R. M. Mattheyses, "A linear-time heuristic for improving network partitions," *Proc. 19th Design Automation Conf.*, pp. 175-181, 1982.

[Fis67]    Fisk, C. J., D. L. Caskey, and L. E. West, "ACCEL: automated circuit card etching layout," in *Proc. of the IEEE*, vol. 55, no. 11, pp. 1971-1982, November 1967.

[Fra86]    Frankle, J., and R. M. Karp, "Circuit placements and cost bounds by eigenvector decomposition," in *Digest of the Intl. Conf. on Computer-Aided Design*, pp. 414-417, November 1986.

[Fuk83]    Fukunaga, K., S. Yamada, H. S. Stone, and T. Kasai, "Placement of circuit modules using a graph space approach," in *Proc. of the 20th Design Automation Conf.*, pp. 465-471, 1983.

[Gar79]    Garey, M. R., and D. S. Johnson, *Computers and Intractability, A Guide to the Theory of NP-Completeness*. San Francisco: W. H. Freeman and Company, 1979.

[Gar68]    Garside, R. G., and T. A. J. Nicholson, "Permutation procedure for the backboard wiring problem," in *Proc. of the IEEE*, vol. 115, pp. 27-30, January 1968.

[Gil62]    Gilmore, P. C., "Optimal and suboptimal algorithms for the quadratic assignment problem," *Journal of the Society for Industrial and Applied Mathematics*, vol. 10, no. 2, pp. 305-313, June 1962.

[Got81]    Goto, S., "An efficient algorithm for the two-dimensional placement problem in electrical circuit layout," *IEEE Trans. on Circuits and Systems*, vol. CAS-28, no. 1, pp. 12-18, January 1981.

[Got86]    Goto, S., and T. Matsuda, "Partitioning, assignment and placement," *Advances in CAD for VLSI, Volume 4: Layout Design and Verification*, T. Ohtsuki, ed., Amsterdam, the Netherlands: North Holland, Chapter 5, 1986.

[Gre84]    Greene, J. W., and K. J. Supowit, "Simulated annealing without rejected moves," in *Proc. of the Intl. Conf. on Computer Design*, pp. 658-663, 1984.

[Gro86]    Grover, L. K., "A new simulated annealing algorithm for standard cell placement," in *Digest of the Intl. Conf. on Computer-Aided Design*, pp. 378-380, November 1986.

[Hal70]    Hall, K. M., "An r-dimensional quadratic placement algorithm," *Management Science*, vol. 17, pp. 219-229, November 1970.

[Han66]    Hanan, M., "On Steiner's problem with rectilinear distance," *SIAM Journal on Applied Mathematics*, vol. 14, no. 2, pp. 255-265, March 1966.

[Han72a] Hanan, M., and J. M. Kurtzberg, "Placement techniques," in *Design Automation of Digital Systems. Volume 1: Theory and Techniques*, M. A. Breuer, ed., Englewood Cliffs, New Jersey: Prentice-Hall, Inc., Chapter 5, pp. 213-282, 1972.

[Han72b] Hanan, M., and J. M. Kurtzberg, "A review of the placement and quadratic assignment problems," *SIAM Review*, vol. 14, pp. 324-342, April 1972.

[Han76a] Hanan, M., P. K. Wolff, Sr., and B. J. Agule, "Some experimental results on placement techniques," in *Proc. of the 13th Design Automation Conf.*, pp. 214-224, 1976.

[Han76b] Hanan, M., P. K. Wolff, Sr., and B. J. Agule, "A study of placement techniques," *Journal of Design Automation and Fault-Tolerant Computing*, vol. 1, no. 1, pp. 28-61, October 1976.

[Han78] Hanan, M., P. K. Wolff, Sr., and B. J. Agule, "Some experimental results on placement techniques," *Journal of Design Automation and Fault-Tolerant Computing*, vol. 2, pp. 145-168, May 1978.

[Han86] Hancock, J. M., and S. DasGupta, "Tutorial on parallel processing for design automation applications," in *Proc. of the 23rd Design Automation Conf.*, pp. 69-77, 1986.

[Hel77] Heller, W. R., W. F. Mikhail, and W. E. Donath, "Prediction of wiring space requirements for LSI," in *Proc. of the 14th Design Automation Conf.*, pp. 32-42, 1977.

[Hel82] Heller, W. R., G. Sorkin, and K. Maling, "The planar package planner for system designers," in *Proc. of the 19th Design Automation Conf.*, pp. 253-260, 1982.

[Hil80] Hillier, F. S., and G. J. Lieberman, *Introduction to Operations Research*. San Francisco: Holden-Day, Inc., 1980.

[Hil86] Hillner, H., B. X. Weis, and D. A. Mlynski, "The discrete placement problem: a dynamic programming approach," in *Proc. of the Intl. Symposium on Circuits and Systems*, pp. 315-318, 1986.

[Hor81] Horng, C., and M. Lie, "An automatic/interactive layout planning system for arbitrarily-sized rectangular building blocks," in *Proc. of the 18th Design Automation Conf.*, pp. 293-300, 1981.

[How86] How, M. M., and B. Y. M. Pan, "AMBER — a knowledge-based area estimation assistant," in *Proc. of the Intl. Conf. on Computer Design*, pp. 180-183, 1986.

[Hua86] Huang, M. D., F. Romeo, and A. Sangiovanni-Vincentelli, "An efficient general cooling schedule for simulated annealing," in *Proc. of the Intl. Conf. on Computer-Aided Design*, pp. 381-384, November 1986.

[Hun80]    Hung, M. S., and W. O. Rom, "Solving the assignment problem by relaxation," in *Operations Research*, vol. 28, no. 4, pp. 969-982, July-August 1980.

[Hwa76]    Hwang, F. K., "On Steiner minimal trees with rectilinear distance," *SIAM Journal on Applied Mathematics*, vol. 30, no. 1, pp. 104-114, January 1976.

[Ios83]    Iosupovici, A., C. King, and M. A. Breuer, "A module interchange placement machine," in *Proc. of the 20th Design Automation Conf.*, pp. 171-174, 1983.

[Joh83]    Johannes, F. M., K. M. Just, and K. J. Antreich, "On the force placement of logic arrays," in *Proc. of the Sixth European Conf. on Circuit Theory and Design*, pp. 203-206, 1983.

[Jon87]    Jones, M., and P. Banerjee, "Performance of a parallel algorithm for standard cell placement on the intel hypercube," in *Proc. of the 24th Design Automation Conf.*, pp. 807-813, 1987.

[Jun81]    Jung, J., S. Goto, and H. Hirayama, "A new approach to the two-dimensional placement problem of wire congestion in master-slice LSI layout design," in *Trans. Inst. of Electronics and Communications Engineers of Japan*, vol. J64-A, no. 1, pp. 55-62, 1981.

[Kam81]    Kamikawai, R., M. Yamada, and T. Chiba, "A critical path delay check system," in *Proc. of the 18th Design Automation Conf.*, pp. 118-123, 1981.

[Kam82]    Kambe, T., T. Chiba, S. Kimura, T. Inufushi, N. Okuda, and I. Nishioka, "A placement algorithm for polycell LSI and its evaluation," in *Proc. of the 19th Design Automation Conf.*, pp. 655-662, 1982.

[Kan83]    Kang, S., "Linear ordering and application to placement," in *Proc. of the 20th Design Automation Conf.*, pp. 457-464, 1983.

[Kan76]    Kani, K., H. Kawanishi, and A. Kishimota, "ROBIN: a building block LSI routing problem," in *Proc. of the Intl. Symposium on Circuits and Systems*, pp. 658-661, 1976.

[Kar84]    Karger, P. G., and M. Malek, "Formulation of component placement as a constrained optimization problem," in *Proc. of the Intl. Conf. on Computer Design*, pp. 814-819, 1984.

[Ker70]    Kernighan, B. W., and S. Lin, "An efficient heuristic procedure for partitioning graphs," *Bell System Technical Journal*, vol. 49, no. 2, pp. 291-307, February 1970.

[Kho77]    Khokhani, K. H., and A. M. Patel, "The chip layout problem: a placement procedure for LSI," in *Proc. of the 14th Design Automation Conf.*, pp. 291-297, 1977.

[Kho81]   Khokhani, K. H., A. M. Patel, W. Ferguson, J. Sessa, and D. Hatton, "Placement of variable size circuits on LSI masterslices," in *Proc. of the 18th Design Automation Conf.*, pp. 426-434, 1981.

[Kir83]   Kirkpatrick, S., C. D. Gelatt, and M. P. Vecchi, "Optimization by simulated annealing," *Science*, vol. 220, no. 4598, pp. 671-680, May 13, 1983.

[Kor72]   Koren, N. L., "Pin assignment in automated printed circuit board design," in *Proc. of the 9th Design Automation Workshop*, pp. 72-79, 1972.

[Koz83]   Kozawa, T., H. Terai, T. Ishii, M. Hayase, C. Miura, Y. Ogawa, K. Kishida, N. Yamada, and Y. Ohno, "Automatic placement algorithms for high packing density VLSI," in *Proc. of the 20th Design Automation Conf.*, pp. 175-181, 1983.

[Koz84]   Kozminski, K., and E. Kinnen, "An algorithm for finding a rectangular dual of a planar graph for use in area planning for VLSI integrated circuits," in *Proc. of the 21st Design Automation Conf.*, pp. 655-656, 1984.

[Kur86]   Kurdahi, F. J., and A. C. Parker, "PLEST: a program for area estimation of VLSI integrated circuits," in *Proc. of the 23rd Design Automation Conf.*, pp. 467-473, 1986.

[Kra86]   Kravitz, S. A., and R. A. Rutenbar, "Multiprocessor-based placement by simulated annealing," in *Proc. of the 23rd Design Automation Conf.*, pp. 567-573, 1986.

[Kri83]   Krishnamurthy, B., and P. Mellema, "On the evaluation of mincut partitioning algorithms for VLSI networks," in *IEEE Intl. Symposium on Circuits and Systems*, pp. 12-15, 1983.

[Kur65]   Kurtzberg, J. M., "Algorithms for backplane formation," in *Microelectronics in Large Systems*, Washington, D.C.: Spartan Books, pp. 51-76, 1965.

[LaP86]   La Potin, D. P., and S. W. Director, "Mason: a global floorplanning approach for VLSI design," *IEEE Trans. on Computer-Aided Design*, vol. CAD-5, no. 4, pp. 477-489, October 1986.

[Lau79]   Lauther, U., "A min-cut placement algorithm for general cell assemblies based on a graph representation," in *Proc. 16th Design Automation Conf.*, pp. 1-10, 1979.

[Law66]   Lawler, E. L., and D. E. Wood, "Branch-and-bound methods: a survey," in *Operations Research*, vol. 14, no. 4, pp. 699-719, July-August 1966.

[Lin65]   Lin, S., "Computer solutions of the traveling salesman problem," *Bell System Technical Journal*, vol. 44, no. 10, pp. 2245-2269, December 1965.

[Loo79]   Loosemore, K. J., "Automated layout for integrated circuits," in *Proc. of the IEEE Intl. Symposium on Circuits and Systems*, pp. 665-668, 1979.

[Lue80]   Luebbert, F., and M. Ulrey, "Gate assignment and pack placement: two approaches compared," in *Proc. of the 17th Design Automation Conf.*, pp. 472-482, 1980.

[Mac87]   Macaluso, E., "Graphical floorplan design of cell-based VLSI circuits," *VLSI Systems Design*, pp. 50-57, April 1987.

[Mag77]   Magnuson, W. G., "A comparison of constructive placement algorithms," in *IEEE 1977 Region 6 Conf. Record*, pp. 28-32, 1977.

[Mal82]   Maling, K., S. H. Mueller, and W. R. Heller, "On finding most optimal rectangular package plans," in *Proceeding of the 19th Design Automation Conf.*, pp. 663-670, 1982.

[Met53]   Metropolis, N., A. W. Rosenbluth, M. N. Rosenbluth, E. Teller, and A. H. Teller, "Equation of state calculations by fast computing machines," *Journal of Chemical Physics*, vol. 21, no. 6, pp. 1087-1092, June 1953.

[Mor78]   Mory-Rauch, L., "Pin assignment on a printed circuit board," in *Proc. of the 15th Design Automation Conf.*, pp. 70-73, 1978.

[Mun57]   Munknes, J., "Algorithms for the assignment and transportation problem," *Journal of SIAM*, vol. 5, pp. 21-38, 1957.

[Mur80]   Murai, S., M. Kakinuma, M. Imai, and H. Tsuji, "The effects of the initial placement techniques on the final placement results − constructive vs. top-down techniques," in *Proc. of the Intl. Conf. on Circuits and Computers*, pp. 80-82, 1980.

[Mur86]   Murakata, A., A. Tanaka, M. Yamada, and T. Mitsuhashi, "A standard cell placement algorithm with predictive row width equalization," in *Digest of the Intl. Conf. on Computer-Aided Design*, pp. 374-377, 1986.

[Nah85]   Nahar, S., S. Sahni, and E. Shragowitz, "Experiments with simulated annealing," in *Proc. of the 22nd Design Automation Conf.*, pp. 748-752, 1985.

[Nai85]   Nair, R., and J. Reif, "Linear time algorithms for optimal CMOS layout," in *VLSI: Algorithms and Architecture*, P. Bertolazzi and F. Luccio, eds. North Holland: Elsevier Science Publishers, pp. 327-338, 1985.

[Nie83]   Niessen, C., "Hierarchical design methodologies and tools for VLSI chips," in *Proc. of the IEEE*, vol. 71. no. 1, pp. 66-75, January 1983.

[Nis78]   Nishioka, I., T. Kurimoto, S. Yamamoto, I. Shirakawa, and H. Ozaki, "An approach to gate assignment and module placement for printed wiring boards," in *Proc. of the 15th Design Automation Conf.*, pp. 60-69, 1978.

[Ng81]    Ng, P., W. Glauert, and R. Kirk, "A timing verification system based on extracted MOS/VLSI circuit parameters," in *Proc. of the 18th Design Automation Conf.*, pp. 288-292, 1981.

[Oda85]   Odawara, G., K. Iijima, and K. Wakabayashi, "Knowledge-based placement technique for printed wiring boards," in *Proc. of the 22nd Design Automation Conf.*, pp. 616-622, 1985.

[Oda87]   Odawara, G., T. Hamuro, K. Iijima, T. Yoshino, and Y. Dai, "A rule-based placement system for printed wiring boards," in *Proc. of the 24th Design Automation Conf.*, pp. 777-785, 1987.

[Oga86]   Ogawa, Y., T. Ishii, Y. Shiriaishi, H. Terai, T. Kozawa, K. Yuyama, and K. Chiba, "Efficient placement algorithms optimizing delay for high-speed ECL masterslice LSI's," in *Proc. of the 23rd Design Automation Conf.*, pp. 404-410, 1986.

[Oht70]   Ohtsuki, T., N. Sugiyama, and H. Kawanishi, "An optimization technique for integrated circuit layout design," in *Proc. of ICCST-Kyoto*, pp. 67-68, September 1970.

[Oht86]   Ohtsuki, T., ed., *Advances in CAD for VLSI, Volume 4: Layout Design and Verification*. Amsterdam, the Netherlands: North Holland, 1986.

[Ott80]   Otten, R. H. J. M., "Complexity and diversity in IC layout design," in *Proc. of the Intl. Conf. on Circuits and Computers*, pp. 764-767, 1980.

[Ott83]   Otten, R. H. J. M., "Efficient floorplan optimization," in *Proc. of the Intl. Conf. on Computer Design*, pp. 499-502, 1983.

[Ott84]   Otten, R. H. J. M., and L. P. P. P. van Ginneken, "Floorplan design using simulated annealing," in *Digest of the Intl. Conf. on Computer-Aided Design*, pp. 96-98, 1984.

[Ott86]   Otten, R. H. J. M., "Annealing applied to floorplan design in a layout compiler," in *Automation '86 High Technology Computer Conf. Proc.*, pp. 185-228, 1986.

[Pal84]   Palczewski, M, "Performance of algorithms for initial placement," in *Proc. of the 21st Design Automation Conf.*, pp. 399-404, 1984.

[Pat84]   Patel, A. M., "A wirability placement algorithm for hierarchical VLSI layout," in *Proc. of the Intl. Conf. on Computer Design*, pp. 344-350, 1984.

[Pay82]  Payne, T. S., and W. M. vanCleemput, "Automated partitioning of hierarchically specified digital systems," in *Proc. of the 19th Design Automation Conf.*, pp. 182-192, 1982.

[Pen81]  Penfield, P., Jr., and J. Rubinstein, "Signal delay in RC tree networks," in *Proc. of the 18th Design Automation Conf.*, pp. 613-617, 1981.

[Per76]  Persky, G, "PRO - an automatic string placement program for polycell layout," in *Proc. of the 13th Design Automation Conf.*, pp. 417-424, 1976.

[Pre77]  Preas, B. T., and C. W. Gwyn, "Architecture for contemporary computer aids to generate IC mask layouts," in *Record of 11th Asilomar Conf. on Circuits, Systems and Computers*, pp. 353-361, 1977.

[Pre78]  Preas, B. T., and C. W. Gwyn, "Methods for hierarchical automatic layout of custom LSI circuit masks," in *Proc. of the 15th Design Automation Conf.*, pp. 206-212, 1978.

[Pre79a]  Preas, B. T., and W. M. vanCleemput, "Placement algorithms for arbitrarily shaped blocks," in *Proc. of the 16th Design Automation Conf.*, pp. 474-480, 1979.

[Pre79b]  Preas, B. T., *Placement and Routing Algorithms for Hierarchical Integrated Circuit Layout*. Doctoral Dissertation, Department of Electrical Engineering, Stanford University, 1979.

[Pre85]  Preas, B. T., and C. S. Chow, "Placement and routing algorithms for topological integrated circuit layout," in *Proc. of the Intl. Symposium on Circuits and Systems*, pp. 17-20, 1985.

[Pre87]  Preas, B. T., "An approach to placement for rectilinear cells," presented at Physical Design Workshop: Placement and Floorplanning, Hilton Head, South Carolina, April 1987.

[Pri57]  Prim, R. C., "Shortest connection networks and some generalizations," *Bell System Technical Journal*, vol. 36, no. 6, pp. 1389-1401, November 1957.

[Qui75]  Quinn, N. R., "The placement problem as viewed from the physics of classical mechanics," in *Proc. of the 12th Design Automation Conf.*, pp. 173-178, 1975.

[Qui79]  Quinn, N. R., and M. A. Breuer, "A force directed component placement procedure for printed circuit boards," *IEEE Trans. on Circuits and Systems*, vol. CAS-26, pp. 377-388, June 1979.

[Rei65]  Reiter, S., and G. Sherma, "Discrete optimizing," in *Journal of the Society for Industrial and Applied Mathematics*, vol. 13, no. 3, pp. 864-889, September 1965.

[Ric84]   Richard, B. D., "A standard cell initial placement strategy," in *Proc. of the 21st Design Automation Conf.*, pp. 392-398, 1984.

[Rom84]   Romeo, F., A. Sangiovanni-Vincentelli, and C. Sechen, "Research on simulated annealing at Berkeley," in *Proc. of the Intl. Conf. on Computer Design*, pp. 652-657, 1984.

[Rom85]   Romeo, F., and A. Sangiovanni-Vincentelli, "Probabilistic hill climbing algorithms: properties and applications," in *Proc. of the 1985 Chapel Hill Conf. on VLSI*, pp. 393-417, 1985.

[Ros86]   Rose, J. S., D. R. Blythe, W. M. Snelgrove, and Z. G. Vranesic, "Fast, high quality VLSI placement on a MIMD multiprocessor," in *Digest of Intl. Conf. on Computer-Aided Design*, pp. 42-45, November 1986.

[Rut86]   Rutenbar, R. A., and S. A. Kravitz, "Layout by annealing on a parallel environment," in *Proc. of the Intl. Conf. on Computer Design*, pp. 434-437, 1986.

[Sch72a]  Schuler, D. M., and E. G. Ulrich, "Clustering and linear placement," in *Proc. of the 9th Design Automation Workshop*, pp. 50-56, 1972.

[Sch72b]  Schweikert, D. G., and B. W. Kernighan, "A proper model for the partitioning of electrical circuits," in *Proc. of the 9th Design Automation Workshop*, pp. 57-62, 1972.

[Sch76]   Schweikert, D. G., "A 2-dimensional placement algorithm for the layout of electrical circuits," in *Proc. of the 13th Design Automation Conf.*, pp. 408-416, 1976.

[Sec84]   Sechen, C., and A. L. Sangiovanni-Vincentelli, "The timberwolf placement and routing package," in *Proc. of the 1984 Custom Integrated Circuit Conf.*, May 1984.

[Sei85]   Seitz, C. L., "The cosmic cube," in *Communications of the ACM*, vol. 28, pp. 22-33, January 1985.

[Sha85]   Sha, L., and R. W. Dutton, "An analytical algorithm for placement of arbitrarily sized rectangular blocks," in *Proc. of the 22nd Design Automation Conf.*, pp. 602-608, 1985.

[Slu80]   Slutz, E. A., *Shape Determination and Placement Algorithms for Hierarchical Integrated Circuit Layout*. Doctoral Dissertation, Department of Electrical Engineering and Computer Science, Stanford University, 1980.

[Smi80]   Smith, V. K., R. J. Smith, and P. A. Preston, "Comet − a fast component placer," in *Proc. of the 17th Design Automation Conf.*, pp. 465-471, 1980.

[Sou81]   Soukup, J., "Circuit layout," *Proc. of the IEEE*, vol. 69, no. 10, pp. 1281-1304, October 1981.

[Spi85]  Spira, P. M., and C. Hage, "Hardware acceleration of gate array layout," in *Proc. of the 22nd Design Automation Conf.*, pp. 359-366, 1985.

[Sta79]  Stabler, E. P., V. M. Kureichik, and V. A. Kalashnikov, "Placement algorithm by partitioning for optimum rectangular placement," in *Proc. of the 16th Design Automation Conf.*, pp. 24-25, 1979.

[Ste61]  Steinberg, L., "The backboard wiring problem: a placement algorithm," *SIAM Review*, vol. 3, no. 1, pp. 37-50, January 1961.

[Sze80]  Szepieniec, A. A., and R. H. J. M. Otten, "The genealogical approach to the layout problem," in *Proc. of the 17th Design Automation Conf.*, pp. 535-542, 1980.

[Tam83]  Tamara, E., K. Ogawa, and T. Nakano, "Path delay analysis for hierarchical building block layout system," in *Proc. of the 20th Design Automation Conf.*, pp. 403-410, 1983.

[Tar83]  Tarjan, R. E., *Data Structures and Network Algorithms*. Philadelphia: Society for Industrial and Applied Mathematics, 1983.

[Ued85]  Ueda, K., H. Kitazawa, and I. Harada, "CHAMP: chip floorplan for hierarchical VLSI layout design," *IEEE Trans. on Computer-Aided Design*, vol. CAD-4, no. 1, pp. 12-22, January 1985.

[Ueh81]  Uehara, T., and W. M. vanCleemput, "Optimal layout of CMOS functional arrays," *IEEE Trans. on Computers*, vol. C-30, pp. 305-312, May 1981.

[Wal75]  Walsh, G. R, *Methods of Optimization*. New York, New York: John Wiley and Sons, Ltd., 1975.

[Wan80]  Wang, P-T., "A placement technique based on minimization and even distribution of crossovers," in *Proc. of the Intl. Conf. on Circuits and Computers*, pp. 87-89, 1980.

[Wat86]  Watanabe, H., and B. Ackland, "Flute — a floorplanning agent for full custom VLSI design," in *Proc. of the 23rd Design Automation Conf.*, pp. 601-607, 1986.

[Whi84]  White, S. R., "Concepts of scale in simulated annealing," in *Proc. of the Intl. Conf. on Computer Design*, pp. 646-651, 1984.

[Wip82]  Wipfler, G. J., M. Wiesel, and D. A. Mlynski, "A combined force and cut algorithm for hierarchical VLSI layout," in *Proc. of the 19th Design Automation Conf.*, pp. 671-677, 1982.

[Won86]  Wong, D. F., and C. L. Liu, "A new algorithm for floorplan design," in *Proc. of the 23rd Design Automation Conf.*, pp. 101-107, 1986.

[Woo86]  Woo, L. S., C. K. Wong, and D. T. Tang, "PIONEER: a macro-based floor-planning design system," *VLSI Systems Design*, pp. 32-43, August 1986.

[Zib74]    Zibert, K., and R. Saal, "On computer aided hybrid circuit layout," in *Proc. of the Intl. Symposium on Circuits and Systems*, pp. 314-318, 1974.

# Chapter 5

# Routing

**Michael J. Lorenzetti**

*Microelectronics Center of North Carolina*
*Research Triangle Park, North Carolina*

**D. Scott Baeder**

*Digital Equipment Corporation*
*Shrewsbury, Massachusetts*

## 5.1  Introduction

Automatic layout of wiring segments for printed circuit boards (PCBs) and integrated circuits (ICs) has been under study for many years. PCBs and many of the application specific integrated circuit (ASIC) design methodologies in use today use a cell-based approach wherein cells, selected from a library, are arranged on a plane and interconnected by wiring. In such design methods, the routing process usually begins after completing placement. This technique leads to an iterative process of placement and routing to optimize certain aspects (performance, layout area, *etc.*) of the design.  In module generation and full custom integrated circuit layout, the design methods do not depend heavily on a cell-based approach.  In these methods, the placement and routing phases have a stronger interaction and are usually integrated into a single phase.  This chapter mainly discusses routing as it applies to cell-based methods, although many of the basic algorithms described are also applicable to module generation and full custom layout as well.

More information about the layout problem for cell-based design styles can be found in [Hig73, Sou81]. Both references are excellent overviews of the placement and routing problem.

## The Routing Problem

Following placement, components are arranged on a plane and the task remains to insert the electrical connections among the components to make them function. For a printed circuit board, the components are IC packages, and the electrical connections are made with a metal etching process. Metal is uniformly deposited on a carrier (typically fiberglass) surface and the unnecessary metal etched away, leaving wire lines. Connections between layers are made by drilling holes through the fiberglass and plating them with metal. In an integrated circuit, wire lines of polysilicon are fabricated to carry electrical signals between circuits. In addition, one or two layers of metal separated by insulating layers of oxide are deposited and etched above the silicon to form wire lines. Holes are left in the oxide to make connections between layers. In both cases, the routing problem is very similar. The router must determine the wire line patterns to be laid out and the positions of the holes, or vias, to interconnect the layers.

In general there are many sets of points (geometric coordinates) that must be electrically connected. The points are either the pins of the packages on the PCB or designated points (usually on the periphery) of a cell in an integrated circuit. The routing problem is to connect all of the points in each set and ensure that the wiring paths of the different sets do not intersect on any surface.

In addition to these requirements, there are often several routing aspects to be optimized. Long wire lengths cause propagation delays, so meandering paths should be avoided. Sometimes, there are critical signals whose delay must be minimized at the expense of others. Available routing space is often variable, in which case overall area should be minimized. The particular feature to be optimized depends on the layout problem to be solved.

The following common terminology will be used throughout the chapter to describe the routing problem:

- *Net* — A set of points to be electrically connected.

  *Layer* — The wiring surface used for interconnection.

- *Segment* — A straight piece of wire on a single layer, which comprises a portion of the wiring path used to connect a net.

- *Via* — A mechanism (hole) for connecting segments on different layers.

## Historical Perspective

In order to better understand the routing problem, it is important to look at the origins of automatic routing algorithms. The first algorithms were originally developed to route PCBs. These algorithms imposed very few constraints on the routing process and operated on a single connection at a time. Many of these routing algorithms have been adapted for routing ICs and continue to play an important role in integrated layout systems. These algorithms are called general-purpose routers in this chapter.

Since these general-purpose routers work on the entire design in a serial fashion, the size of the problems they reasonably attempt is limited. However, they were adequate for the small to medium PCB problems. They are much more efficient than doing the task manually and enjoyed a widespread popularity.

Later, restricted routing algorithms were developed. These require some constraints on the routing problem, such as empty rectangular areas with all of the pins on the periphery. Such problem restrictions allow these routers to consider how routing one net in a particular fashion impacts the other nets to be routed. Because of their limited scope, these routers can do a better job of modeling the contention of the nets for the routing resources and, therefore, can be viewed as routing the nets in parallel.

Restricted routing algorithms are employed as part of a three-step process. Large problems (a PCB or IC layout) are broken into smaller tasks that meet these restrictions. Decomposition of routing problems in this manner is a substantial task in itself, but this "divide and conquer" approach makes automatic routing of today's large VLSI circuits possible. The steps in such a routing system are:

(1)  Divide the routing surface into routing areas which meet the restrictions of the router to be employed.

(2)  Use a *global router* to assign each net to a subset of the routing areas. Global routers do not define wire segments and vias but rather take the original net information and carrier topology and define a set of restricted routing problems.

(3)  Use a restricted *detailed router* to calculate actual wiring paths. A detailed router is one which reads a routing problem and produces wire segments and vias to realize the required connections.

## A Router Taxonomy

To provide a perspective on the various routing algorithms and to aid in organizing the chapter, a routing taxonomy is presented in Figure 5.1. It is important to note that any such taxonomy is somewhat subjective. It should, however, aid in understanding the material. The reader may wish to refer to Figure 5.1 while reading the later parts of this chapter.

At the first level of the hierarchy are the global and detailed routers introduced above. A third category of *specialized routers* has been added. This category refers to routers which have been designed to solve specific problems and are not as generally applicable as those in the previous two categories.

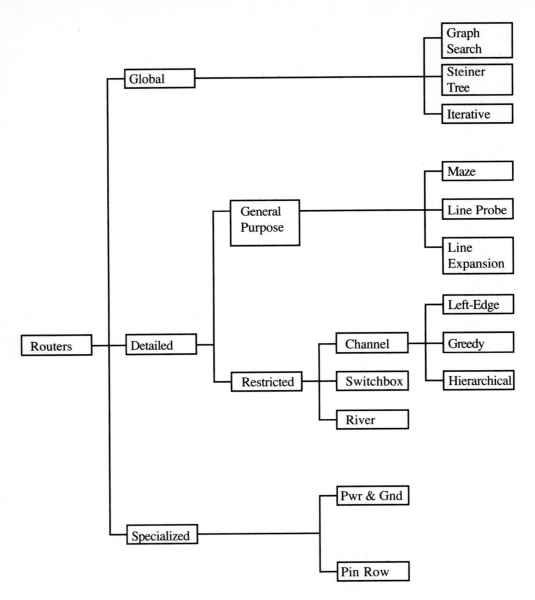

**Figure 5.1** Router Taxonomy

Detailed routers further divide into general-purpose and restricted as discussed above. This chapter presents three main types of restricted detailed routers: channel, switchbox and river routers. The rightmost column of the diagram refers to algorithms which will be described in this chapter.

**Routing Chapter Overview**

Since many detailed and global routing algorithms rely heavily on graph theory (in particular, the problem of finding the shortest path between two vertices in a weighted graph), this chapter begins with an overview of shortest path techniques as they apply to the routing problem. Enough material is presented to provide an adequate foundation for the routing algorithms presented.

A focus of the chapter is on detailed routing algorithms, so they are presented before the others. Both general-purpose algorithms and restricted algorithms are discussed. Following the detailed routers, global routers are presented, along with algorithms used to refine or improve existing layouts. Finally, some examples of the types of routers used in actual layout systems, along with topics of current research and references to further reading material, are given.

An alternative chapter ordering might be to present the routers in order of their use within a routing system: global routers, specialized detailed routers, and then general-purpose detailed routers. However, we chose our ordering for the following reasons:

- Our ordering is that in which the algorithms first appeared in the literature and thereby gives a better historical perspective.

- Our ordering allows us to describe specialized detailed routers immediately after the general-purpose ones so they can be compared and contrasted.

- Understanding the strengths and limitations of the restricted detailed routers is important in developing global routing techniques that exploit them.

Since the area of automatic routing algorithms is fairly broad, this chapter provides an overview of the various techniques. References to the original descriptions of the algorithms are given for readers interested in the details of a particular technique.

## 5.2 Shortest Path Algorithms

Several automatic routing algorithms are based on finding the shortest path through a weighted graph. It is appropriate to review the theoretical work in this area before examining any specific routing algorithms.

All techniques presented in this section determine the shortest path between two vertices by selecting one of the vertices as the starting point and exploring paths outward until the other vertex is reached. Such exploration emanates in the form of a tree, with the source vertex at the root of the tree. Once the target vertex is encountered along one of the paths, the task remains to retrace, and mark, the path back toward the root. This retracing is easy if the appropriate information regarding parentage is available at each vertex along the way. The main distinction between these techniques is the method of steering the exploration.

## Breadth-First Search

One of the earliest published solutions to the shortest path problem appeared in [Moo59]. Moore's algorithm starts by marking the source vertex with a cost of zero. The next step marks all of the adjacent vertices with the cost associated with the weight of the edge connecting the vertex to the source. In subsequent steps, the neighbors of the vertices marked in the previous step are marked with the cost on the previous vertex plus the cost of the intervening edge. Whenever a vertex is encountered that is already marked, its cost is remarked only if the new cost is less than the cost already on the vertex. This process is repeated until a step occurs in which no mark is changed.

Upon completion, every vertex in the graph is labeled with the cheapest cost to it from the original source vertex. Retracing from the target (or any other vertex for that matter) back to the source can be done by repeatedly finding the neighboring vertex which is marked with a cost equal to that of the current vertex minus the cost of the intervening edge. The existence of more than one such neighbor indicates that there is more than one lowest cost path back to the source.

Edge weights in the graph can be either positive or negative, as long as there are no negative cycles. Worst case runtime for Moore's algorithm is $O(n^3)$ where $n$ is the number of vertices in the graph [Hul78].

## Label Setting Methods

An alternative solution to the problem was proposed by Dijkstra in 1959 [Dij59]. Dijkstra labels the vertices in the same manner as Moore. The main difference between the algorithms is in the order in which the vertices are marked. Label setting methods do not immediately label the neighboring vertices of the source, but rather make a list of these vertices along with their accumulated cost. The lowest cost vertex is then selected from this list and labeled, and its neighbors are added to the list, along with the accumulated cost (calculated in the same way as in Moore's algorithm). Such a technique is referred to as a "uniform cost search" in some references (for example, [Bar81]).

When a vertex is labeled, the label is known to be the cost of the shortest path to that vertex and need never be changed. Thus the designation as a "label setting method" by [Hul78], as opposed to the "label correcting method" of Moore. Because of this, the process can stop when the target vertex is labeled (*i.e.*, pulled off the list as the lowest cost vertex). The exploration process does not need to mark the entire graph as in Moore's algorithm.

Each edge of the graph is examined only once; therefore, the worst case is a complete graph exploration which has $n(n-1)$ edges and a runtime of $O(n^2)$. Average runtime can be much shorter since few layout problems have edges between all pairs of vertices. Edge weights are restricted to being non-negative (rather than just free of negative cycles as in the breadth-first search technique).

## 5.3  Detailed Routers

In this section several algorithms for detailed routing are presented. The earliest algorithm for automatic routing, the maze router, is presented first, followed by the line-probe and line-expansion algorithms. These are general-purpose routing algorithms. These algorithms are followed by the more restrictive channel, switchbox, and river routers.

### Maze Routers

The term *maze router* refers to a set of general-purpose routing algorithms which operate on a gridded model and route a single connection at a time within that model. In a gridded model, the entire routing surface is represented as a rectangular array of cells. The size of the cells is defined such that wires can be routed through adjacent cells without violating the width and spacing rules for the wires. In this model, two points are connected by finding a sequence of adjacent cells between the cells containing the two points. Cells which are unusable, because they already contain wiring or some other obstruction, are marked as blocked. Figure 5.2 shows a small section of a printed circuit board in which blocked cells are shaded. The pattern of shaded and unshaded cells forms a maze, from which the router gets its name.

Maze routers are applications of finding the shortest path through a weighted graph. The weighted graph is formed by modeling each cell in the gridded model as a vertex in the graph and inserting an edge between vertices whose cells are adjacent. This forms a very regular graph, in which each vertex has degree four (except those on the perimeter of the routing surface). This regularity should be exploited in the data structures for these algorithms. A simple two-dimensional array will suffice, rather than adjacency lists or other more complex structures required for more general graphs.

**The Basic Maze Router.**   The *Lee algorithm* is the most widely known technique for finding a path between two points in a rectangular gridded model [Lee61]. In terms of the graph theory techniques presented earlier, the Lee algorithm is a refinement of Moore's algorithm. It is a breadth-first search technique and so always finds a path if one exists and guarantees that the path found is of minimum length.

The algorithm operates by maintaining a single list which contains the leaf nodes of the exploration tree. In each successive step, the neighbors of each entry on the original expansion list go onto a new, updated copy of the list,

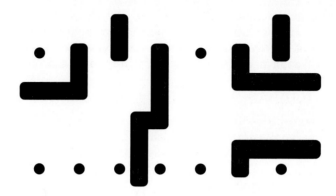

(a) Section of a Printed Circuit Board

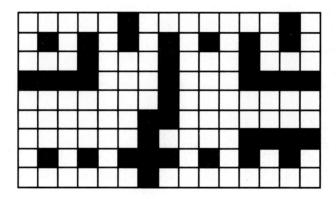

(b) Corresponding Gridded Model

**Figure 5.2**  Gridded Model Example

along with the accumulated cost to that vertex. The exploration phase stops when the target vertex first appears in the list. Stopping the search in this manner is a refinement of Moore's algorithm (which continues until all the vertices are explored and no relabelling is possible). This refinement is made possible by the additional restriction that no edge weight can be negative. On the regular grid structure with uniform costs on the edges, the exploration emanates as a diamond shape wavefront as in Figure 5.3.

The Lee algorithm is important because of its shortest path guarantee. Most maze router papers describe their routers as "modified Lee routers." The major weak point of the Lee algorithm is that it does one connection at a time and has no provision for ensuring that an early connection will not prevent the routing of connections not yet processed. In practice, isolating a point of an as

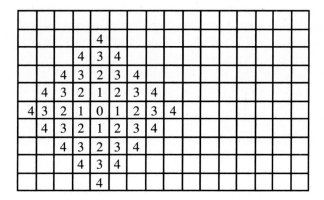

**Figure 5.3**  Symmetric Expansion

yet unrouted connection is the main failure mechanism of Lee routers. Other disadvantages include large memory requirements (one storage element for every cell on the routing surface) and long search times (search times increase as the square of the lengths of the connections). The following sections present modifications which address these weaknesses.

**Reduction of Memory Requirements.**  In a breadth-first search such as the Lee algorithm, only two types of neighbors to a cell need to be distinguished during the retrace phase: cells toward the target and cells toward the source of the exploration. This information can be coded in a single bit for each cell [Ake67]. Add to these retrace markings the additional states of "unexplored" and "blocked" and there are a total of four states, or two bits per cell for a single-layer problem. Assignment of these four states greatly reduces the amount of storage for the Lee router, since each cell otherwise requires a counter large enough to hold the maximum distance in grids that would be encountered. For a large design (on the order of 2000 grids on a side), Lee's algorithm requires at least 12 bits per cell.

It is important to note that Akers' coding scheme is only applicable because of the breadth-first nature of Lee's maze router. Other types of cost functions and exploration techniques allow a third type of neighbor: one that has been explored but does not lie on a least cost path through the cell of interest. Also, coding schemes such as Akers' do not change the fact that the storage requirements increase as the square of the length of a side of the routing surface; they only reduce the storage by a constant factor.

**Speed Improvements.**  The basic maze router explores symmetrically from the source point going equally far away from the target as it goes toward it. Soukup [Sou78] proposed a technique which provides a 10 to 50 times speed improvement over the original algorithm. Instead of a single list containing the leaf nodes of the exploration tree, the algorithm maintains two lists which are designated as the old reach set and the new reach set. The primary difference between Soukup's technique and a basic maze router is as follows. When a cell being entered into the new reach set is found to be in a direction toward the target cell, the building of the new reach set is halted. The contents of the new reach set is appended to the old reach set after the last unused entry, and the program enters a new loop. In this loop, the program repeatedly explores the next cell in the same direction for as long as it continues to make progress toward the target. The exploration becomes a "bee-line" toward the target. When no progress is made, it returns to the process of exploring cells out of the old reach set.

There are two things to note about Soukup's approach. First, it is a breadth-first search only until the "bee-line" feature is invoked. Second, it is essentially a hill climbing technique when the "bee-line" feature is operating and therefore offers no guarantee of finding the shortest path. The result is especially obvious if a large amount of congestion is present on the wiring layer.

Another speed improvement can be realized by exploring out from more than one source point at a time. By using these "multiple expansion units" [All76], the minimum cost path can be found which will connect one of the sources with any other source in the same net. This approach means that it is not necessary to decide in advance which pair of nodes in a large net are to be connected first. Multiple expansion units are simply a bidirectional search taken a step further. This approach cuts the number of cells searched by more than half. It also allows the expansion process itself to determine which pairs of pins to connect, rather than forcing a presumptive decision.

Given two points to be connected, search time is affected by careful selection of which point is the source and which is the target. Exploration begun at points nearer the boundary limits the symmetric expansion and reduces the number of cells examined [Rub74a]. The search space can be further restricted by imposing an artificial boundary on the area to be explored [Tad80].

**Increased Parallelism.** The Lee algorithm is inherently serial, since connections are made one at a time. Several modifications to the Lee router have been proposed to alleviate the router's tendency to block unrouted connections.

One such idea is to implement a flexible means to incorporate different edge weights in the graph to cause the router to favor cells away from unconnected pins and thereby avoid blocking them. In such a scheme, edge weights around connection points are high and edge weights in the middle of empty routing regions are low. The primary difficulty here is that this is a very large graph and entering costs individually is prohibitive. Hightower [Hig83] proposed taking advantage of the repeated structure of gate array masterslices and devised a language to describe the costs in small areas of the chip. Allen [All76] took a similar approach to identify regions of the surface to have a horizontal or vertical bias.

**Summary of Maze Routers.** Because the entire routing surface is stored in the form of a grid, memory increases as the routing area (square of the length of one side), and runtime increases as the square of the lengths of the individual paths. However, maze routers are still important components in most production layout systems because of the guarantee of finding a path if one exists. It is important to note that this guarantee applies only to the connection currently being routed. Maze routers do not do well on an empty surface requiring a large number of connections, because they tend to block connections not yet routed. In modern systems, the main application of maze routers is as a final cleanup phase to route any failures left by the more specialized routers. The guarantee of finding a path if one exists is important in completing the last few connections on a crowded routing surface.

Another drawback to maze routers is that all features (pins, wires, vias) must be on a regular grid to allow efficient routing. Cells are marked as either blocked or available; therefore, a feature that lies between cells (off-grid) must be modeled as blocking all the cells on which it impinges. In order to avoid this inefficiency, a layout system that uses a maze router for final cleanup must be designed so that all routers in the system produce wiring that is on-grid.

### Line-Probe Routers

Both Mikami [Mik68] and Hightower [Hig69] independently proposed an alternative to routing algorithms based on maze routing. These alternative routers, called *line-probe routers*, do not store the entire routing surface in memory in the form of a grid. Instead, they store only those features already placed onto the wiring surface (terminals, wiring segments, *etc.*). Line-probe routers model the routing surface as lists of lines.

The basic idea of line-probe routers is to project line-probes from a point until an obstruction is encountered. The algorithm starts by determining the two points to be connected. From each point, potential wiring segments (or *probes*) are projected as far as possible in both the horizontal and vertical directions. If the probes intersect, the connection is complete and will have at most one bend. However, if the probes are stopped by some routing obstruction (for example, already placed wiring segments), the algorithm must choose a new point (called an *escape point*) along the current probes. Additional probes are sent out from the escape point. The process continues until the probes emanating from one point intersect the probes emanating from the other. At this point, there is a sequence of lines which connect the two points. Figure 5.4 from [Hig69] illustrates the sequence of probes generated to connect two points ($A$ and $B$ in the figure). Solid lines represent existing wire segments and dashed lines are the line-probes generated. Points identified as $E$ are escape points.

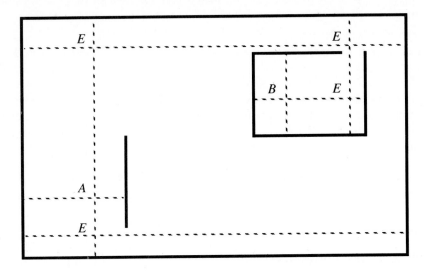

**Figure 5.4** Line-Probe Router Example from [Hig69]

Organization of the data to allow efficient search for obstructions to a probe is a key consideration for the surface model of these routers. A typical organization is two lists for each layer: one for horizontal lines and one for vertical lines. Two lists are required because a vertical probe is only limited by horizontal lines and, similarly, a horizontal probe is limited by vertical lines. Separate lists allow lines parallel to the direction of probe to be ignored without slowing down the search time. The horizontal list is sorted by $y$ coordinates and the vertical list by $x$ coordinates. Implementation can be as linear arrays (allowing binary searches) or as linked lists of lines for each $y$ value (in the case of horizontal lines), which allows easy insertions.

The process of choosing escape points is the difference between the two original line-probe algorithms. For each existing line segment at a given level, Mikami generates additional perpendicular lines at each "grid" intersection. This approach is, in essence, a complete breadth-first search. This approach will guarantee a solution, if one exists.

Hightower's algorithm adds only a single escape point to each existing probe. In the simple case of a probe parallel to a wiring segment, the escape point is placed just past the endpoint of the segment. Hightower describes three such escape processes, designed to allow the router to find a path around different types of obstacles. Figure 5.5 depicts a simple example (first presented by Ohtsuki [Oht86]) which illustrates the difference between the algorithms. Note that the Mikami algorithm generates 17 escape points, while the Hightower approach uses only 3.

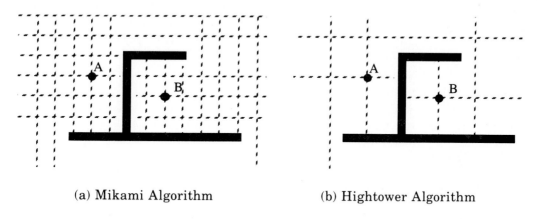

(a) Mikami Algorithm                    (b) Hightower Algorithm

**Figure 5.5**    Mikami and Hightower Algorithms Applied to a Simple Problem

While the number of escape probes generated by Hightower's algorithm is significantly less than that of Mikami, it may not produce a successful connection, even if a path exists. In some cases this potential lack of completion may actually be an advantage over maze algorithms. If a very complex path with a large number of segments and bends is placed on the wiring surface, it may block many additional paths and lower the overall completion rate. It may be better for the algorithm to fail to find a path in such cases and defer the connection to a cleanup phase rather than to define a circuitous path which blocks subsequent connections.

When compared to the grid based models used in maze routers, line-probe routers have a major advantage in the use of memory. Since only the wiring segments and obstructions are stored, the amount of memory is proportional to the number of segments, not the size of the routing region. Line-probe routers also offer a memory advantage related to the degree of precision that may be obtained. Since only the coordinates of the wiring segments are stored, the precision is limited only by the precision of the computer. In a grid-based system, the size of the grid (the precision available) has a dramatic effect on the amount of memory required.

Mikami's algorithm guarantees to find a solution if one exists, but does not guarantee to find the shortest path. It is also an inherently gridded approach. Hightower's algorithm is not gridded but does not guarantee to find a path even though one may exist. An extension has been proposed [Hey80] to do a more exhaustive approach in a line-probe router, but the problem remains that these routers, like the maze router, are serial in nature. Connections routed early in the layout process tend to block those which are routed later. Line-probe routers are used predominantly in layout systems that have interactive routing editors [Hig85]. These editors allow the designer to complete the last few connections, usually with the aid of a puck or mouse and a color graphics display. The advantages of using a line-probe router in such an interactive system is that it can allow features to be off-grid, and the user can steer the algorithm by helping in the selection of escape points and moving wires aside when necessary.

### Channel Routers

The majority of modern automatic IC routing systems are based on *channel routers*; they are the workhorse of IC routing. These systems employ a "divide and conquer" approach in which the layout problem is divided into several channel routing problems which are each solved separately. Decomposing the entire routing problem into several channel routing problems (channel decomposition and global routing) is a separate topic and is covered later in this chapter.

In descriptive terms, a *channel* typically consists of a rectangular space between two parallel rows of pins or *terminals*. The locations of these terminals are fixed along two opposite sides of the channel, while the location of any pins on the remaining two sides is determined during routing. The discussion in this chapter will presume the channels are oriented horizontally (*i.e.*, the fixed terminals are on the top and bottom sides). They can, of course, be oriented vertically without loss of generality, but adopting this convention avoids confusion in the description of the algorithms. Furthermore, some channel routing algorithms allow contours along the top and bottom edges of the channels, so the channel need not be strictly rectangular. We will assume rectangular channels in the descriptions of the algorithms and point out which algorithms can be readily modified to handle non-rectangular channels.

Channel routers impose restrictions on net topology; this allows them to consider connections in parallel and thereby avoid the inherent problems of serial approaches such as maze and line-probe routing. Most channel routers assume there are two layers available for routing. Instead of approaching the routing problem one connection at a time, channel routers emphasize the allocation of the *routing resources* (the area available for defining segments and vias). The task of the channel router is to route all of the connections, while minimizing the amount of routing resources consumed.

To describe channel routing, this chapter uses the terminology presented in [Sat80]. Generally, vertical wires extending into the channel from the top and bottom terminals are on the same layer. The horizontal wires which connect these wires are on the opposite layer. The horizontal wires are called *trunks*, and the lines along which they are placed are called *tracks*. The vertical wires connecting the trunks to the top and bottom of the channel are called *branch*es. Figure 5.6 illustrates the net topology and the nomenclature employed. Since the length of the channel is fixed, the problem becomes one of assigning the required trunks to the minimum possible number of tracks.

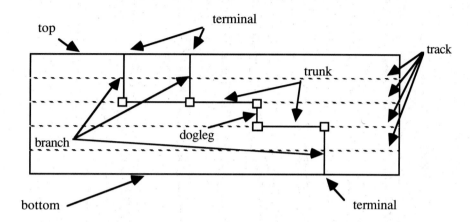

**Figure 5.6**  Channel Routing Terminology

There are three fundamental approaches to the channel routing problem. The left-edge algorithm (LEA) and its extensions proceed one track at a time. The greedy channel router proceeds one column at a time. The hierarchical channel router uses an innovative approach of repeatedly splitting the area and refining the solution after each split.

**The Basic Left-Edge Algorithm.**  The original channel routing algorithm, proposed by Hashimoto and Stevens [Has71], was designed to be used on two-layer PCBs where the majority of the components were dual inline packages. The packages were placed in rows, with the area underneath the packages and between the rows divided into rectangular routing channels.

The basic *left-edge algorithm* imposes the restrictions that each net consists of a single trunk section and that the convention that trunks be routed on one layer and branches on the other is strictly enforced. Subject to these conditions and in the absence of vertical constraints (a term defined in the next subsection), the algorithm produces a solution with the the minimum possible number of tracks. This number is referred to as *density* and is found by determining the maximum number of nets which span any $x$ position along the channel.

First, all of the required trunks are sorted in increasing order of their leftmost endpoints. The algorithm proceeds one track at a time, starting at the left-edge of the channel (hence, the name left-edge algorithm), and scans the sorted segments. The first unplaced segment encountered is placed into the first track. Then, in order to fill the track as much as possible, the algorithm

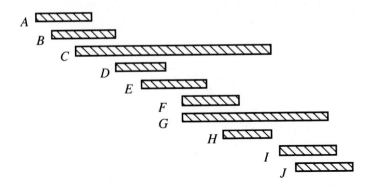

(a) Segments Sorted by Starting Point

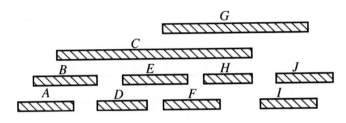

(b) Segments Assigned to Tracks by LEA

**Figure 5.7**  Left-Edge Example

continues scanning for the next segment that will also fit into the first track. This scan continues until the end of the track is reached. The algorithm repeats for as many additional tracks as are required. Figure 5.7 shows a sample problem, with the trunks sorted and then assigned to tracks by the left-edge algorithm.

After completing the assignment of the trunk layer, the branches can be easily added. However, if there are terminals at both the top and bottom at a given location along the channel, these branch layer wiring segments may overlap, causing a short circuit. This overlap occurs if the trunk associated with the top terminal is placed in a track below the trunk connected to the bottom terminal. The overlap is easily corrected in the PCB environment for which this router was invented, since there is always additional space between adjacent pins on a package to "jog" one of the conflicting segments, as shown in Figure 5.8. Spacing between terminals was wide enough to allow at least one extra branch section between terminals.

The basic left-edge algorithm produces a solution with the minimum number of tracks subject to the constraint that each net consists of one trunk section and that branch-layer conflicts can be resolved as in Figure 5.8.

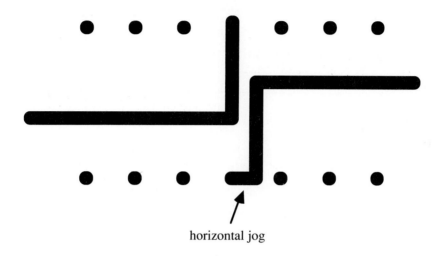

horizontal jog

**Figure 5.8**  Horizontal Jogs to Avoid Conflicts

**Constrained Left-Edge Algorithms.**  Simple resolution of these branch layer conflicts is not possible in the fine line technology of modern ICs. There is generally insufficient room between adjacent terminals to jog one of the branch wires. In addition to ensuring that none of the trunk segments overlap, channel routers for modern ICs must also ensure that the branch segments do not overlap. If two terminals are within via-to-line spacing in the $x$ direction (one on the top edge of the channel, the other on the bottom edge) and are in different nets, a *vertical constraint* exists. Such constraints are resolved by placing the trunk segment connected to the top terminal in a track above the trunk connected to the bottom terminal.

Since the positions of the terminals are fixed before beginning the routing process, many of these constraints may need to be satisfied. In fact, the constraints may form chains of arbitrary length (for example, $A$ above $B$, $B$ above $C$, $C$ above $D$, *etc.*) or even form constraint loops. The simplest constraint loop is formed between two connections (for example, $A$ is above $B$ at one terminal position, while $B$ is above $A$ at another). This information is usually represented in the form of a directed graph as depicted in Figure 5.9.

Optimal solutions can be found for this constrained left-edge problem using branch-and-bound methods [Ker73]. However, such approaches consider only solutions in which there is a single trunk section per net. This restriction is

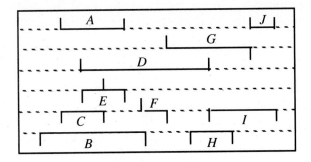

(a) Channel Routing Problem

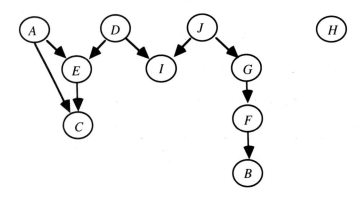

(b) Graphical Representation of Constraints

**Figure 5.9**  Constraints and Their Graph Representation

unnecessary and often results in a solution with more tracks than necessary. Consider the situation in Figure 5.10. The constraint in column 1 requires net $A$ to be placed above net $B$, while the constraint in column 4 requires net $B$ to be above $C$. If there is sufficient space, a *dogleg* can be added to net $B$. This dogleg reduces by one the number of tracks required to route these three connections.

The insertion of a dogleg in the region of maximum density will prohibit the channel from being routed in density since the connection with the dogleg uses two tracks instead of one. Notice that in Figure 5.10, the density is two, but with the dogleg the minimal solution requires three tracks. However, without the dogleg, routing one of the connections causes the other to fail. Finding the optimal number and location of doglegs to break constraint loops and reduce channel height has been shown to be NP-complete [Szy85].

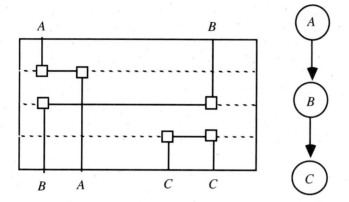

(a) Three-Track Solution without Doglegs

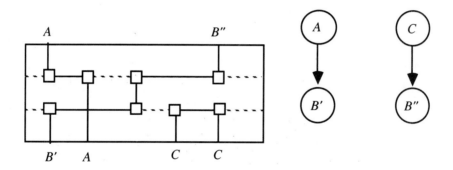

(b) Two-Track Solution with One Dogleg

**Figure 5.10**  Using Doglegs to Break Constraint Chains

By studying the examples where doglegs were not used, Deutsch [Deu76] observed the following: the primary reason the solution required many tracks in excess of the theoretical minimum or density was due to a few crucial nets. These nets, usually clock lines, were heavily connected to both sides of the channel and generated a long constraint chain. Since these connections have numerous terminals, it does not overly restrict the solution to limit doglegs to terminal positions. Furthermore, doglegs at terminal positions introduce one extra via, whereas those at other positions require two additional vias. This observation allows the algorithm to divide all of the long connections into a series of two-pin connections or trunk segments. When placing a trunk segment into a track, the next trunk section of the same net may be placed in the same track. This procedure works reasonably well but can introduce more doglegs than are necessary. In order to minimize this, and reduce the number of vias added, Deutsch introduced a concept called *range*. Range represents the number of consecutive trunk segments that must be assigned to the same track. By increasing the range parameter, the router can be made to put in fewer doglegs.

Despite the name "left-edge algorithm," it is possible to employ these same concepts starting on the right edge and working left. By controlling the starting corner (top/bottom, left/right), different results may be obtained. There are numerous combinations for ordering the packing of the tracks. Instead of going from bottom to top or from top to bottom, the algorithm can alternate between topmost and bottommost tracks. The starting corner gives one of four possibilities. When changing to an alternate side, there is a choice of two corners for starting the packing. This simple scheme yields eight combinations of routing. There are, of course, numerous other combinations, each resulting in a potentially different solution. Since LEA is fast, good results are obtained in reasonable runtime by routing several combinations of packing orders and range values and keeping the one with the best results.

A different approach to handling constraints was implemented in a router called YACR2 (for "Yet Another Channel Router, the Second") [Ree85]. This router employs a two-phase approach. A modified LEA, which allows vertical constraint violations to occur, is followed by a branch layer routing phase that can define more complex wiring than strictly vertical lines.

After completing the track-assignment phase, the branch layer segments are placed for all columns that do not violate vertical constraints. YACR2 then attempts to connect all of the remaining trunk layer segments using what is referred to as "maze routing techniques." These techniques are not based on Lee's algorithm, as the name might imply, but use a pattern router. While a Lee algorithm could be used, the authors observed that many of the conflicts can be resolved with simple connections. A *pattern router* is one which searches for solutions that match a particular topology. These patterns avoid the overhead of a complete maze search and give better control over the number of vias used.

YACR2 uses three patterns: *maze*1, *maze*2 and *maze*3. The key to the *maze*1 pattern is in its use of non-preferred direction wire segments. It attempts to route one of the constrained branch connections using a horizontal "jog." Figure 5.11 shows an example. This jog resolves the constraint, and the other connection can be completed normally. The other two patterns attempt to "dogleg" one (*maze*2) or both (*maze*3) of the branch connections using the existing space on the trunk layer.

If all three patterns fail, YACR2 adds an additional track. This extra track creates additional empty space that may be used to resolve constraints during the track assignment. (The heuristics involved will assign the trunk segments to different tracks if the number of tracks is increased.) Before discarding all of the previous work, YACR2 makes a last-ditch attempt to complete the connection

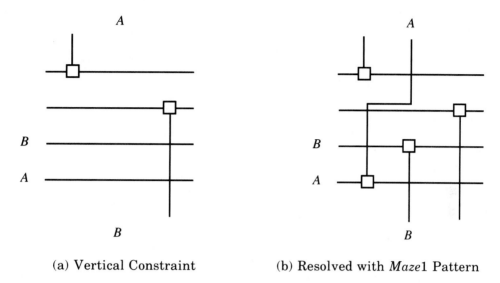

(a) Vertical Constraint          (b) Resolved with *Maze*1 Pattern

**Figure 5.11** *Maze*1 Pattern for Resolving Constraint for YACR2

using a single dogleg (*maze*2) by inserting an empty track at each possible location across the channel and checking whether this allows completion of all remaining branches. If this attempt fails, YACR2 starts over with the LEA phase using one additional track.

YACR2 also has a unique mechanism for removing cyclic constraint loops. The vertical constraints that form the "loops" in the constraint graph are identified and minimized. If all of the "maze" patterns have failed to resolve a constraint that is part of a loop, additional tracks alone will not be enough to allow successful routing. In this situation, YACR2 will add an additional column to one end of the channel along with one or more tracks. Then, the connection can be completed by using a process similar to *maze*2.

**Greedy Channel Router.**   While the above-mentioned channel routers are all based on the LEA which routes the channel one track at a time, Rivest and Fiduccia [Riv82a] developed an algorithm which works column by column. This algorithm routes the wire segments by walking across the channel from left to right, placing all of the segments of a given column before proceeding to the next. In each column, the router applies a set of simple, "greedy" heuristics to maximize the utility of the wiring produced. Instead of performing doglegs at terminal locations, the greedy router allows any horizontal connection to change tracks, which causes it to insert a large number of vias. In some cases it requires that additional columns be added at the end of the channel to complete the wiring.

The greedy router uses five steps in routing each column. The first step is to connect any terminals to a trunk segment. This connection is done using a minimum of branch layer wiring. Each terminal is connected to the first empty track, or to the track that already contains the net.

In the second step, the greedy router attempts to collapse any nets with more than one track using additional branch layer segments, as depicted in Figure 5.12. This collapsing may bring a terminal connection up to the correct track if it had stopped on an earlier empty track. Since there may still be nets in more than one track, the third step is to reduce the range, or distance between these tracks, as shown in Figure 5.13. This reduction is done by the addition of branch layer segments.

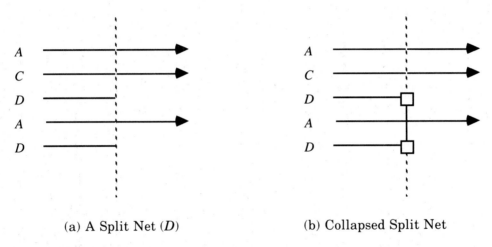

(a) A Split Net (*D*)          (b) Collapsed Split Net

**Figure 5.12**  Collapsing Split Nets (step 2) of Greedy Router

The fourth step tries to move all nets closer to the edge they prefer, based on the edge of the next pin in the net. If the next pin is on the top of the channel, the greedy router tries to move the net to a higher track. In this stage, the router tends to favor longer distance moves, since they are effective in preventing later constraint conflicts. However, no explicit consideration of these potential conflicts is made.

The final step is used only if a terminal failed to enter the channel in step 1. In this situation, a new track is added and the pin connected to the empty track. After processing the entire column, the router extends the trunks of the nets, which continue to the next column, and repeats the above procedure.

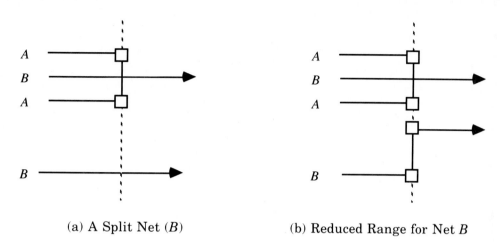

(a) A Split Net (*B*)        (b) Reduced Range for Net *B*

**Figure 5.13**  Reducing Range of Split Nets (step 3)

In some cases, the algorithm uses heuristics that select which of the many possible paths are accepted. In other cases, the router can exhaustively try all possible alternatives, since the number of tracks at any given column is limited.

Because the greedy channel router contains fewer assumptions about the topology of the connections, it is more flexible in the placement of doglegs than the left-edge approaches. It does not impose the restriction of one trunk section per net or one trunk section per terminal pair. The greedy router often results in fewer tracks than a left-edge solution but also uses more vias and more wire bends in the solution.

**Hierarchical Channel Router.**    A third approach to channel routing was proposed by Burstein [Bur83a]. This approach looks at the problem in a hierarchical fashion. Initially the channel is split horizontally, forming two routing regions. Using a Two-By-N (TBN) algorithm, where *N* is the length of the channel, Burstein allocates each of the connections to one of the two regions. The TBN algorithm routes each connection serially, but performs random rerouting of connections after all connections are routed. This rerouting helps smooth out the congestion and often allows the process to continue without adding additional tracks. Additional tracks are added only when it is impossible to route a connection without violating the current channels' capacities.

The process of splitting the routing region is repeated in a hierarchical fashion until each channel section contains only a single track. This decomposition of the channel allows the generation of very robust Steiner points during the routing and tends to produce results very close to channel density.

**Channel Router Summary.**   Channel routers are the workhorse of IC routing. Most commercial routing systems are based on channel routers. Because of their widespread use, they have also received widespread publication, making the channel routing problem well understood.

"Deutsch's difficult example," first published in [Deu76], has become the benchmark by which channel routers are measured. It is a single channel from an actual design which has been shown to be very difficult to route in density (19 tracks). Table 5.1 summarizes the results obtained on this example using various routers. The basic LEA (one track per net, starting at the lower left corner) requires 31 tracks to route this channel. When extended to use doglegs, and the eight track packing combinations obtained by alternating top and bottom sides, LEA produces a 21-track solution. It has recently been discovered by the author that a 20-track solution can be obtained using the dogleg approach and packing each track from left to right beginning at the bottom (the original LEA track packing order). This solution, which is presented in Table 5.1 as "New Dogleg," occurs when the range value is set to four. Greedy and hierarchical algorithms achieve better solutions in terms of tracks but at the expense of many extra vias and bends, which can cause manufacturing problems.

| Router | Tracks | Vias | Wire Length | Source |
|--------|--------|------|-------------|--------|
| LEA | 31 | 290 | 6526 | * |
| Original Dogleg | 21 | 346 | 5331 | [Deu76] |
| Greedy | 20 | 403 | 5381 | [Riv82a]** |
| New Dogleg | 20 | 329 | 5078 | * |
| Hierarchical | 19 | 336 | 5023 | [Bur83] |
| YACR2 | 19 | 287 | 5020 | [Ree85] |

*LEA and New Dogleg results were obtained by the author using a version of LEA developed at the Microelectronics Center of North Carolina.

**Original Greedy results reported in [Riv82a] were in error. Greedy results reported here were obtained by private communication with David Deutsch of Bellcore.

**Table 5.1**  Published Results for Deutsch's Difficult Example

Another criterion for comparing channel routing algorithms is their adaptability to variations in the routing problem. Such variations as terminals not on-grid or channels which are not rectangular are more easily accommodated by some algorithms than by others. Greedy routers and hierarchical routers require that all terminals be on-grid. In fact, the spacing between vertical grid lines must be at least as large as the via-to-via spacing requires. LEA-based routers can be readily modified for off-grid terminals (see [Deu76]), and the terminal spacing can be as small as the via-to-line spacing. LEA-based routers are also more easily adaptable to non-rectangular channels than the other approaches. Greedy routers, on the other hand, can route problems with fixed terminals on three sides of the channel, which the other approaches cannot.

With the advent of routing compaction techniques (see Section 5.5), the channel routing problem has been cast in a new light. The problem is no longer to produce a routing with the fewest tracks, but to produce one which compacts to the smallest overall height [Deu85, Roy87].

A problem in channel routing is how to apply this technique to newer technologies that offer more than two layers of interconnect. While there has been some promising work done in this area [Bru85, Bra86, Con87, Jou87], more effort is required to put these algorithms into practical, widespread use. These approaches assume that the design rules for all three layers are the same, or at least close enough to allow a three-layer grid to be superimposed on the channel. This is not always the case in practice.

### Switchbox Routing Algorithms

The switchbox routing problem consists of a rectangular routing area with fixed pin locations on all four sides of the region. Unlike channels, which have adjustable height, the size of the switchbox routing region is fixed. Switchboxes, often referred to as "four-sided channels," are usually formed at the intersection of two or more channel routing regions and occur in general cell layout systems.

**Extensions to Channel Routing Algorithms.** Both the hierarchical and greedy channel routing algorithms have been extended to work with the switchbox problem. In the hierarchical algorithm, Burstein [Bur83b] extended the original algorithm by alternating the direction of the "cutlines" used to partition the routing area.

The switchbox router called "Detour" [Ham84] is essentially an extension of the greedy channel router mentioned above. Included as a part of the Magic VLSI layout system [Ous84], Detour extends the notion of "rising" and "falling" nets (those that prefer the top or bottom track, respectively) to include nets with pins at the end of the switchbox region. Detour also includes mechanisms to "split" a net into two tracks to accommodate nets with multiple pins on the "far" side of the region (such as net $C$ in Figure 5.14). Nets that contain pins on only the right edge of the region (such as net $D$ in Figure 5.14) also present a problem. These nets must be introduced at some column prior to reaching the right end of the switchbox. The router adds these nets near the end of the routing region, at a column where two trunk wires can be added.

Another interesting feature of the Detour router is that it begins at the point of maximum density and operates outward in both directions. This variation of the greedy router was suggested (but not implemented) by Rivest and Fiduccia [Riv82a]. The routing region is split into two pieces at the column of maximum density, and tracks are assigned for this column, creating two

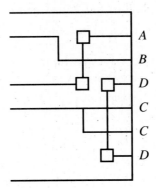

**Figure 5.14**  Problem Nets for a Greedy Switchbox Router

independent switchbox problems. The intent of splitting the channel (or switchbox) into two regions is to improve the routability of each subchannel. This improvement ensures that there is enough space to route all of the connections, especially if the width of the routing area is close to the density of the channel. It also helps to remove some of the vertical constraints.

**An Artificial Intelligence-Based Approach.**  A totally different approach was taken by Joobbani and Siewiorek [Joo86] in the development of a router called "Weaver." Weaver combines algorithmic approaches like the vertical constraint graph with simple deductive and expert knowledge. Since there are many metrics that need optimization, Weaver provides an "expert" on each: constraint propagation, wire length, congestion, *etc.* These experts observe the problem and make suggestions based on their own area of concern. A scheduler, an expert system itself, then decides on the best application of all suggestions received.

For example, one of Weaver's expert systems, the Via expert, suggests alternatives to the routing of a completed net. It attempts to remove unnecessary vias by changing the layers of some wire segments. This expert not only reduces the number of vias required, but also frees up some routing area on the segment's original layer. In addition to its "staff" of ten experts, Weaver provides the user with the ability to act as an additional resource. The user can override any system decision and either pre-route or delete wiring segments. Weaver is also free of the constraints of keeping all of the horizontal (or vertical) wiring on a single layer. Combined with the global view made possible by the expert systems' implementation, Weaver is able to route Burstein's difficult switchbox automatically [Bur83b] using fewer vias and less wiring than a manually guided solution obtained by the greedy approach outlined above. Additional details on Weaver are contained in Chapter 9 of this book. The concepts of least commitment and graceful retreat in switchbox routing are also discussed in [Ho85, Mar85].

## River Routing

The *river routing* problem is a special case of the channel routing problem. In a river routing problem, all nets have exactly two terminals, one on the top of the channel and the other on the bottom. Furthermore, the nets occur in the same left to right order on the top layer as they do on the bottom. Because of this topology, all wiring is to occur on a single layer, and the goal (as in channel routing) is to minimize the height of the channel.

River routing was initially studied for its applications in the PCB area, but is applicable to today's IC designs, particularly in dataflow architectures where multi-bit buses flow from one block of logic to the next. River routing is also applicable to symbolic IC design systems. Here, after mapping from symbolic to actual dimensions and compacting, the terminals of different modules may no longer directly connect. There is already an implicit ordering of the connections, and routing in a single layer may be desirable.

River routing is a special case of the more general *planar routing problem*. Planar routing is a problem in which the interconnection topology of the nets is planar. That is, all connections can be realized on a single layer.

A small example of a planar routing problem is shown in Figure 5.15, which is from [Pin83]. Notice that the terminals occur on the periphery of a rectangular routing region. River routing is to channel routing as planar routing is to switchbox routing. The algorithms for both are quite similar. We will concentrate on the more general planar routing problem.

The first step in the process is to determine if the problem can be routed in a single layer. While there are general-purpose algorithms for determining the planarity of a graph, in this case it is only necessary to check if the terminals along the boundary of the switchbox match without intersecting.

This check is easily done in the following manner:

(1)    Choose a starting terminal, and place it onto a stack.

(2)    Scan the list of terminals around the boundary. For each terminal encountered, if the net of the terminal is the same as the terminal at the top of the stack, pop the stack. If they are not in the same net, push the terminal onto the stack.

(3)    If, after scanning all terminals, the stack is empty, the routing problem is planar. Otherwise, it cannot be realized in a single layer.

This technique is only applicable to two-terminal nets. Generalizing it to multi-terminal nets is left as an exercise for the reader.

After determining that the problem can be routed in a single layer, the actual routing is done. This routing proceeds in a fashion similar to the algorithm presented above, but whenever a net is "popped" off the stack, it is routed. This routing is done by contouring along the outer boundary, with the routing path staying as close as possible to the boundary formed by the edges of the routing region or the wiring paths already defined.

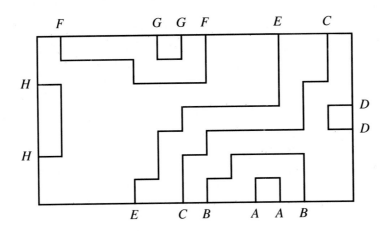

**Figure 5.15** A Planar Routing Example

In a similar algorithm, Hsu [Hsu83a] provides additional heuristics to reduce the number of bends and to shorten wiring paths without impairing the final solution.

## 5.4  Global Routing

The purpose of a global router is to decompose a large routing problem into small, manageable problems so that the previously discussed detailed routers can be applied. This decomposition is done by assigning each net to a set of routing regions (channels, switchboxes or river-route areas). The objective of this assignment is to plan the routing globally to reduce chip size, shorten connection length (and, therefore, delay), improve routability, and balance congestion across the layout surface. In some references, this process is also called loose routing or topological routing.

Before global routing can begin, channels must be defined on the routing surface. Channel definition is straightforward for standard cells and gate arrays because they are usually laid out in rows or columns with intervening empty areas for routing. Figure 5.16 illustrates a typical floorplan for gate array or standard cell designs. Channel definition is less obvious for general cell layouts.

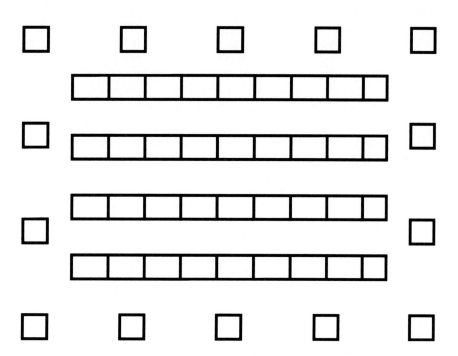

**Figure 5.16** Typical Gate Array or Standard Cell Floorplan

A good explanation of the considerations involved is given in [Per81]. Channel definition for general cells is highly dependent on placement. It has been proven that order constraints for channel routing are acyclic if, and only if, the channel structure corresponds to a slicing of the chip [Sup83]. Partitioning-based placement techniques described in the previous chapter have the desirable property of creating such a structure as a natural side effect. They are, there-fore, well suited for general cell layouts which use channel routers. Much work has been done to automatically generate channel structures for general cell lay-outs [Cho85a].

### Global Routing Models

There are two models used in global routing, those which model the channel structure in the form of a channel-graph and those which model the routing sur-face as a coarse routing grid. Both of these models ultimately view the problem as finding paths in a weighted graph, and many of the same algorithms apply to both models.

**Channel-Graph Global Routing Models.**    Preas [Pre79] defined a *channel intersection graph* in which each vertex represents a channel intersection and each edge represents a channel or channel subsection. Figure 5.17 shows a general cell structure and the corresponding channel intersection graph. By assigning edge weights to model wire length and channel capacity, such a graph can be used to select a channel sequence for each net. Preas employs the shortest path algorithm of Dijkstra (described in Section 5.2), but any of a number of path finding algorithms can be used on this graph.

In gate arrays and standard cells, it is necessary to model the availability of feedthroughs. *Feedthrough*s are wiring paths which cross the cell rows and thereby provide paths between two channels which do not intersect. This means that connections can leave and enter the channels at places other than channel intersections, and therefore a finer resolution is required. To model this feature, the channel intersection graph must be refined by subdividing channels into *minichannels*. Figure 5.18 shows a channel broken into minichannels. Vertices appear at not only the channel intersections but also at the boundaries of the minichannels. Edges are added to the graph to represent feedthroughs which connect adjacent channels. The effect of minichannels is to model each channel not as a single edge but rather as a string of edges. The result is longer search time but more accuracy of the edge weights and more accurate modeling of a wire that does not span the entire length of the channel.

An alternate channel-graph model is arrived at by representing channels as vertices and channel adjacencies as edges (the opposite of what is done in the channel intersection graph) [Hig80]. The use of such a graph is the same as for the channel intersection graph. It is a weighted graph, and the edge weights model the length and capacity of the routing channels. The differences between these two models are subtle. For the channel intersection graph where the edges correspond to channels, the notion of density can be used to calculate the portion of the cost due to congestion. For the other model, edges represent the boundaries between adjacent channels. Remaining capacity can be readily determined at these boundaries, but the area of maximum density of a channel usually occurs near the middle rather than at the channel intersections. Representing the channel intersections as edges, therefore, may fail to model the most critical portion of the channel in terms of density. Employing minichannels alleviates this problem by adding edges at intervals along the length of the channel.

**Coarse Grid Global Routing Models.**    It is possible to take greater advantage of the regularity of standard cell and gate array structures by confining the graph to be a regular grid. Such an approach was taken by [Pat85], in which a row-column structure is imposed on the routing surface. The wiring capacity of the cells formed by this structure is readily computable. This structure is modeled as a grid graph much like that used in Lee routers, except that the wiring capacity in each direction is no longer limited to a single segment (see

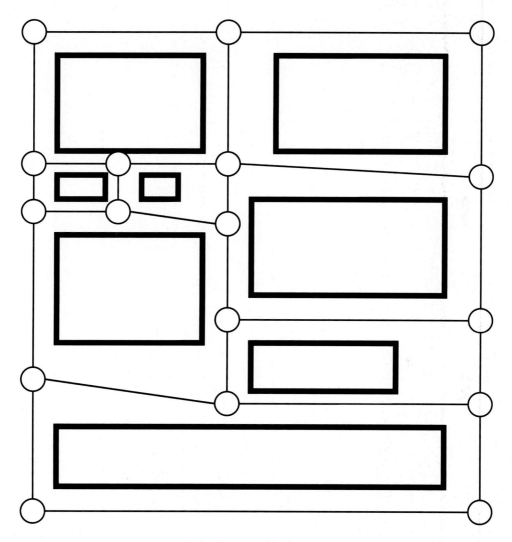

**Figure 5.17** Channel Intersection Graph

Figure 5.19). A cost is supplied to each edge in the graph based upon the remaining track capacity at the cell boundary. Note that exact positioning of the wires along the cell edge is not modeled; that is the task of the detailed router. During the global route stage, only the capacity of the cell boundary (*i.e.*, the number of wires that can cross it) and the number of wires that are currently assigned to cross that boundary need to be known. Wire length is modeled implicitly by the fact that each cell is the same size as all the others.

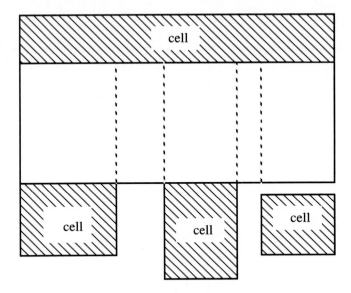

**Figure 5.18**  Horizontal Channel Broken into Minichannels

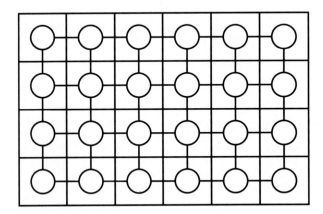

**Figure 5.19**  Coarse Grid and Corresponding Grid Graph

This consistency simplifies determination of minimum length paths. However, using such a model in the general cell case is complicated by the fact that cells are not usually permeable, and cell boundary capacities must reflect that.

## Global Routing Algorithms

Whichever model is selected, the task of the global router is to find a suitable path for each connection through the weighted graph. Ideally, a minimal cost Steiner tree would be found for each net, but this problem is NP-complete in the general case. Furthermore, since edge weights model wiring congestion, the optimal wiring for each net depends on the wiring for all the other nets. Consequently, heuristic techniques must be used which yield acceptable results that allow successful detailed routing of the circuit.

Many systems employ two phases: an initial global route followed by an iterative improvement phase.

**Initial Global Route Techniques.**   The first technique to be discussed is modeled after the Lee router presented earlier. Nets are sequentially selected for routing, and the edge weights are updated after each net is completed. Within each net, one pin is selected to be the source point, and symmetric expansion is performed until any one of the other pins in the net is encountered. This subtree is then used as the source of the next expansion, and the process continues until all pins in the net are in the same subtree. The global routing models contain many fewer vertices than the maze routing model; therefore, goal-directed searching is not usually necessary. The shortest path guarantee (see Section 5.3) is still viable.

As in the Lee router, this net-at-a-time routing is order dependent. There is a great deal of inconclusive study on connection ordering [Abe72], but it is certain the results are different for different connection orderings of the same problem. An advantage of such an approach is that nets designated as critical (because of delay considerations) can be favored by scheduling them early in the routing process.

Another approach to initial global routing is to construct an initial layout by determining the minimum-distance Steiner tree for each net, ignoring other nets in the process [Pat85]. By ignoring wiring capacities in this step, the connection ordering dependencies discussed previously are avoided. This approach requires the use of an improvement phase of the global router to alleviate congestion. In other words, some routing regions may be grossly oversubscribed, while others are left nearly empty. However, this technique gives a good starting point for improvement, because each net is known to have its minimum length path.

A third approach to global routing is to employ a least-commitment technique. For some simple nets, there is only one choice (or one obvious choice) for routing. For example, a connection that goes straight across a single channel has only one feasible alternative to consider.

This situation suggests a three-step approach [De86]:

(1)    Route the simple nets: those which have only two- or three-pin nets which occupy a single row. These comprise approximately 75% of the nets in a typical design.

(2)    Use a "close-to-optimal" Steiner tree algorithm to route nets of intermediate length.

(3)    Route remaining "big" nets (four or more terminals) using maze routing procedures.

The observation that 75% of the nets fall into the "simple" category is important. By laying them out first, we have a fairly accurate model of the channel congestion before beginning the longer connections (which usually have several alternative paths) in steps 2 and 3.

A least-commitment approach that is somewhat more dynamic is proposed in [Li84]. It is based on the realization that some cells (a vertex in the coarse grid model) have capacity for only the local connections through their boundaries. Such cells are designated as non-pass-through (NPT) cells. A cell is NPT if the number of unconnected pins inside of it is equal to or one less than the remaining capacities on its four boundaries. Nets with pins in the vicinity of NPT cells are sometimes forced to take unique routes.

The algorithm proceeds as follows:

(1)    Connect any nets that have only two pins which are located in two adjacent cells.

(2)    Identify NPT cells and other barriers (such as a string of zero capacity edges).

(3)    Find unique routes forced by the NPT cells and barriers. Make connections for nets which have one pin in an NPT cell and another pin in an adjacent cell.

(4)    Connect pin pairs, trying to keep to the outside of the grid structure. See [Li84] for details of this step.

(5)    Update the cell graph and create new NPT cells. Repeat step 3.

(6)    Remove redundant or dangling connections.

(7)    Go back to step 4 until no more connections can be embedded.

**Iterative Improvement Global Route Techniques.** The search-type, initial global routers are highly dependent on net ordering. The Steiner-type, initial global routers often result in channels which are oversubscribed. Both of these problems can be reduced by following the initial global routing with an improvement phase.

The basic idea is to start with a feasible global routing solution and improve it by repeatedly selecting connections, removing them, and reconnecting them by some alternate path. The various approaches are distinguished by how the connections are selected for removal and how they are reconnected.

One selection approach is to identify critical nets. Critical nets are those nets which, if removed, would make the design smaller [Pre79]. These nets can be readily identified by finding a position of maximum density along a critical channel. Nets which have segments within this area are potentially critical nets. These nets are then rerouted with a graph search technique. Because the graph is updated to reflect the current capacities (with all nets now routed), the path chosen is likely to be different from the initial routing solution arrived at with less complete congestion information.

An alternate approach, applicable to a coarse grid model, starts by selecting a rectangular grid graph which contains all of the pins of the given net and repeatedly removes the most expensive edge which does not disconnect the net [Pat85]. The grid structure makes the selection of an appropriate subgraph a straightforward procedure while still considering a large portion of the feasible solutions for the net. Of course, the boundary capacities must be updated after each iteration.

A final approach to iterative improvement global routing is to apply the technique of simulated annealing [Vec83]. (See Chapter 4 for more information on simulated annealing.) By considering L-shaped (and later Z-shaped) paths for each two-point connection, the annealing technique can be used to distribute the connections uniformly [Kir83].

### Global Router Summary

This chapter has presented two different weighted graph models for global routing and discussed several techniques for defining global routes within them. Unfortunately, there are no benchmark problems (such as Deutsch's difficult example for channel routers) which can be used to compare approaches. The quality of the global router result depends heavily on the detailed router employed and therefore is very difficult to measure.

Data structures for the "C" implementations of some of these techniques are given in [Xio86]. Also recommended for further reading are papers which employ very similar techniques without explicitly building the graph models described here [Ham85, Clo84].

## 5.5 Routing Refinement Techniques

### Routing Compaction

Channel routers inherently try to minimize the number of routing tracks required. They assume, however, that the routing tracks are straight lines and do not bend horizontal wires to reduce overall channel height. Deutsch [Deu85] has shown that by applying compaction techniques to completed routing channels, a 15%-20% reduction in size can be realized. (For comparison, an innovative channel routing algorithm that results in a two-track reduction for Deutsch's difficult example represents only a 10% size reduction.)

The details of this technique will not be elaborated here except to say that it is similar to the one-dimensional, grid-based compaction scheme presented in Chapter 6, with the further simplification that only wires and contacts appear in the channel. There are no transistors or wells to worry about. In the channel routing realm, this technique corresponds to bending each track (starting at the bottom) to fit into the contour of the track and wiring below it. This bending is followed by a step which attempts to remove the unnecessary bends. Figure 5.20 shows a sample channel section at the end of each of these steps.

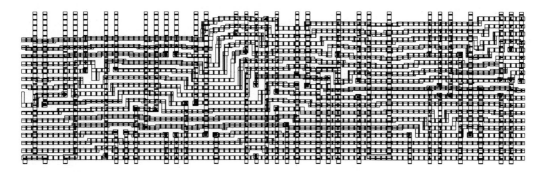

(a) A Compacted Channel

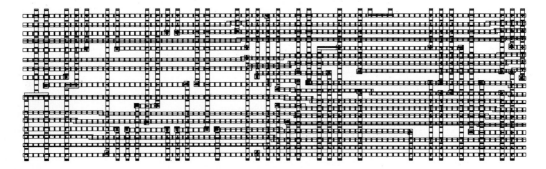

(b) The Straightened Version

**Figure 5.20**  Channel Routing Compaction

### Rip-Up and Shove-Aside

One hundred percent completion can be guaranteed for standard cell and general cell layouts due to the expandability of the routing channels. One hundred percent completion cannot be guaranteed for gate arrays and PCBs, in which the size of the routing space is fixed. Modifying existing routes to get in the last few connections has intuitive appeal; this is the way a manual designer would approach the problem. The idea is not to start over from placement and discard all the work done so far, but, rather, *shove-aside* existing wires to make room for new ones (Figure 5.21) or *rip-up* the wires that are in the way and route them differently (Figure 5.22). This process can be automated.

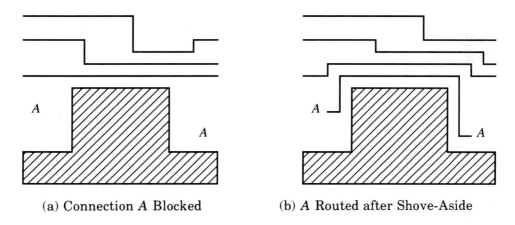

(a) Connection *A* Blocked          (b) *A* Routed after Shove-Aside

**Figure 5.21**  A Shove-Aside Operation

**Rip-Up and Reroute.**  The basic problem in ripping-up and rerouting connections is the selection of wires to remove. The key is to select wires which allow failures to be routed and which can be rerouted themselves by some alternate path.

One approach [Bol79] is to use a maze router which ignores blockage by other nets and removes those wires which interfere. Although this approach offers no guarantee of improvement, such iteration almost always outperforms a noniterative approach.

Dees and Karger [Dee82] made the observation that a large number of failures result from what they call "node isolation." That is, the endpoints of the connection become isolated from the rest of the routing surface. Based on this observation, they proposed a rip-up strategy which simply probes in the vicinity of the nodes of a failed connection and rips-up the wires which isolate it. This procedure results in reasonably good connectivity improvements and greatly reduces runtime because fewer wires are ripped-up at each step and less time is

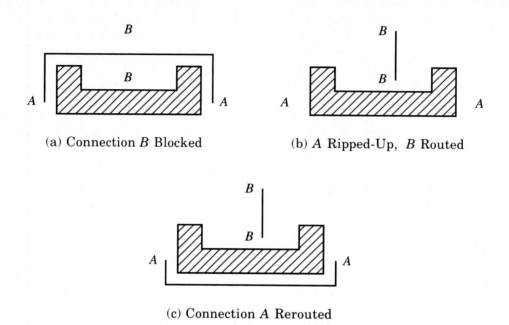

(a) Connection B Blocked          (b) A Ripped-Up,  B Routed

(c) Connection A Rerouted

**Figure 5.22**  A Rip-Up and Reroute Operation

spent locating them.  The trade-offs involved in different orderings of connections for rerouting are discussed in [Dee81].

**Shove-Aside.**    The rerouting problem can be avoided if wires can be moved aside without breaking existing connections.  The problem again is to identify the wires to be moved which will allow completion of failed connections.  A maze router can again be used for this task.  By defining "dummy" grid lines between the existing ones, a maze router can find paths which circumnavigate existing barriers [Rub74b].  Figure 5.23 illustrates the basic idea.  The actual shoving of the wires is a recursive process.  A vertical wire can be moved to the right if the grid line to the right is empty or occupied by another vertical wire which can be moved to the right.  Figure 5.24 illustrates the basic idea.  Automatic identification and movement of wires in this fashion performs effectively what would be several complicated steps of rip-up and retry iterations.

Shove-aside techniques are also employed in interactive systems such as the "plow" feature in Magic [Ous84].  In such a system the user performs the function of identifying the wires to be moved, and the program need only be concerned with the shoving-aside [Sco84].  The user can then route the connections through the newly available space.

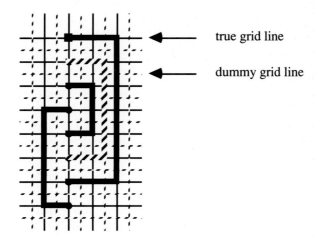

**Figure 5.23**  Dummy Wiring Grid

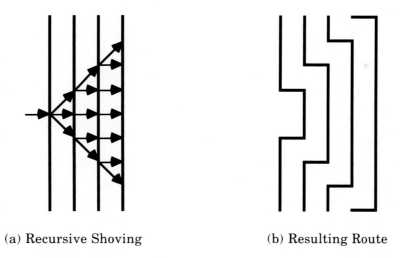

(a) Recursive Shoving        (b) Resulting Route

**Figure 5.24**  Recursive Shoving of Vertical Wires

## 5.6  Specialized Routers

This section discusses routing techniques that have been devised to solve very specialized problems.  These techniques are not as generally applicable as those previously presented.

### Power and Ground Routing

Routing of power and ground wiring requires special attention for two reasons:

- Due to added parasitics of contacts, they are usually defined on a single layer.

- They are usually much wider than other wires because they must carry more current.

For gate array designs, the power and ground wiring is designed by the designer of the masterslice and is the same for all personalizations of that masterslice. For standard cell designs the power and ground bus routing is somewhat straightforward. Power and ground wiring is defined as part of the cells and connected to the pad ring by predefined patterns. Normal wiring routed in the vicinity of the patterns must avoid the power wiring, but little additional special handling is required.

For general cell layouts, the situation is considerably more complex. In order to satisfy the first constraint mentioned above, these nets are usually routed using a tree-based approach. One such approach is to bring the ground wires in from one side of the chip and the power from the other and interdigitate the two patterns onto a single layer [Rot81].

The key to this technique of finding a planar solution is twofold: By selecting a routing order based on the horizontal distance of each connecting point from the edge of the chip, both nets are grown simultaneously. In addition, the following two conditions are imposed on the topology of the macro cells being connected:

- Each macro cell must have only a single connection point for each of the power and ground nets.

- The cells must be spaced such that both nets may be routed next to any macro cell edge.

Figure 5.25 shows a sample layout produced using this method.

Other approaches to finding a planar solution are based on congestion and "traffic rules" to prevent the nets from crossing [Sye82] or application of graph theory by finding a short Hamiltonian cycle that passes through all of the modules [Mou83].

Once a planar solution has been found, the wire widths need to be adjusted to satisfy the electrical demands of the circuit. In addition to metal migration and total current flow, the resistance (and corresponding voltage drop) must be considered. Since the methods mentioned above ensure a planar, or tree-structured, solution, it is possible by simple tree analysis to calculate the current flow in each branch.

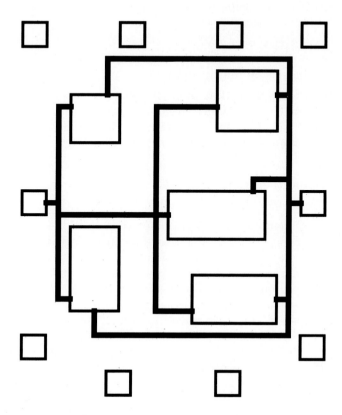

**Figure 5.25** Power and Ground Routing Example

This analysis can be used to determine the required width of each segment. Just meeting the current and voltage requirements may not give the best solution, since there are many choices [Cho85b]. To further constrain the solution, a mathematical formulation of the problem can be created that includes minimization of the total area taken up by the power and ground nets. Using a set of linear equations, an objective function which minimizes the total power and ground wiring area while satisfying all of the current and voltage drop conditions is imposed.

## Pin-Row Routers

PCBs present a slightly different environment for routing than integrated circuits. The space available for routing is fixed (as it is in gate arrays), and the surface is generally free of blockages (there are no transistors or other circuitry to be avoided). Furthermore, the pins of the devices tend to line up in an orderly pattern to form a two-dimensional matrix. Finally, there are often more than two layers available for interconnect.

Channel routers can be employed for two-layer PCBs; in fact, they were invented for such an environment. Other techniques have been developed which take advantage of the special features of PCBs.

Based on the observation that pins on a PCB are usually lined up in rows, techniques have been devised which weave a set of connections along a row. Effective algorithms [Tsu81, Car83, Tar84, Rag82, Han84] were developed to solve this unique problem and this led to a multi-step approach:

(1)    Allocate nets to sets of (horizontal and vertical) pin-rows.

(2)    Assign the connections to layers.

(3)    Apply a pin-row algorithm to the individual rows on each layer.

Doreau [Dor78] made several significant contributions in the development of pin-row routing. First, he devised a transformation of the pin-row problem to a permutation of the nets involved, with some constraints to model the capacity between pins. He was then able to apply a branch-and-bound technique to search through the permutations. Furthermore, he generalized the single-row problem to a double-row problem to model the intervening channel capacity more accurately. These techniques were applied in TWIGY [Dor81], a production routing system.

## 5.7  Routing in Layout Systems

This section will present a brief overview of how the algorithms discussed earlier in this chapter are put to practical use. While the choice of algorithms is somewhat subjective, this overview gives the reader a sense of the types of algorithms that may be used in an actual production layout system.

### Gate Array Routing Systems

Many current gate arrays are characterized by fixed rows or columns of cells. This arrangement easily lends itself to the use of a channel routing approach. However, since the routing space between cells is fixed, the channel router may fail to make all the connections in the given number of tracks. Additional routes may be needed to achieve one hundred percent of the connections.

**Channel Generation.**    This step is usually not required in a gate array system. The layout of the channels is fixed by the designer of the basic array.

**Global Routing.**    Since the placement of the rows or columns of cells defines the channel boundaries, the first step is to do the global routing. As previously shown in Figure 5.18, the channels can be divided into "minichannels" before building the channel intersection graph. This technique allows effective use of any feeds that may be present on the array. Also, when computing the edge weights of the weighted graph, the number of tracks in the channel is known and the goal of the global router is not to exceed that capacity for any channel.

**Power and Ground Routing.** This step is usually not required in a gate array system. The bussing is done by the designer of the basic array. It is only necessary to avoid this pre-determined power wiring during the detailed routing phases which define the signal wiring.

**Channel Routing.** After completing the global routing phase, any of the channel routing algorithms may be used. However, they are slightly modified, since they are constrained to a fixed area [Dee85]. This constraint also means that a cleanup router may be necessary.

**Cleanup Routing.** The cleanup router is usually based on one of the maze routing algorithms for two reasons: they guarantee to find a solution if it exists, and they are topologically less restrictive than the channel router. However, even a maze router may not be able to complete all of the connections. In these cases rip-up and reroute approaches may be used.

**Interactive Routing.** Most layout systems also allow the designer to adjust the wiring manually. Several systems use a manually directed, line-probe router to assist the designer in determining the wiring paths that may be inserted [Hig85].

In addition to the standard gate arrays described above, "sea-of-gates" style arrays are becoming more popular. In these arrays, the placement of cells onto the routing surface defines the channel areas. Even though the widths of the routing channels are not pre-determined, the amount of space available is still fixed, and the routing algorithms mentioned above still apply.

### Standard Cell Routing Systems

As in the gate array problem outlined above, simple standard cell systems are characterized by fixed height rows of cells and lend themselves to the use of a channel routing approach. However, unlike gate arrays, the size of the layout surface is not pre-determined. As opposed to gate arrays, the problem is not to make one hundred percent of the connections in a given amount of space, but to take the smallest amount of space to make one hundred percent of the connections. While this is a subtle distinction, it is the key difference between the routers for these design styles.

**Channel Generation.** Since the placement of the rows may be flexible, many systems employ some type of channel generating step. This step is especially important in systems that allow combinations of large functional blocks along with the simple, standard-height cells.

**Global Routing.** Since there may not be sufficient paths across the cell rows after placing the standard cells, most global routing systems include the addition of simple *feedthrough* cells. Feedthrough cells are simple cells that contain only a vertical wire plus the power buses for the row. They are inserted between the cells in a given row. Their job is to provide additional feedthroughs while maintaining the power and ground connections for the cell row. As with gate arrays, the channels can be divided into "minichannels" before building the channel intersection graph. The goal of the global router is to keep the design as small as possible. This goal translates into distributing connections so channels are fully utilized. If the local density of a channel is very high in a small region and very low elsewhere, there will be wasted space and an unnecessarily tall design. Edge weights, therefore, are usually based on the local density of the channel compared to the maximum density for the channel.

**Power and Ground Routing.**    After global routing, the first step in detailed routing is to connect the power and ground connections. This connection may involve routing the rings around the I/O section of the chip, as well as the connections of the individual rows of cells. However, it is a much simpler task than in the general macro cell case. This phase must also take into consideration the power requirements and size the buses accordingly.

**Channel Routing.**    Once the power and ground routing has been completed, any of the channel routing algorithms may be used. Again, the key is to minimize the amount of space used in completing one hundred percent of the connections. Depending on the topology of the channels generated, a switchbox router may also be included.

**Cleanup Routing.**    Since a complete solution is guaranteed by the channel routing phase, no cleanup router is needed. This removes many of the restrictions on the system, since no gridded model need be used.

**Interactive Routing.**    Most standard cell layout systems still allow the designer to adjust the wiring manually. Here, the intent is to provide tools to help minimize the area, or to "pre-route" critical signals such as clock lines.

### General Cell Routing Systems

As in standard cell systems, the objective of a general cell layout system is to complete the layout process using the minimum area.

**Channel Generation.**    Since the placement of general cells is more flexible, all cell systems employ some type of channel generator. In fact, cell generation may be combined with a floorplanning tool to aid the designer in the physical organization of his or her design.

**Global Routing.**    Since most general cells are not row based, there is no need to insert additional feedthroughs. However, some systems may use connection paths that have been built into the cell during the global routing phase. The goal of the router is to keep the chip area as small as possible. This is done by calculating edge weights based on local density versus maximum density for the channel. Additionally, the global router must keep track of which channels can and cannot be expanded without increasing chip area [Pre79].

**Power and Ground Routing.** As in standard cell systems, the power and ground connections need to be completed as outlined in Section 5.6. These connections may be done either before or in parallel with other detailed routing.

**Channel Routing.** Again, any of the channel routing algorithms may be used. Since macro cell layouts are much more likely to include three- or four-sided routing regions, a switchbox router or modified channel router is a necessity.

**Cleanup Routing.** As in the standard cell system, no cleanup router is needed.

**Interactive Routing.** Most macro cell layout systems also allow the designer to adjust the wiring or cell placement manually. Again, the intent is to provide tools to help minimize the chip area of the final design.

### Summary of Routing in Layout Systems

Integration of different routing tools into a routing system is a large problem in itself. The quality of the global router depends heavily on the channel router used: the channel router must create wires on-grid if a gridded cleanup router is to be used, and so on. Several system overview papers for gate array [Tin83, Hig80] and standard cell [Anw85, Hsu85, Per76] and general cell layout [Lau85, Che83, Pre78, Kim83, Riv82b] are available.

## 5.8 Current Research Topics in Routing

It is appropriate to conclude this chapter with some speculation on the next breakthroughs in routing research. There are two areas which have been receiving a great deal of attention lately but which have not as yet been employed in production routing systems.

### Hardware Accelerators

Since routing is a computationally intensive task, it is an appropriate domain to explore special-purpose computing architectures. Investigations so far have concentrated on architectures for maze routers. The two-dimensional grid structure of maze routers leads naturally to an approach using a two-dimensional array of interconnected processors. Schemes for mapping a routing surface onto such an array have been proposed [Bla81, Suz86, Nai82, Ios86].

The two-dimensional nature of the regular grid structure also suggests that special-purpose engines for image processing tasks might be used for maze routing. Rutenbar [Rut84] has reported on such an approach.

### Topological Routers

Another interesting area of exploration is that of routers which attempt to model only the relative positioning of wires during interconnection and delay transformation to the realm of geometric coordinates [Hit69, Lor83, Kes86].

Such an approach involves dividing the routing surface into regions, much like channel routing, but without the assumptions about one direction per layer. Layering is done during the global route phase, during which wires remain very pliable and are pushed aside, when necessary, to make room for new connections. Techniques for routing the individual regions have been devised by several researchers [Hsu83b, Mar83].

The convention of designating one layer for horizontal wires and the other for vertical wires was devised to facilitate the task of (human and automatic) routers. It does, however, represent a major (and somewhat arbitrary) restriction on the solution space of the general routing problem. Topological routing is a promising approach, but still needs some development work to be useful in a production system.

## Acknowledgements

We would like to thank Bryan Preas, James Cohoon, Bill Dees, Dave Deutsch, and the other co-authors of this book for their comments and critiques of early drafts of this chapter. In addition, we would like to thank the many people with whom we have spent countless hours discussing various aspects of automated routing over the years, especially Rob Smith, Dave Hightower, Bill Dees, and Bryan Preas. While the opinions expressed in this chapter are not necessarily those of our employers, we would like to thank them for their support of this project.

## Exercises

5.1    Explain the advantages and disadvantages of using a maze router compared to a line-probe router.

5.2    Describe two enhancements to Lee's algorithm and their benefits/drawbacks.

5.3    Describe the three basic channel routing algorithms.

5.4    Compare the advantages and drawbacks to using a line-probe router for final cleanup, as opposed to a maze router. Why is this "cleanup" phase necessary?

5.5    Why do designers use channel and switchbox routers rather than maze or line-probe routers?

5.6†   What restrictions are usually placed on general cell power/ground routing algorithms?

**5.7†**  Route the channel in Figure 5.26 using the following:

  **a)**  The left-edge algorithm. First, build a complete constraint graph. Constraint loops (if any) may be broken by inserting doglegs.

  **b)**  The greedy channel router.

  **c)**  The hierarchical channel router.

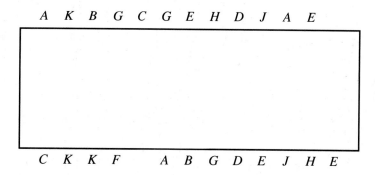

*A   K   B   G   C   G   E   H   D   J   A   E*

*C   K   K   F        A   B   G   D   E   J   H   E*

**Figure 5.26**  A Channel Routing Problem

**5.8†**  Apply one-dimensional compaction to the results obtained in Exercise 5.7a.

**5.9†**  Construct an instance where the depth-first search heuristic of Soukup improves the performance of a basic maze router algorithm. Construct an instance where this heuristic does not improve the performance of a basic maze router. Under what conditions is it appropriate to apply the heuristic?

**5.10‡**  Design and implement a simple, two-layer maze router. Experiment with different connection lengths and ordering schemes.

**5.11†**  Write a program which interconnects the terminals of a net within a gridded model using:

  **a)**  A minimal Steiner tree with trunk (allow the trunk to be either horizontal or vertical).

  **b)**  A minimal spanning tree.

  **c)**  A minimal chain.

  See Figure 4.2 for descriptions of these topologies.

**5.12‡** Design and implement a program to determine the vertical constraint graph of a channel. Develop heuristics to remove loops from the graph by inserting doglegs.

**5.13‡** Design and implement a program to construct a channel graph given the description of the cells and their locations.

**5.14‡** Design and implement a simple general cell global router.

## References

[Abe72]   Abel, L., "On the ordering of connections for automatic wire routing," *IEEE Trans. on Computers*, pp. 1227-1233, November 1972.

[Ake67]   Akers, S. B., "A modification of Lee's path connection algorithm," *IEEE Trans. on Electronic Computers*, pp. 97-98, February 1967.

[All76]   Allen, J. R., "A topologically adaptable cellular router," in *Proc. 13th Design Automation Conf.*, pp. 161-167, June 1976.

[Anw85]   Anway, H., G. Farnum, and R. Reid, "PLINT layout system for VLSI chips," in *Proc. 22nd Design Automation Conf.*, pp. 449-452, June 1985.

[Bar81]   Barr, A., and E. A. Feigenbaum, *The Handbook of Artificial Intelligence, Volume I*, William Kaufmann, Inc., Los Altos, CA, 1981.

[Bla81]   Blank, T., M. Stefik, and W. vanCleemput, "A parallel bit map processor architecture for DA algorithms," in *Proc. 18th Design Automation Conf.*, pp. 837-845, June 1981.

[Bol79]   Bollinger, H., "A mature DA system for PC layout," in *Proc. 1st Intl. Printed Circuit Conf.*, pp. 85-99, 1979.

[Bra86]   Braun, D., J. Burns, S. Devadas, H. K. Ma, K. Mayaram, F. Romeo, and A. Sangiovanni-Vincentelli, "Chameleon: a new multi-layer channel router," in *Proc. 23rd Design Automation Conf.*, pp. 495-502, June 1986.

[Bru85]   Bruell, P., and P. Sun, "A 'greedy' three layer channel router," in *Digest Intl. Conf. on Computer-Aided Design*, pp. 298-300, November 1985.

[Bur83a]  Burstein, M., and R. Pelavin, "Hierarchical channel router," in *Proc. 20th Design Automation Conf.*, pp. 591-597, June 1983.

[Bur83b]  Burstein, M., and R. Pelavin, "Hierarchical wire routing," *IEEE Trans. on Computer-Aided Design*, vol. CAD-2, no. 4, pp. 223-234, October 1983.

[Car83]   Carter, H. W., and M. A. Breuer, "Efficient single-layer routing along a line of points," *IEEE Trans. on Computer-Aided Design*, vol. CAD-2, no. 4, pp. 259-266, October 1983.

[Che83]   Chen, N. P., C.-P. Hsu, E. S. Kuh, C. C. Chen, and M. Takahashi, "BBL: a building-block layout system for custom chip design," in *Digest Intl. Conf. on Computer-Aided Design*, pp. 40-41, 1983.

[Cho85a]  Chow, C.-S., "Phoenix: An Interactive Hierarchical Topological Floorplanning Placer," Master's Thesis, Department of Electrical Engineering and Computer Sciences, Massachusetts Institute of Technology, June 1985.

[Cho85b]  Chowdhury, S., and M. A. Breuer, "The construction of minimal area power and ground nets for VLSI Circuits," in *Proc. 22nd Design Automation Conf.*, pp. 794-797, June 1985.

[Clo84]   Clow, G. W., "A global routing algorithm for general cells," in *Proc. 21st Design Automation Conf.*, pp. 45-51, June 1984.

[Con87]   Cong, J., D. F. Wong, and C. L. Liu, "A new approach to the three layer channel routing problem," in *Digest Intl. Conf. on Computer-Aided Design*, pp. 378-381, 1987.

[De86]    De, V. K., "A Heuristic Global Router for Polycell Layout," Master's Thesis, Department of Electrical Engineering, Duke University, 1986.

[Dee81]   Dees, W. A., Jr., and R. J. Smith II, "Performance of interconnection rip-up and reroute strategies," in *Proc. 18th Design Automation Conf.*, pp. 382-390, June 1981.

[Dee82]   Dees, W. A., Jr., and P. G. Karger, "Automated rip-up and reroute techniques," in *Proc. 19th Design Automation Conf.*, pp. 432-439, June 1982.

[Dee85]   Dees, W. A., Jr., and M. J. Lorenzetti, "Channel routing in a gate array environment," in *Proc. IEEE Intl. Conf. on Computer Design*, pp. 458-462, October 1985.

[Deu76]   Deutsch, D. N., "A 'dogleg' channel router," in *Proc. 13th Design Automation Conf.*, pp. 425-433, June, 1976.

[Deu85]   Deutsch, D. N., "Compacted channel routing," in *Digest Intl. Conf. on Computer-Aided Design*, pp. 223-225, November 1985.

[Dij59]   Dijkstra, E. W., "A note on two problems in connexion with graphs," *Numerische Mathematik*, vol. 1, pp. 269-271, 1959.

[Dor78]   Doreau, M. T., and L. Abel, "A topologically based non-minimum distance routing algorithm," in *Proc. 15th Design Automation Conf.*, pp. 92-99, June 1978.

[Dor81]  Doreau, M. T., and P. Koziol, "TWIGY — a topological algorithm based routing system," in *Proc. 18th Design Automation Conf.*, pp. 746-755, June 1981.

[Ham84]  Hamachi, G. T., and J. K. Ousterhout, "A switchbox router with obstacle avoidance," in *Proc. 21st Design Automation Conf.*, pp. 173-179, June 1984.

[Ham85]  Hamachi, G. T., and J. K. Ousterhout, "Magic's obstacle-avoiding global router," in *1985 Chapel Hill Conf. on Very Large Scale Integration*, pp. 145-164, Computer Science Press, Inc., Rockville, Maryland, 1985.

[Han84]  Han, S., and S. Sahni, "Single-row routing in narrow streets," *IEEE Trans. on Computer-Aided Design*, vol. CAD-3, no. 3, pp. 235-241, July 1984.

[Has71]  Hashimoto, A., and J. Stevens, "Wire routing by optimizing channel assignment within large apertures," in *Proc. 8th Design Automation Workshop*, pp. 155-163, 1971.

[Hey80]  Heyns, W., W. Sansen, and H. Beke, "A line-expansion algorithm for the general routing problem with a guaranteed solution," in *Proc. 17th Design Automation Conf.*, pp. 243-249, June 1980.

[Hig69]  Hightower, D. W., "A solution to the line routing problem on a continuous plane," in *Proc. 6th Design Automation Workshop*, pp. 1-24, 1969.

[Hig73]  Hightower, D. W., "The interconnection problem — a tutorial," in *Proc. 10th Design Automation Conf.*, pp. 1-21, June 1973.

[Hig80]  Hightower, D. W., and R. Boyd, "A generalized channel router," in *Proc. 17th Design Automation Conf.*, pp. 12-21, June, 1980.

[Hig83]  Hightower, D. W., "The Lee router revisited," in *Proc. Intl. Conf. on Computer Design*, pp. 136-139, October 1983.

[Hig85]  Hightower, D. W., "A manually driven line search router in a line editor," in *Digest Intl. Conf. on Computer-Aided Design*, pp. 276-278, November 1985.

[Hit69]  Hitchcock, R. B., "Cellular wiring and the cellular modeling technique," in *Proc. 6th Design Automation Workshop*, pp. 25-41, June 1969.

[Ho85]  Ho, W. P.-C., D. Y. Y. Yun, and Y. H. Hu, "Planning strategies for switchbox routing," in *Proc. Intl. Conf. on Computer Design*, pp. 463-467, October 1985.

[Hsu83a]  Hsu, C.-P., "General river routing algorithm," in *Proc. 20th Design Automation Conf.*, pp. 578-583, June 1983.

[Hsu83b] Hsu, C.-P., "Minimum-via topological routing," *IEEE Trans. on Computer-Aided Design*, vol. CAD-2, no. 4, pp. 235-246, October 1983.

[Hsu85] Hsu, C.-P., B. N. Tien, K. Chow, R. A. Perry, and J. Tang, "ALPS2, a standard cell layout system for double-layer metal technology," in *Proc. 22nd Design Automation Conf.*, pp. 443-448, June 1985.

[Hul78] Hulme, B. L., and J. A. Wisniewski, "A comparison of shortest path algorithms applied to sparse graphs," *Technical Report at Sandia Laboratories*, vol. NUREG/CR-0293, SAND78-1411, RS, National Technical Information Service, Springfield, VA 22161, August 1978.

[Ios86] Iosupovici, A., "A class of array architectures for hardware grid routers," *IEEE Trans. on Computer-Aided Design*, vol. CAD-5, no. 2, pp. 245-255, April 1986.

[Joo86] Joobbani, R., and D. Siewiorek, "WEAVER: a knowledge-based routing expert," *IEEE Design and Test of Computers*, vol. 3, no. 1, pp. 12-23, February 1986.

[Jou87] Jou, J. M., J. Y. Lee, and J. F. Wang, "A new three-layer detailed router for VLSI layout," in *Digest Intl. Conf. on Computer-Aided Design*, pp. 382-385, 1987.

[Ker73] Kernighan, B. W., D. G. Schweikert, and G. Persky, "An optimum channel-routing algorithm for polycell layouts of integrated circuits," in *Proc. 10th Design Automation Workshop*, pp. 50-59, 1973.

[Kes86] Kessenich, J., and G. Jackoway, "Global forced hierarchical router," in *Proc. 23rd Design Automation Conf.*, pp. 798-802, June 1986.

[Kim83] Kimura, S., N. Kubo, T. Chiba, and I. Nishioka, "An automatic routing scheme for general cell LSI," *IEEE Trans. on Computer-Aided Design*, vol. CAD-2, no. 4, pp. 285-292, October 1983.

[Kir83] Kirkpatrick, S., C. D. Gelatt, Jr., and M. P. Vecchi, "Optimization by simulated annealing," *Science*, vol. 220, no. 4598, pp. 671-680, 1983.

[Lau85] Lauther, U., "Channel routing in a general cell environment," in *VLSI 85, VLSI Design of Digital Systems*, ed. E. Horbst, pp. 393-403, Elsevier Science Publishers (North Holland), Amsterdam, the Netherlands, 1986.

[Lee61] Lee, C., "An algorithm for path connections and its applications," *IRE Trans. on Electronic Computers*, VEC-10, pp. 346-365, September, 1961.

[Li84] Li, J-.T., and M. Marek-Sadowska, "Global routing for gate array," *IEEE Trans. on Computer-Aided Design*, vol. CAD-3, no. 4, pp. 298-307, October 1984.

[Lor83]    Lorenzetti, M. J., "Algorithms and Models for a Topologically Based Interconnection Routing System," Ph.D. Dissertation, The University of Texas at Austin, May 1983.

[Mar83]    Marek-Sadowska, M., and T. T.-K. Tarng, "Single-layer routing for VLSI: analysis and algorithms," *IEEE Trans. on Computer-Aided Design*, vol. CAD-2, no. 4, pp. 246-259, October 1983.

[Mar85]    Marek-Sadowska, M., "Two-dimensional router for double layer layout," in *Proc. 22nd Design Automation Conf.*, pp. 117-123, 1985.

[Mik68]    Mikami, K., and K. Tabuchi, "A computer program for optimal routing of printed circuit connectors," *IFIPS Proc.*, vol. H47, pp. 1475-1478, 1968.

[Moo59]    Moore, E. F., "The shortest path through a maze," *Annals of the Computation Laboratory of Harvard University*, vol. 30, pp. 285-292, Harvard University Press, 1959. (Also appears in *Proc. of International Symposium on Theory of Switching*, 1957.)

[Mou83]    Moulton, A. S., "Laying the power and ground wires on a VLSI chip," in *Proc. 20th Design Automation Conf.*, pp. 754-755, June 1983.

[Nai82]    Nair, R., S. J. Hong, S. Liles, and R. Villani, "Global wiring on a wire routing machine," in *Proc. 19th Design Automation Conf.*, pp. 224-231, June 1982.

[Oht86]    Ohtsuki, T., "Maze-running and line-search algorithms," in *Layout Design and Verification*, ed. T. Ohtsuki, pp. 99-131, North Holland, Amsterdam, 1986.

[Ous84]    Ousterhout, J., G. Hamachi, R. Mayo, W. Scott, and G. Taylor, "Magic: a VLSI layout system," in *Proc. 21st Design Automation Conf.*, pp. 152-159, June 1984.

[Pat85]    Patel, A. M., N. L. Soong, and R. K. Korn, "Hierarchical VLSI routing — an approximate routing procedure," *IEEE Trans. on Computer-Aided Design*, vol. CAD-4, no. 2, pp. 121-126, April 1985.

[Per76]    Persky, G., D. N. Deutsch, and D. G. Schweikert, "LTX — a system for the directed automatic design of LSI circuits," in *Proc. 13th Design Automation Conf.*, pp. 399-407, June 1976.

[Per81]    Persky, G., C. Enger, and D. M. Selove, "The Hughes automated layout system — automated LSI/VLSI layout based on channel routing," in *Proc. 18th Design Automation Conf.*, pp. 22-28, June 1981.

[Pin83]    Pinter, R. Y., "River routing: methodology and analysis," in *The Third Caltech Conference on VLSI*, pp. 141-163, March 1983.

[Pre78]    Preas, B. T., and C. W. Gwyn, "Methods for hierarchical automatic layout of custom LSI circuit masks," in *Proc. 15th Design Automation Conf.*, pp. 206-212, June 1978.

[Pre79]    Preas, B. T., "Placement and Routing Algorithms for Hierarchical Integrated Circuit Layout," Ph.D. Dissertation, Stanford University, August 1979.

[Rag82]    Raghavan, R., and S. Sahni, "Optimal single row router," in *Proc. 19th Design Automation Conf.*, pp. 38-45, June 1982.

[Ree85]    Reed, J., A. Sangiovanni-Vincentelli, and M. Santomauro, "A new symbolic channel router: YACR2," *IEEE Trans. on Computer-Aided Design*, vol. CAD-4, no. 3, pp. 208-219, July 1985.

[Riv82a]   Rivest, R. L., and C. M. Fiduccia, "A 'greedy' channel router," in *Proc. 19th Design Automation Conf.*, pp. 418-424, June 1982.

[Riv82b]   Rivest, R. L., "The 'PI' (placement and interconnect) system," in *Proc. 19th Design Automation Conf.*, pp. 475-481, June 1982.

[Rot81]    Rothermal, H., and D. Mlynski, "Computation of power supply nets in VLSI layout," in *Proc. 18th Design Automation Conf.*, pp. 37-42, June 1981.

[Roy87]    Royle, J., M. Palczewski, H. VerHeyen, N. Naccache, and J. Soukup, "Geometrical compaction in one dimension for channel routing," in *Proc. 24th Design Automation Conf.*, pp. 140-145, June 1987.

[Rub74a]   Rubin, F., "The Lee path connection algorithm," *IEEE Trans. on Computers*, vol. c-23, no. 9, pp. 907-914, September 1974.

[Rub74b]   Rubin, F., "An iterative technique for printed wire routing," in *Proc. 11th Design Automation Workshop*, pp. 308-313, 1974.

[Rut84]    Rutenbar, R. A., T. N. Mudge, and D. E. Atkins, "A class of cellular architectures to support physical design automation," *IEEE Trans. on Computer-Aided Design*, vol. CAD-3, no. 4, pp. 264-278, October 1984.

[Sat80]    Sato, K., H. Shimoyama, T. Nagai, M. Ozaki, and T. Yahara, "A 'grid-free' channel router," in *Proc. 17th Design Automation Conf.*, pp. 22-31, June 1980.

[Sco84]    Scott, W., and J. Ousterhout, "Plowing: interactive stretching and compaction in Magic," in *Proc. 21st Design Automation Conf.*, pp. 166-172, June 1984.

[Sou78]    Soukup, J., "Fast maze router," in *Proc. 15th Design Automation Conf.*, pp. 100-102, June 1978.

[Sou81]    Soukup, J., "Circuit layout," *Proc. of the IEEE*, vol. 69, no. 10, pp. 1281-1304, October 1981.

[Sup83]   Supowit, K. J., and E. A. Slutz, "Placement algorithms for custom VLSI," in *Proc. 20th Design Automation Conf.*, pp. 164-170, June 1983.

[Suz86]   Suzuki, K., Y. Matsunaga, M. Tachibana, and T. Ohtsuki, "A hardware maze router with application to interactive rip-up and reroute," *IEEE Trans. on Computer-Aided Design*, vol. CAD-5, no. 4, pp. 466-476, October 1986.

[Sye82]   Syed, Z., and A. El Gamal, "Single layer routing of power and ground networks in integrated circuits," *Journal of Digital Systems*, vol. 6, no. 1, pp. 53-63, 1982.

[Szy85]   Szymanski, T. G., "Dogleg channel routing is NP-complete," *IEEE Trans. on Computer-Aided Design*, vol. CAD-4, no. 1, pp. 31-41, January 1985.

[Tad80]   Tada, F., K. Yoshimura, T. Kagata, and T. Shirakawa, "A fast maze router with iterative use of variable search space restriction," in *Proc. 17th Design Automation Conf.*, pp. 250-254, June 1980.

[Tar84]   Tarng, T. T.-K., M. Marek-Sadowska, and E. S. Kuh, "An efficient single-row routing algorithm," *IEEE Trans. on Computer-Aided Design*, vol. CAD-3, no. 3, pp. 178-183, July 1984.

[Tin83]   Ting, B. S., and B. N. Tien, "Routing techniques for gate array," *IEEE Trans. on Computer-Aided Design*, vol. CAD-2, no. 4, pp. 301-312, October 1983.

[Tsu81]   Tsui, R., and R. J. Smith II, "A high-density multilayer PCB router based on necessary and sufficient conditions for single row routing," in *Proc. 18th Design Automation Conf.*, pp. 372-381, June 1981.

[Vec83]   Vecchi, M. P., and S. Kirkpatrick, "Global wiring by simulated annealing," *IEEE Trans. on Computer-Aided Design*, vol. CAD-2, no. 4, pp. 215-222, October 1983.

[Xio86]   Xiong, J. G., "Algorithms for global routing," in *Proc. 23rd Design Automation Conf.*, pp. 824-830, June 1986.

# Chapter 6

# Symbolic Layout and Compaction

**Wayne H. Wolf**

**Alfred E. Dunlop**

*AT&T Bell Laboratories*
*Murray Hill, New Jersey*

## 6.1 Introduction

Anyone who has designed a chip with graph paper and pencil or a computer-aided drafting system knows that layout design is a complicated, tedious, and error-prone process. The layout designer is driven by two conflicting demands: to produce as small a layout as possible, to minimize manufacturing cost and maximize circuit performance; and to create the layout quickly, to minimize design time and cost. The designer is responsible for both *topological design* — the relative placement of components and wires — and *geometric design* — the physical positions of the layout elements. Geometric design is particularly difficult because the layout must completely satisfy the many design rules of the manufacturing process.

If we can improve layout productivity, designers can explore more alternatives and create better chips. Because designers work by first producing a sketch of the layout's topological design and then refine the sketch to include geometric detail, it is natural to automate the two phases separately. *Symbolic layout* and *compaction* are two closely related design methodologies that encourage the separation of topological and geometric design and help to automate geometric design. Symbolic layout allows the topological designer to work with transistors, wires, and cells as primitives rather than manipulating individual polygons; compaction changes the geometry of the topological design to produce a small layout while enforcing the design rules. Although symbolic layout and compaction can be seen as distinct methodologies, the two used in combination make for more powerful design methods for both the CAD engineer and the chip designer.

We will first discuss important issues in symbolic layout: representing a mask layout with symbols, languages for symbolic layout, techniques for separating processing technology information from the symbolic description, and algorithms that manipulate symbolic layouts. In Section 6.3 we will use our knowledge of symbolic layout to understand layout compaction, studying the following: different styles of compaction, leaf cell compaction, hierarchical compaction, and methods for optimizing layouts during compaction.

## 6.2  Symbolic Layout

### Overview

Symbolic layout simplifies layout specification by allowing designers to work with symbols that represent primitive elements — components and wires — in the layout rather than the polygons actually used in fabrication. The symbolic layout is easier to draw because it contains fewer elements. A symbolic layout can be translated into a *mask layout* made entirely of rectangles suitable for fabrication. Figure 6.1 shows symbolic and mask layouts for a CMOS inverter. Transistors and vias are represented as symbols connected by symbolic wires. Symbols may represent one or more polygons (typically rectangles) in the mask layout. Some symbols may be parameterized — the transistors in this layout have different channel lengths and widths. The symbolic layout also includes *ports*, which indicate locations where external connections can be made to the cell. A mask layout is made from a symbolic layout by expanding each symbol into the rectangles that implement it, using the coordinates of the symbols to guide the placement of the rectangles.

Symbolic layouts can be specified hierarchically by encapsulating groups of symbols into *cells*. A *leaf* symbol (or leaf cell) contains no other cells — only primitive symbols. A non-primitive cell contains instances of other cells, called *subcells*. A subcell is typically drawn as a boundary and its ports; the designer may be able to describe the cell boundary, or the system may use a simplified boundary such as the cell's bounding box. Hierarchical layout design offers many advantages common to other forms of hierarchical design. As shown in Figure 6.2, once a cell is defined, it can be reused; encapsulation into cells also helps organize the design.

Some workers have chosen to take symbolic layout as meaning any description that uses cells to define a design hierarchy. But because many algorithms are greatly simplified by having components and wires described as primitive symbols, we define a symbolic layout to be one built from primitive symbols — wires, transistors, vias, *etc.* — and from cells built from primitives and/or cells. For convenience, we use these additional terms: *symbol* for any

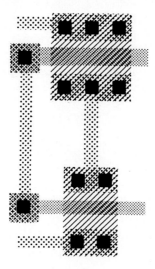

(a) Symbolic

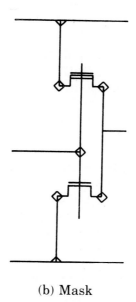

(b) Mask

**Figure 6.1** Symbolic and Mask Layouts

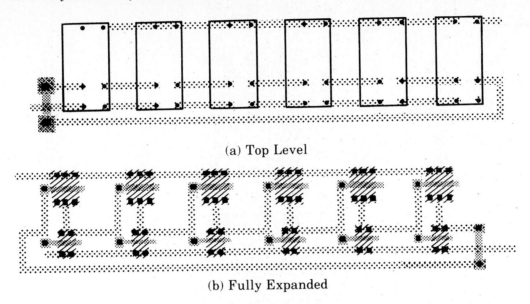

(a) Top Level

(b) Fully Expanded

**Figure 6.2**  A Hierarchical Symbolic Layout

primitive element, such as a transistor, and *component* for any primitive symbol that is not a wire.

Symbolic layout offers the designer a number of advantages. Because primitive symbols can describe several rectangles in the mask layout, the designer has fewer objects to manipulate. The mask layout implementation of those symbols is correct-by-construction. Reusing cells in a layout saves designer time. The symbolic cell representation reduces CPU time requirements by minimizing the storage space and plotting time needed. If the symbolic layout system allows cells to take parameters — like the parameters taken by procedures in programming languages — the designer can use one cell description to produce several different cell layouts. Encapsulating the layouts for components and wires into symbols simplifies a number of synthesis and analysis algorithms. And as Section 6.3 will show, the additional information provided by the layout symbols also makes it easier to create layout compaction programs that can ensure design rule correctness.

### History and Taxonomy of Symbolic Layout

Symbolic layout systems can use symbols to represent mask layout elements in several different ways. It may be tempting to view symbolic and mask layouts as fundamentally different — drawn on different grids with some complex mapping from symbolic to mask layout. But this approach creates more problems than it solves. Considering the translation process as a simple macro expansion makes symbolic layouts and algorithms that operate on them easier to understand. We can, however, define the relation between symbolic and mask components and wires in different ways that trade off expressive simplicity and power: we can change the display of cells or the relative sizes of symbolic and mask cells.

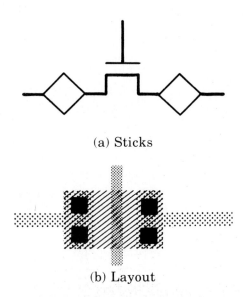

(a) Sticks

(b) Layout

**Figure 6.3**  Sticks and Layout Views of a Symbolic Layout

The alternative display methods are shown in Figure 6.3. A symbolic layout can be drawn as a *stick diagram* (a term introduced by Williams [Wil78]), with wires drawn as line segments and components as symbols, or as a *layout* display with wires and devices drawn as rectangles similar or identical to those used in the layout. (The layout display may not exactly correspond to the final mask layout to emphasize certain features for the designer.) The stick diagram may clarify the cell topology, while the layout display gives the designer a feel for the relative sizes of layout elements. Many systems support both views.

A symbolic layout system can translate symbols parameterized by size, like transistors, into mask layouts in one of two different ways. In a *true-size* representation the symbols may be assigned a size proportional to their size in the mask layout. Alternatively, in a *unit-size* representation all primitive symbols may be displayed at the same size, regardless of their size parameters. Unit-size representation simplifies changing the channel sizes of transistors: as shown in Figure 6.4, wires may have to be added to the layout for connections between symbols with differing mask sizes to produce a legal layout. However, a unit-size representation necessarily allows only one port on the source or drain of a transistor, while a true-size representation may allow several ports distrib-

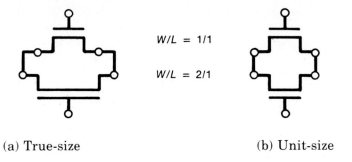

(a) True-size                                    (b) Unit-size

**Figure 6.4**  Unit- and True-Size Representations of Symbols

uted along a transistor's width; designers can use multiple connection points along the source or drain to optimize wiring.

The history of symbolic layout began in the late 1960s as designers started to produce layouts with significant numbers of components and computers became cheap enough to be used as design tools. Systems like those of Larsen [Lar71, Lar73, Lar78], the LSI CAD system [Bar75], and SLIC [Gib76] increased layout productivity by combining two ideas: using symbols to reduce the drawing time for a layout and placing symbols on a grid to eliminate design rule errors. The LSI CAD system was a *coarse-grid* system — the grid spacing was equal to the maximum design rule distance, so that any layout produced would be design rule correct. Other systems used a *fine-grid*, in which the grid spacing was less than the maximum design rule distance to allow more flexibility in layout design. The finer grid meant that the layouts had to be design rule checked, but the symbols and grid could be used to aid checking.

Both coarse- and fine-grid systems do not allow the designer to make maximum use of the layout area. Later symbolic layout systems generally worked in coordinate systems with a resolution equal to the smallest distance that could be fabricated and relied on independent design rule checking to ensure correctness. These systems are often called *gridless* although they use a grid whose spacing is equal to that of the minimum allowed distance; we prefer to call them *fine-resolution* systems. Symbolic layout systems designed for use with a compactor, which will be discussed in Section 6.3, are all fine-resolution systems. Examples of fine-resolution symbolic layout systems not expressly designed for use with a compactor are the i language [Joh80], the HP-IGS editor [Inf78], SIDS [Sap79], and IMAGES [Hil87]. The icon editor [Hil84] was designed to produce i language descriptions using an interactive graphics editor.

Several other languages are *nearly symbolic* — they can describe rectangles but their styles are oriented toward leaf cells built from wires and a hierarchical design style. Nearly symbolic languages include Earl [Kin82], CLL [Sax83], SILT [Dav83], ALI [Lip83], ICL [Ayr83], and Allende [Mat85a]. EDIF [Com85] is a design interchange language that includes elements of symbolic layout languages. The EDIF language supports the description of a design at a number of levels of abstraction, including both mask and symbolic layout. The language also supports procedural description and many other features to increase the language's utility.

The interested reader should also compare symbolic layout systems to other hardware description languages, like SCALD [McW78], SDL [van77], and VHDL [Int85], and mask layout systems like Icarus [Row78] and Magic [Ost84]. Many of the issues encountered in the representation of schematics and mask layouts — scoping of names, declaration of global nets, access to libraries, parameterization — are also important in symbolic layout.

## Symbolic Layout Languages

Symbolic layout languages and graphical editors complement each other. An interactive graphic editor reads and writes symbolic language descriptions; some design tasks, like leaf cell design, are best done with interactive graphics, while others — like the design of regular, parameterized structures — may be best done textually. A textual description language is also necessary to create programs that generate symbolic layouts. Beyond describing hierarchies of cells, other features — parameterized cells, array constructs, and global nets — can be added to the system to simplify design.

The construction of a symbolic layout language translator is much like the design of the front end of a compiler or interpreter. The standard phases and tools of compiler writing — lexical analysis, parsing, symbol tables, and action routines — are equally useful here and have been covered elsewhere [Aho86]. Similarly, other books [New79, Fol82] describe issues in the construction of interactive graphic editors. We will concentrate on features that can be implemented and trade-offs between the power of the language and the difficulty of its implementation.

Simple symbolic layout languages are *declarative*, not *procedural* — they describe static, unchanging layouts, not procedures to construct the layouts. Because they are declarative, the languages are simpler to translate than procedural programming languages. We can see from examples of symbolic layout descriptions that the symbolic language statements have direct translations into elements in the mask layout.

Figure 6.5 shows the description of an inverter in the i language and the resulting layout. The description starts with a declaration of the cell name, *inv*. In the body of the cell description, the external ports (called *xdefs* in i) are declared first by giving them a name and a position. Next, the transistors and vias in the cell are declared with names, positions, and parameters ($\%\mathbf{W} = 1$ for a width of one in the case of the transistors). Components may be instantiated with rotations and reflections as well. Symbolic points — simply names for positions in the cell — are declared to simplify the descriptions of wires. Finally, the wires that connect these points are declared. Connections are made to primitive components with declarations like *t1.ds* for t1's southern diffusion connection. If the cell contained subcells, they would be declared similarly to primitive components; a connection to a port on a subcell is described as *subcell.port*, just as with primitive components.

We can extend the language with more powerful constructs that make the layout easier to describe and modify. Points can be manipulated as complex numbers, allowing the user to write arithmetic expressions on points. Expressions can add or subtract points, or operate on the x and y components of a point separately, for instance, $\{3, 5\} + point1 - \{\mathbf{Xof}(point2) + 1, \mathbf{Yof}(point3) - 1\}$.

A wire's path can be specified with relative coordinates. Once the starting point of the wire is set, the wire's path can be described as a motion from the last position along the wire. If the language allows the designer to change the wire's layer (inserting a contact at the point where the layer changes), a complex layout can be described in a single wire statement. An example of relative wiring from CLL is shown in Figure 6.6.

The language can be extended to include conditional inclusion (**if**) and repeated instantiation (**for**) of symbols. Although the syntax of these statements makes them look like procedural programming language constructs, they are strictly declarative. Because the values of the expressions controlling the **if** and **for** statements are known at translation time, they can be treated as macros to be expanded. The **if** statement can be used to add features to a cell, like a clear line for a register. **For** loops are typically used to build arrays of components or wires, with point arithmetic on the iteration variable used to place the symbols. For example, an array of wires can be built with the statement **for** $i$ **from** *0* **to** $n-1$ **do** *wire* **from** *start_x, i\*width* **to** *end_x, i\*width*. (Trimberger [Tri81] and Mayo [May86] have explored methods for parameterizing graphic layout designs.)

Closely related extensions are constant declaration and cell parameterization. Declared constants clarify a cell description and allow it to be more easily modified. Most languages support both scalars and points as constants and have scope rules as in programming languages to determine the range of validity of a declaration. Parameterized cells can change their contents based on the values of the parameters, so one cell definition can describe a number of different cell instances. Parameters are most often used to control **if** and **for** statements.

The language may allow components and wires to be grouped into structures that are syntactically similar to structures in C [Ker78] or records in Pascal [Jen74]. The symbols in a structure may then be referred to as a group. ALI has a fairly general structure declaration facility similar to Pascal's; Lava [Eic86] has a structure-definition facility for wires.

Beyond these basic features, newer languages are experimenting with more advanced features to facilitate the description of very large designs. IMAGES has a package construct similar to Ada: a cell is declared to be part of a

```
SYM inv {
/* ports */
XDEF POLY vdd_l (-20,16); XDEF POLY in (-20,0);
XDEF POLY vss_l (-20,-16); XDEF POLY vdd_r (12,16);
XDEF POLY out (12,0); XDEF POLY vss_r (12,-16);
/* transistors */
TP %W = 3.000 tp1 (-4,8);
TN %W = 2.000 tn1 (-4,-8);
/* vias */
MP mp3 (-12,8); MP mp4 (-12,-8);
MDP mdp5 (-4,12); MDP mdp6 (-4,4);
MDN mdn7 (-4,-4); MDN mdn8 (-4,-12);
/* wires */
METAL (-4,-4) UP = 8; METAL (-4,0) RIGHT = 16;
METAL (-12,-8) UP = 16; METAL (-20,0) RIGHT = 8;
METAL (-20,-16) RIGHT = 32; METAL (-20,16) RIGHT = 32;
METAL (-4,16) UP = 4; METAL (-4,-16) UP = 4;
IDEF point1 (-12,-8); /* intermediate point */
POLY point1 RIGHT = 4; POLY (-12,8) RIGHT = 4;
}
```

(a) i Language Description

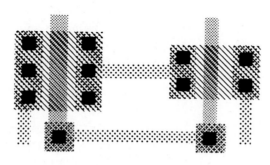

(b) Layout

**Figure 6.5** An Inverter Layout in the i Language

**#define** *point1 10,12* /* start of wire */
**#define** *point2 3,18*
**#define** *point3 7,13*
**#define** *const1 15* /* scalar constant */
**wire metal** *point1* /* starting point, wiring in metal */
        **y** *const1* **l** *2* /* go to y = const1, then left 2 */
        **poly l** **Xof***(point2)-3* /* switch to poly layer, then left */
        **d** **Yof***(point3)* /* down */
        **Xof***(point2)-1,***Yof***(point3);* /* construct new point from x, y values */

(a) CLL Description

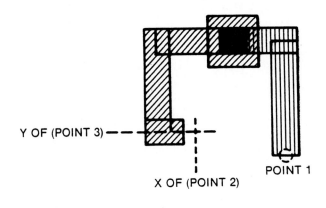

(b) Layout

**Figure 6.6** Wiring Using Relative Location

package and is known to other cells only if its name is exported to other packages. Packages allow the designer to control the name space of symbols so that, for instance, two designers on a project can create cells called *nand* without conflict. Several languages also support the declaration of global signals; all instances of a global signal are assumed to be electrically connected. Global declarations for power, ground, and clock signals simplify extraction and simulation of designs.

A powerful extension of a symbolic layout language comes from combining it with a programming language. Both the ic language [Joh81] and CHISEL [Ull84] allow a designer to embed layout language statements in a C program, calculating the layout parameters with C expressions or functions. A preprocessor translates the source into a pure C program with the layout language statements replaced by function calls or print statements; when the compiled program is executed, it creates the layout description as its output. Embedded languages allow the procedural description of a layout — the designer can use a complex algorithm to build a layout. Embedded languages have limitations: the program that is producing the layout description must work blind — it cannot examine the layout as it is built and modify it to satisfy design requirements; and the user can debug only the modified program written by the preprocessor, which may look very different from the user's original program.

The final powerful method for symbolic description is design by constraint. Languages like i, Earl, ALI, and IMAGES allow the designer to specify the positions of subcells as linear inequalities involving other positions: $point1.x >$ $point2.x + 5$. The system collects the constraints and solves them to find a consistent assignment of positions for all subcells. Constraints allow a designer to position symbols in a cell and to allow the cell to stretch, if positions in the constraint are parameters to the cell. Because the solution of these constraints is closely related to compaction, we will postpone describing the algorithms until Section 6.3.

### Technology Independence and Technology Databases

To interpret a symbolic layout, we need several different types of information about the primitive symbols: their electrical connectivity and their mask layout realization, to name just two. Because the symbolic layout system will likely be used for a variety of fabrication technologies, it is good software engineering practice to keep the description of the primitives separate from the code that builds layouts. Furthermore, the program can read technology fabrication information at execution time to allow designers to update a symbolic layout by binding it to a new technology. The primitive descriptions are kept in the *technology database*.

The technology database defines a number of different *views* of a primitive symbol: its electrical connectivity description, mask layout implementation, *etc.* Because several views all look at mask layout layers, it also defines a set of layers and their attributes. Figure 6.7 illustrates how the symbolic layout system uses the technology database. Externally, the database appears as a set of routines that provide information about the primitives. The code making the requests provides parameters to these routines, including the type of symbol, any parameters for the symbol instance, and the view required. Alternatively, the caller may request information about a layer in one of the views. Internally, the database procedures compute the view from the allowed layers, the description of the primitive symbol, and the given parameter values. Some of the layer information and views may be kept in the database as simple tables, while other data may be computed by procedures using the supplied parameters. The database is initialized from an external file that describes the layers and views; the designer can bind a symbolic layout to different technologies by loading different technology files into the system.

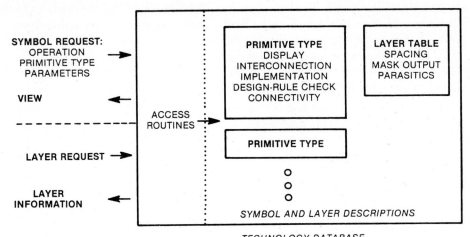

**Figure 6.7** The Technology Database of Primitive Symbols

Although many technology databases have been implemented for symbolic layout, relatively few descriptions of them have been published; one description is the technology database for the VIVID system [Smi85]. The design of a technology database should be well thought out because it can limit a symbolic layout system in two ways: by running so slowly as to make application programs like compaction too slow to use or by not being general enough to describe the important properties of real design rule sets.

The technology database must support these views of a primitive:

- *Display* — How the symbol is to be drawn.

- *Interconnection* — Where to make electrical connections to the cell's ports.

- *Implementation* — How to translate a symbol into a pure layout.

- *Design rule* — How to check that a symbol will satisfy all design rules when the layout is translated into pure form.

- *Electrical* — How the various ports on the symbol are electrically connected.

Figure 6.8 shows views for an unparameterized via and a parameterized transistor. Views of unparameterized symbols can be easily generated from tables. Parameterized symbols require procedures to compute a view from the parameters, which are usually driven by tables indexed by the parameters.

The display view for a symbol is generated procedurally in virtually all symbolic layout systems — the system executes special-purpose code to draw a

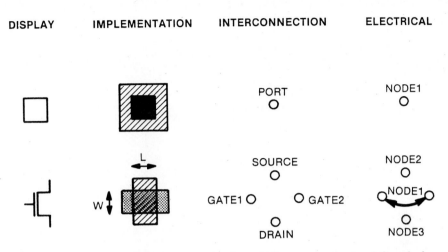

**Figure 6.8** Views of Parameterized and Unparameterized Symbols

symbol directly on an output device. Users demand fast drawing times, so optimizing the display view procedures is important.

Parameters slightly complicate the generation of the other views, but a few assumptions about the form of primitives can greatly simplify and speed the generation of views. As shown in Figure 6.9, the geometric views can be described in terms of linear equations that depend on a single parameter and are relative to a single origin on the symbol. A point of a feature in a view is described as $p_i = <s_{ix}a_{ix}+c_{ix}, s_{iy}a_{iy}+c_{iy}>$, where $a_{ix}$ and $a_{iy}$ are parameters for the symbol, $s_{ix}$ and $s_{iy}$ are the scales for those parameters, and $c_{ix}$ and $c_{iy}$ are constant offsets. For example, in the implementation view of a transistor shown in Figure 6.9, each rectangle is described by two points and parameterized by the channel length ($L$) and width ($W$). The transistor's active area is bounded by $<-.5L,-.5W>$ and $<.5L,.5W>$. The source region is bounded by $<.5L,-.5W>$ and $<.5L+d_{source},.5W>$, where $d_{source}$ is the extension of the source beyond the active region. A table like the one in the figure can be used to select the parameters and constants to be substituted into the expressions that describe these points. More complicated components can be described by allowing multiple origins for a component.

The symbol's electrical connectivity view must allow the system to derive how wires are connected to the symbol's ports. The view consists of *abstract node numbers* for each port. The next section will describe how to use the abstract node numbers to determine circuit connectivity.

The SPARCS compactor [Bur87] is unique among symbolic layout systems in using a technology database without built-in component types. The external model for a cell is a *protection frame*, which defines what regions in the cell

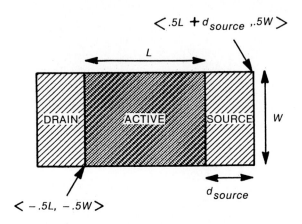

(a) Partial Implementation View

| point | x | | | y | | |
|---|---|---|---|---|---|---|
| | $a_x$ | $s_x$ | $c_x$ | $a_y$ | $s_y$ | $c_y$ |
| active-lower | $L$ | $-.5$ | $0$ | $W$ | $-.5$ | $0$ |
| active-upper | $L$ | $.5$ | $0$ | $W$ | $.5$ | $0$ |
| source-lower | $L$ | $.5$ | $0$ | $W$ | $-.5$ | $0$ |
| source-upper | $L$ | $.5$ | $d_{source}$ | $W$ | $.5$ | $0$ |
| drain-lower | $L$ | $-.5$ | $-d_{drain}$ | $W$ | $-.5$ | $0$ |
| drain-upper | $L$ | $-.5$ | $0$ | $W$ | $.5$ | $0$ |
| gate-lower | $L$ | $-.5$ | $0$ | $W$ | $-.5$ | $-d_{gate}$ |
| gate-upper | $L$ | $.5$ | $0$ | $W$ | $.5$ | $d_{gate}$ |

(b) Implementation View Generation Table

**Figure 6.9**  Generating Views of Primitive Symbols

cannot be overlapped by other cells, and a *terminal frame*, which identifies the cell's ports. The internal representation of a cell is a collection of rectangles on layers. Transistors, vias, *etc.*, are implemented as cells, which allows the designer to create non-rectangular transistors and other special-purpose primitives. SPARCS shows that a well-designed technology database can support queries on generalized cells quickly enough to support applications like compaction that make many database queries.

**Algorithms for Symbolic Layout**

The structure inherent in the symbolic description of a layout greatly simplifies some common layout operations and makes possible some new ones. Extraction of electrical connectivity and design rule checking are both made simpler because components and wires are encapsulated into symbols. Transistor sizing — determining the widths of transistors to optimize some function of circuit delay and layout area — would be much harder to use with mask layouts because of the difficulty of recognizing transistors and the loads they drive.

The algorithms described in this section rely on two basic properties of a symbolic layout: that all components are encapsulated in primitive symbols or cells and that all connections are made between ports. These requirements mean that the designer cannot generate the layout using tricks common in rectangle-based editing: creating new connections or components by overlaying new geometry on an existing cell or splitting components between cells. Parameterized cells can describe virtually all the structures that designers create using ill-structured techniques and make the designer's intent more clear.

In all these applications there may be additional work — tub smoothing, small feature elimination, *etc.* — that cannot easily be done directly on the symbolic layout. Tasks that are difficult to perform on symbolic layouts are best tackled by performing them directly on the mask layout using the algorithms described in Chapter 8.

**Connectivity Extraction.** Connectivity extraction assigns a unique number to each electrical node in a cell. All points in the symbolic layout with the same node number are at equipotential, ignoring parasitic resistance. Circuit extraction is both driven by and produces an *electrical abstraction* of the cell in which each port is assigned a node number. The electrical abstraction of a cell can be used as a representation for the cell's instance in a larger cell to perform hierarchical connectivity extraction.

The extraction algorithm, described in Figure 6.10, uses the union-find algorithm [Aho74] to build maximal sets of connected ports. First, the electrical abstractions of the cell's components are used to build initial sets of connected ports. Each set of disconnected ports on a component is assigned to a separate set. In the transistor electrical abstraction of Figure 6.8, the source and drain ports would be put in separate sets, and the *gate1* and *gate2* ports would be put together in a third set. Wire segments direct the merging of sets of connected ports: first, find all the ports and already processed wire segments that intersect the given wire segments, then merge all the sets to which these objects belong. When all wire segments have been processed, all electrically connected ports in the circuit are in a single set.

This simple algorithm has two restrictions. First, the algorithm as given does not detect the connection of two coincident ports that are not connected by a wire segment. Although the algorithm could be easily modified to handle this

```
procedure extract_connectivity(var nodes: port_set);
begin
{ initially, place each port in a separate set }
nodes := build_empty_union_structure();
for comp in components do
    for port in ports(comp) do
        add_set(nodes);
{ find wire_segments that connect distinct sets }
for wire_seg in wire_segments do
    begin { merge node sets connected by this wire segment }
    node1 := find(connected_to(wire_seg,end_1));
    { set of components that touch end 1 of wire_seg }
    node2 := find(connected_to(wire_seg,end_2));
    { set of components that touch end 2 of wire_seg }
    if node1 <> node2 then
        { these nodes are really connected }
        union(node1,node2);
    end; { of merging nodes connected by a wire segment }
end; { of extract_connectivity }
```

**Figure 6.10**  Algorithm for Connectivity Extraction

case, the system probably does not want to allow implicit connections of components. Second, it allows connections between wires only at their ends — wires that cross to form a T or X are not considered connected. If we allow connections at any point along the wire segment, we simplify the specification of the design but slow down the search for connected objects. Many systems compromise by allowing connections to be made anywhere along a wire segment and in a preprocessing step break the wires to create only endpoint connections.

The extraction algorithm is simple because we assume that connections occur only at the ports declared for cells. If cells are allowed to overlap arbitrarily, connections may be made at places other than ports. Chapter 8 discusses the problems in hierarchical connectivity extraction with arbitrary overlap.

**Design-Rule Checking.**  Virtually all design rules can be put into one of two categories. Some rules are *composition rules*, which describe how to construct components and wires from the layers that can be fabricated. Other rules are *spacing rules*, which describe how far apart objects in the layout must be for them to be reliably built. Checking a mask layout requires verification of adherence to both the spacing and composition rules. In a symbolic layout the primitive symbols are correct-by-construction, so we have to check only conformance to the spacing rules. Design rule checking is further simplified because most spacing rules are pairwise, depending on the interaction of only two symbols. (A reflection rule, described in Chapter 8, is one rule that depends on three symbols.) Therefore, we can check these simple rules by comparing pairs of symbols.

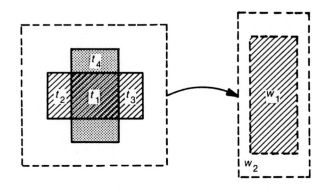

**Figure 6.11**  Checking Spacing Between Two Symbols

Spacing rules are checked by expanding the components and wires into rectangles as specified by their design rule views.  The spacing requirements between the rectangles are checked by referring to the layer-layer rules stored in the technology database.  First consider checking the spacing between two primitive symbols, as shown in Figure 6.11.  The design rule view of the wire includes two rectangles: $w_1$ is diffusion and $w_2$ is tub.  The transistor's design rule view includes five rectangles: $t_4$ for the gate, $t_5$ for the tub, $t_2$ and $t_3$ for the source and drain, and $t_1$ for the channel.  Although the implementation view of the transistor includes only one diffusion rectangle, the design rule view must separate the source, drain, and active regions because they are on separate electrical nodes.  Spacing is checked by comparing each rectangle in the transistor to every rectangle in the wire, by asking the technology database for the spacing required between the rectangles (based on their electrical connectivity) and then checking that the closest points of the rectangles are at least as far apart as required.  In this case the spacing is determined by the tub-tub spacing.

Most design rules are relations between pairs of symbols, so the most common checking algorithms for leaf cells compare only pairs of primitive cells.  Not all pairs need be fully checked: construct a bounding box for each cell, equal in size to the cell's size plus one-half maximum design rule distance in each direction; only cells whose bounding boxes intersect need be considered further.  An algorithm for finding all pairs of intersecting rectangles [Pre85] finds the pairs of cells that require a complete analysis.

Hierarchical design rule checking for symbolic cells can be done in one of two ways: by flattening the hierarchy and checking the entire symbolic layout at once or by building an abstraction of the cell to represent instances of it during checking. Chapter 8 describes the pros and cons of each method.

Symbolic specification of the layout does not simplify the checking of all design rules. Tub boundaries are usually modified after mask layout generation, as described in the next section, for electrical and fabrication reasons. Therefore, the tub inclusion rules cannot be fully checked in the symbolic layout. Many specialized rules depend on the interaction of more than two symbols; while one might be able to formulate algorithms to check them, the algorithms may be too slow to use. Reflection rules, as mentioned previously, depend on the interaction of three symbols. Small layout features can be created by an arbitrary number of layout elements, making rules governing them even harder to check. Verification of these rules is best done in the mask layout domain using the algorithms of Chapter 8.

**Performance Optimization.**    High-performance circuits demand optimization over area, speed, and power. Logic optimization and transistor sizing are the two main types of optimization that affect digital circuits. Transistor sizing is particularly easy to apply to symbolic layouts, in which transistor sizes are parameters of the transistor symbols and can be easily changed. Since optimization is beyond the scope of this book, we will give only a brief overview of transistor sizing.

Performance optimization by transistor sizing can reduce the delay through a circuit by as much as an order of magnitude. Figure 6.12 shows the importance of doing a good job of sizing. The graph shows circuit delay *vs.* transistor width for a simple circuit. When transistors are too small, the total circuit delay is larger than necessary because the transistors cannot drive their loads quickly. Increasing the transistor widths to increase their output current speeds up the circuit. However, if the transistors are made too large, their gate capacitance becomes too great for the driving circuit, and the circuit is made both slower and larger. A transistor sizing program selects transistor sizes that minimize circuit delay without wasting layout area.

Several researchers [Rue77, Mat85b, Hed85b, Mar86, Nye83] have used mathematical optimization techniques like Newton, quasi-Newton or feasible direction methods to solve the performance optimization problem for both PLA and random logic. One drawback of these systems is that they can usually handle only a small circuit (a few hundred transistors will consume too much CPU time and/or memory). Optimizing an entire VLSI chip is out of the question.

A simple and effective program by Fishburn and Dunlop called TILOS [Fis85] uses heuristics to solve the random logic transistor optimization problem. By restricting circuits to be static logic, synchronous and CMOS designs, they have shown that the problem is convex for the Penfield-Rubinstein distributed RC delay model [Rub83]. A steepest descent algorithm searches the convex

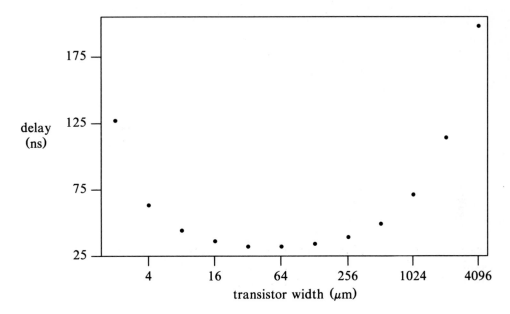

**Figure 6.12** Transistor Size *vs.* Delay

solution space to find the transistor sizes that satisfy an optimization program: minimize $AT^n$, minimize $A$ subject to $T < bound$, or minimize $T$ subject to $A < bound$, where $A$ is total transistor active area and $T$ is delay. The inner loop of TILOS uses sensitivities to find the transistor that improves the worst failing path by the largest amount. For typical circuits TILOS runs in $n\log(n)$ time, where $n$ is the number of transistors. It has optimized circuits with 30,000 transistors in 40 minutes of CPU time on a VAX/11-780.

**Mask Layout Generation.** The symbolic layout must be translated into rectangles for submission to fabrication. Because virtually all mask layout languages (like CIF [Mea80], EDIF [Com85], and XYMASK [Fow69]) allow hierarchical description, the translation to mask layout is straightforward. The implementation of primitive symbols is taken from the technology database, and cells in the symbolic layout are translated directly into cells in the mask layout. The procedure becomes slightly more complicated if cells can be parameterized. Very few mask layout languages permit parameterized cells, so a different version of the mask layout cell must be generated for each unique set of parameter values. The translator should avoid making unnecessary duplicates of parameterized cells called with the same values; it will also have to make unique names for each mask layout version of the cell.

Some mask generation operations cannot be performed directly on the symbolic layout. A common example is tub insertion for CMOS layouts. Symbols are modeled with minimum required amount of tub surrounding them. This produces a ragged tub edge when expanded directly. Each tub rectangle must be electrically connected to a $V_{dd}$ (for n-tub) or $V_{ss}$ (for p-tub) contact known as a *tub tie*. Tub boundaries must also be smoothed to eliminate small features. Tubs are therefore generated in two steps: after producing the minimum required tub, a separate procedure ensures that each tub rectangle is properly connected to the power supply. A crude but effective technique is to require the symbolic layout to include the tub ties, then stretch the tub rectangles in each symbol to overlap the nearest tub tie. Stretching the tub rectangles also helps smooth the tub boundary and eliminate small features on the tub mask. The program must check that no design rule violations were created by stretching a tub rectangle over a transistor of the opposite polarity; this rarely happens in practice.

Some processes require layers, like the complement of the tub region, that are difficult to derive from the symbolic description. Mask shops may also require that the masks contain no overlapping rectangles, notches, or mask features which are closer than the minimum design rule even if they are part of the same electrical net. Problems like these are best solved using the techniques of Chapter 8.

## Applications of Symbolic Layout

Symbolic layouts are simpler to specify than mask layouts because they contain fewer objects. Just as important, because the primitive components are correct-by-construction, the designer is less likely to make mistakes in the layout. Symbolic layout is therefore well-suited to semi-automatic layout compilation. The Plex compiler [Bur83] produced layouts for custom microprocessors by executing a set of ic language programs that generated a complete chip layout based on parameters determined from the microprocessor's architecture.

Adding parameters to cells, arithmetic on points, and positioning by constraint to a language makes symbolic layouts even easier to design. For example, several authors [Ng84, Rab86, Hil86] have described routers that generate flexible routing cells in symbolic form.

Unfortunately, the addition of features for stretchable cell design also exposes the fundamental weaknesses of a simple symbolic layout system. Point arithmetic and constraints were introduced into languages to support both design rule correctness (so that the designer could express the positions of components relative to the design rules) and *pitchmatching* (or stretching cells so that their ports can be connected by abutment). Writing a description of a cell that always stretches as desired while making sure that the cell always remains design rule correct is a difficult task. An exercise for the inquisitive reader would be to write a description for a simple shift register cell that stretches correctly − maintaining compactness and design rule correctness − as its ports are moved in x and y. Even this small example can take a long time to properly design. The designer needs help on geometric design to be able to take advantage of the simplified topological design that symbolic layout provides; we can provide this help by applying compaction to symbolic layout.

## 6.3 Compaction

### Overview

Compactors speed layout design by automating geometric design. The designer gives the compactor a preliminary layout. The compactor moves components and wires in the plane to optimize for two goals: that the layout be small and that it be design rule correct. The compactor moves subcells only in the plane, preserving the designer's topology for the cell. The designer can therefore have a great deal of control on the layout without performing the tedious work required to turn a sketch of a layout into a correct, space-optimized design.

Most of the steps in the compaction process are made much simpler by working from a symbolic description of the layout. The symbolic layout, expressed as components and wires, shows which parts of the layout can stretch (for example, wires) and which cannot (components) and captures the electrical connectivity that would have to be extracted from a mask layout. The compactor runs faster interpreting a symbolic layout, and it is better able to deduce the designer's intent than a program that derives structure and connectivity from a mask layout.

Combining compaction with symbolic layout creates advantages for the CAD developer as well. Because the compactor is responsible for assigning locations to components, the system's symbolic layout language can be made simpler — relative wiring and point arithmetic become less important, though they are still useful when generating symbolic layouts from programs. And programs that generate symbolic layouts are freed from the need to design the cell's geometry for themselves.

### History and Taxonomy of Compaction

Compaction algorithms can be classified along two axes. The first describes how components move during compaction. As shown in Figure 6.13, compaction can be either *one-dimensional (1-D)*, so that components are moved only in x or in y, or *two-dimensional (2-D)*, where a single step can move a component in both x and y. In 1-D compaction the cell is alternately compacted in x and y, while in 2-D compaction components are selected to move as required to improve the layout.

The second axis covers the algorithm used to position the components. The two major types of algorithms, as shown in Figure 6.14, are *constraint-graph* and *virtual grid*. The constraint-graph method describes the required

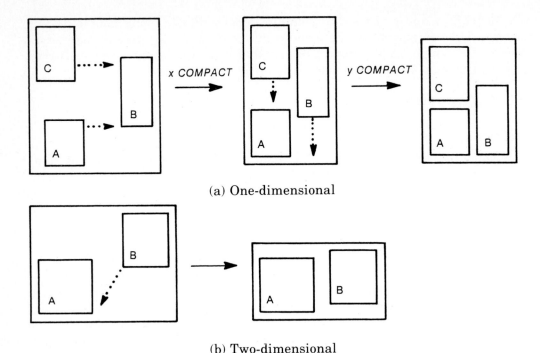

(a) One-dimensional

(b) Two-dimensional

**Figure 6.13**  Styles of Compaction

connections and separation rules as linear inequalities, which can in turn be modeled as a weighted, directed graph. The constraint graph is used to find new positions for the components, and the result is applied back to the layout. The virtual grid method finds the subcell positions by considering the layout to be drawn on a grid; it moves all components on a grid line together so that adjacent virtual grid lines are as close as possible while satisfying all required separations between the symbols on the two virtual grids.

*Hierarchical compaction* works on cells that are constructed from other cells as well as primitive layout symbols. Any of these techniques can be made hierarchical by applying the compaction algorithm to a leaf cell and then using the compacted cell in larger cells. Hierarchical compaction is also called *cell assembly*. A variety of hierarchical compaction algorithms have been developed for both constraint-graph and virtual grid compaction.

The first compaction algorithm, *shear-line compaction*, was proposed by Akers *et al.* [Ake70]. The shear-line algorithm removes excess space in connected regions so that the layout will grow smaller in one dimension. Their work discussed only applications to mask layout. Although the shear-line algorithm is historically important as the first compaction algorithm, modern systems have abandoned it for more powerful methods. A major problem in shear-

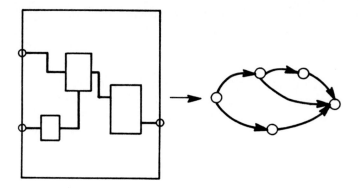

(a) Constraint-graph

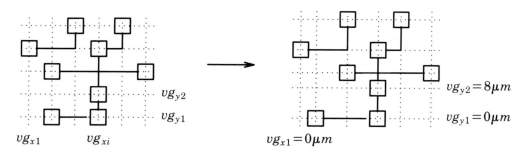

(b) Virtual Grid

**Figure 6.14**  Compaction Algorithms

line compaction is finding connected regions of white space that can be removed from the layout; the solution is a special case of constraint-graph compaction.

Symbolic layout and compaction were first combined in the STICKS system [Wil78]. Williams discussed a number of ideas that were to appear in later symbolic layout and compaction systems, but the methods he describes are largely *ad hoc*. Dunlop combined symbolic layout and shear-line compaction in the SLIP system [Dun78].

The next algorithm to be introduced for compaction was constraint-graph compaction. This method was first used in the FLOSS system [Cho77], though the constraint-graph algorithm was not discussed in the original publication. CABBAGE system [Hsu79] and SLIM system [Dun80, Dun81] independently applied constraint-graph methods to symbolic layout compaction and also explored a variety of optimization techniques.

The virtual grid compaction algorithm was introduced by the MULGA system [Wes81]. MULGA was also the first cell assembly system. MULGA was succeeded by other virtual-grid compactors, including VIVID [Rog85] and SYMPLE [Sza84].

A separate line of development applied constraint-graph methods to hierarchical compaction. Python [Bal82] used the constraint-graphs to hierarchically compact fixed cells. The *port abstraction* method for stretching cells was first used in Earl and first applied to compaction in Lava [Mat82, Ull84]. Constraint-graph hierarchical compaction has since been used in a variety of systems, including SPARCS [Bur86], PSI [Fie84], SLS [Pos86], and MACS [Lo87, Cro87].

The plowing subsystem in Magic [Sco84] directly compacts mask layouts. Plowing moves an edge through the mask layout, pushing other layout elements along as necessary to satisfy the design rules. The algorithm must perform some component extraction to determine what parts of the layout can stretch and what polygons on different layers must move together. Its technique for determining how far polygons must move is very similar to CABBAGE's compaction algorithm.

Most recent work on compaction has focused on extending 1-D techniques to produce better layouts. One line of work focused on methods for optimizing layout during compaction: using the critical path to find local transformations [Wol84, Wol88], other local transformation methods [Shi86], and several methods for automatic jog introduction [Mal85, Var85].

Another line of work explored integer programming-based, two-dimensional compaction algorithms. Extensions of the constraint-graph model for optimal, two-dimensional compaction have been described by several authors, including Schlag *et al.* [Sch83b] and Kedem and Watanabe [Ked84]. Optimal, two-dimensional compaction is NP-hard [Sas82], but Kedem and Watanabe have developed heuristics for two-dimensional compaction. Mosteller *et al.* [Mos87] developed a Monte Carlo method for 2-D compaction.

### Algorithms for Compaction

One-dimensional constraint-graph and virtual grid compaction both follow the same general procedure in compacting a cell, as shown in Figure 6.15. The first step is to determine the physical connections in the layout — which ports need to remain physically coincident during compaction. (Virtual grid compaction does not require this step, because physical connectivity is maintained by the virtual grid during compaction.) The cell then enters the main loop in which compaction takes place. Each iteration through the loop is a *compaction step*. First, the layout is analyzed to determine the spacing rules that must be obeyed. Then the layout is compacted to satisfy those constraints. Finally, a *wire length minimization* algorithm is applied to the compactor's solution to adjust the positions of non-critical wires in the layout. (Virtual grid compactors generally do

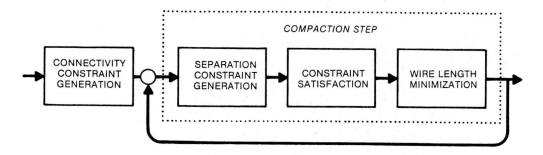

**Figure 6.15** The Compaction Process

not apply wire length minimization because the virtual grid keeps wires from moving far away from their connections.) A cell must go through at least one x and one y compaction step before the compactor can guarantee that the layout is correct. In constraint-graph compaction further compaction steps may shrink the layout size. As will become clear when we discuss separation constraints, performing two x or two y compactions in a row gives no further improvement of the layout.

The result of this process is a list of the $<x, y>$ positions for all components and wires in the layout. The positions can be used to create a new, compacted version of the input symbolic layout which can be viewed and edited symbolically or recompacted. However, some compactors change the layout in ways that cannot be expressed in the compactor's input language: they may make connections off-center to the symbolic ports or convert transistors from unit-size to true-size to take advantage of multiple source and drain ports, for example. These compactors must write their results as mask layouts. The designer of the compactor must trade off ease of specification in the input language, power of the compaction transformations, and the need for symbolic output when choosing a symbolic description and a compaction method.

Our discussion of compaction algorithms starts with a description of the constraint-graph model for layouts, which we will use to understand both constraint-graph and virtual grid compaction. We next introduce algorithms for 1-D leaf cell compaction for both the constraint-graph and virtual grid methods and briefly consider 2-D compaction. We will then extend the leaf cell algorithms to hierarchical compaction. Next, we consider two optimization methods for compaction: wire length minimization and critical path analysis. Then we consider the tools designers require to analyze the results of a compaction for manual improvement. Finally, we analyze and compare the performance of these algorithms.

**The Constraint-Graph Model.**    In Section 6.2 we distinguished composition and spacing rules. The compactor's job is to recognize and enforce spacing rules while minimizing area. Spacing rules can be written as constraints, similar to the constraints one writes in a symbolic layout language. The compactor must also maintain the connections described in the input symbolic layout, which can be described as *physical connectivity constraints*.

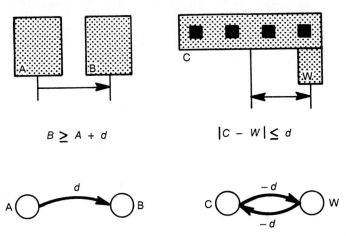

(a) Separation Constraints    (b) Physical Connectivity Constraints

**Figure 6.16**  Representing Linear Inequalities as Graphs

As shown in Figure 6.16 the inequalities that describe separation and physical connectivity constraints can themselves be represented as weighted, directed graphs. Because the compaction algorithm is one-dimensional, an inequality will never mix variables representing x and y positions; in this case the constraints describe x positions. The positions of the components, represented as variables in the inequality, are represented by vertices in the graph.

Maintaining separation of two components requires only one linear constraint and one edge in the constraint graph. If two cells are required to be at least $d$ units apart, the constraint $B \geq A + d$ is represented in the graph as an edge from $A$ to $B$ with weight $d$. (Because the constraint is linear, it determines the order of $A$ and $B$ as well as their separation.)

Maintaining a connection requires two constraints. A connection between a wire and a component's port may allow *slop*, as in the connection of a small wire to a large via. The condition on their relative positions is $|C - W| \leq d$ where $d$ is the amount of slop in each direction; it can also be written as two linear inequalities, $C \geq W - d$ and $W \geq C - d$. The negative weights on the graph edges indicate that each component is allowed to be to the left of the other: $C$

can be at least $-d$ units to the right of $W$ (or $d$ units to the left), and *vice versa*. The edges representing these constraints form a cycle in which the sum of the weights around the cycle is non-positive.

A fixed connection is a special case of the loose connection in which the sum of the weights around the cycle is zero. If the wire and via must be exactly $d_{fix}$ units apart, the constraints $W \geq C + d_{fix}$ and $C \geq W - d_{fix}$ reduce to the equality constraint $W = C + d_{fix}$.

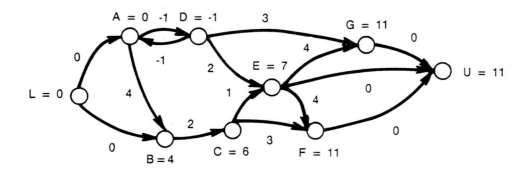

**Figure 6.17**  A Compaction Constraint Graph

The physical connectivity and separation constraints can be assembled into a graph that represents the 1-D compaction problem. The graph represents a 1-D and not a 2-D problem because all the variables in the constraints represent positions in the same dimension. A constraint graph for a 1-D compaction step is shown in Figure 6.17. The graph includes two artificial vertices, $L$ and $U$, which represent the lowest and highest positions in the cell. $L$ is a *source* of the constraint graph because all other vertices are explicitly or implicitly required to be above it; similarly, $U$ is a *sink* of the graph. Values assigned to the vertices represent positions of the cells in the dimension of compaction.

The two most important problems to be solved during compaction are *longest path* (also called *constraint solving* or *solving the graph*) and *transitive closure*. The role of these problems in compaction will become clear in the next sections, and we will defer describing algorithms for solving these problems until then. For now, we simply define the problems.

Constraint solving assigns values to the vertices that satisfy the constraints. Solving a system of constraints produces a complete placement in one dimension for the cells in the layout. We want to find values for the vertices that minimize the distance from $L$ to $U$. The minimal $L \rightarrow U$ distance is determined by the longest path of constraints from $L$ to $U$. In finding this longest path (also called the *critical path*), we can also assign values to the other vertices, thus producing a placement for everything in the cell. Figure 6.17 shows a solution that minimizes the distance from $L$ to $U$; this solution is not unique. In the figure, the paths $L \rightarrow A \rightarrow B \rightarrow C \rightarrow E \rightarrow F \rightarrow U$ and $L \rightarrow A \rightarrow B \rightarrow C \rightarrow E \rightarrow G \rightarrow U$ have the same length, so each determines the $L \rightarrow U$ distance. (In an abuse of graph theory terminology, we will refer to the set of all paths that limit the $L \rightarrow U$ distance as the critical path.) The critical path is a valuable diagnostic tool that helps the user determine what constraints limit the size of a cell.

We can translate the longest path problem into a shortest path problem by negating the weights of the edges. The operations research literature contains a number of analyses of various shortest path problems; good surveys of solution methods include Dreyfus [Dre69], Tarjan [Tar83], and Gallo and Pallottino [Gal86].

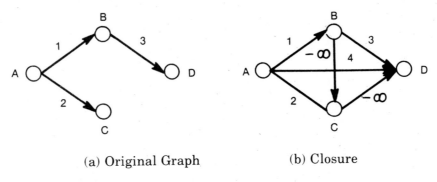

(a) Original Graph          (b) Closure

**Figure 6.18**  A Constraint Graph and Its Closure

The transitive closure of a constraint graph is the set of all constraints implied by the constraints represented by the original graph. Because the transitive closure gives the minimum allowable distances between all pairs of cells, it completely describes how cells can move during compaction. A graph with $n$ nodes has $\dfrac{n(n-1)}{2}$ edges in its transitive closure, one edge between each pair of nodes. (An edge from $A$ to $B$ of weight $d$ implies an edge from $B$ to $A$ of weight $-d$. Because of the obvious redundancy of this edge, we define our transitive closure graph to include only one of the two edges.) The transitive closure is therefore represented as a directed graph with one edge between every node pair. The weight of each edge is equal to the length of the longest path through the graph from the edge's source to its sink. Computing the transitive closure is equivalent to finding the longest path between all pairs of edges in the graph. Figure 6.18 shows a simple constraint graph and its transitive closure. The given graph has four nodes so its closure has six edges. A weight of $-\infty$ implies that there is no path between the two nodes connected by the edge.

The constraint graph is one intuitive model of the rules that govern layout design. Operations research gives us two alternative interpretations of compaction constraints, each of which leads us to new useful terminology and algorithms. We can view 1-D compaction as a *linear programming* problem:

$$
\begin{array}{llllll}
\textbf{minimize} & U & -L \\
\textbf{subject to} & & -L & +A & & \geq & 0 \\
& & -L & & +B & \geq & 0 \\
& & & -A & +B & \geq & 4 \\
& etc.
\end{array}
$$

The *objective function* of the linear program is $U-L$, the distance between the lower and upper edges of the graph. This function's value is to be minimized subject to the constraints expressed in the constraint graph edges. A linear program has an optimal solution to its objective function which lies at the simultaneous solution to two or more of the constraints. Because a linear program's solution space is convex [Dan63], an optimum can be found by a monotonic search of these simultaneous solution values. The most common linear programming algorithm, the *simplex* algorithm, works by performing such a search. (For a geometric interpretation of linear programs and the simplex method, see an operations research textbook, like Hillier and Lieberman [Hil80].) The average number of iterations required for the simplex method is very nearly linear in the number of constraints [Wol63].

We can also see compaction as a specialization of a general linear programming problem that takes advantage of the simple nature of the constraints (all of which have exactly two variables) and of the objective function (which depends on only two variables). The *network-flow* model interprets the edges as carrying some commodity from the graph's source ($L$) to its sink ($U$) with the capacity of an edge set to the negative of its weight. To find the longest path from $L$ to $U$, we force one unit of the commodity to flow from $L$ to $U$. It will flow through the highest capacity path, which, because capacities are the negatives of the constraints' minimum distances, is the longest path. The network flow is the dual of the longest path problem. This formulation allows the use of a number of interesting algorithms for the solution of network-flow problems. The network-flow approach to compaction has been pioneered by Eichenberger [Eic86]. A detailed discussion of these algorithms is beyond the scope of this chapter, but Bazaraa and Jarvis [Baz77] describe efficient algorithms for solving network flows: the simplex algorithm performed directly on the constraint graph and the out-of-kilter algorithm, which can solve the linear program without finding an initial feasible solution.

**Constraint-Graph Leaf Cell Compaction.**    Remember from Figure 6.15 that a compactor first builds the physical connectivity constraints, then finds the separation constraints and solves the combined constraints at each step. In this section we will develop constraint-graph-based algorithms for each of these steps.

Connectivity constraints are found by examining all legal connections in the symbolic layout. Different types of connections (wire-wire, wire-via, wire-source, *etc.*) may require different amounts of slop, which can be looked up in a table. Each connection creates sets of constraints, one for x and one for y.

If the distance between two vertices is fixed, they can be collapsed into a single vertex. Collapsing equality constraints eliminates the one-to-one mapping from layout cells to graph vertices but speeds constraint solving by reducing the size of the graph. To collapse two vertices, we choose one to keep and rewrite all constraints on the other vertex to refer to the kept vertex, adjusting each edge weight by the distance between the two vertices. We must also keep a table relating components to vertices, with the offset from the component's origin to the vertex representing it. The union-find algorithm can be used to find sets of vertices that can be collapsed.

Separation constraint generation is performed once per compaction step, so an efficient algorithm is important. The constraint generation method used should ideally generate an irredundant set of constraints since the cost of solving the constraint graph is proportional to the number of edges in the graph, or the number of constraints. However, in practice, we must trade increased constraint-generation time to generate the smaller constraint set for smaller constraint solving times. The first algorithm developed to generate succinct constraint graphs was *shadowing*. Although better algorithms now exist, we will describe shadowing in detail to illustrate the problems of constraint generation. We will also discuss two newer, more efficient algorithms: the *intervening groups* method and the *perpendicular-plane-sweep* algorithm.

The shadowing algorithm examines the positions of cells to determine what constraints are redundant. As shown in Figure 6.19, we can imagine the region that contains cells that must be constrained against $A$ in x as falling under a *shadow* cast from $A$. The shadow's maximum height is the height of $A$ extended by the maximum design rule distance in each direction. If the shadow falls on a cell, that cell must be constrained against $A$; thus, we must generate a constraint from $B$, $C$, and $D$ to $A$. But because $E$ will be constrained against $C$, we do not need to add a constraint from $E$ to $A$. We represent this by making $C$ block the shadow of $A$. Since the shadow does not fall on $E$, no constraint from it to $A$ is generated.

The algorithm of Figure 6.20 describes how to generate the constraints to a cell (*sink*) from all the cells below it or to its left. The algorithm keeps the bounds of the shadow's left edge in *shadowfront*. The shadow is a list of edges, each of which has a position in the constraint-generation dimension and an

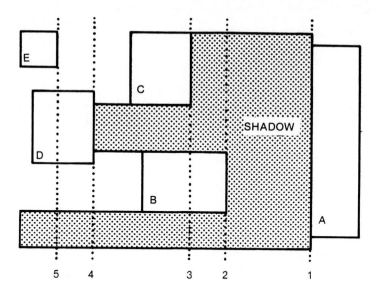

(a) Layout

| position | intervals |
|:---:|:---:|
| 1 | [shadow.bottom,shadow.top] |
| 2 | [shadow.bottom,B.bottom] [B.top,shadow.top] |
| 3 | [shadow.bottom,B.bottom] [B.top,C.bottom] |
| 4 | [shadow.bottom,B.bottom] |
| 5 | [shadow.bottom,B.bottom] |

(b) Shadow Range *vs.* Position

**Figure 6.19**  Shadowing to Eliminate Unnecessary Constraints

upper and lower bound in the opposite dimension. The shadow edge set can be efficiently implemented as a tree of edges. In the example of Figure 6.19, the shadow will always cover the y region, extending one maximum design rule distance above and below $A$ and to the right of the edges maintained in *shadow-front*. The algorithm starts with the shadow extending from $A$ to the cell's left edge, then searches from right to left to find cells that block portions of the shadow. The function *in_shadow* returns true if a cell's right edge is in the shadow, which implies that it must both overlap at least one edge in *shadowfront* in y

**procedure** shadow_cell( *graph*: graph; *sink* : cell; *dim* : dimension);
{ Generate all constraints to sink in dimension dim from all
  cells to its left. }
begin
{ initialize shadow front − extends from cell edge to sink
  in dimension dim and height of sink in opposite dimension }
*shadowfront* := initialize_shadow(0,*sink.ll*[opposite(*dim*)],
                              *sink.ur*[opposite(*dim*)]);
{ check each cell to sink's left, starting at sink and moving
  to the left }
**for** *source* **in** sort_by_position(cells,in_range(0,*sink.ll*[*dim*]),*dim*) **do**
  **begin**
  **if** in_shadow(*shadowfront*,*source.ur*[*dim*],
     *source.ll*[opposite(*dim*)],*source.ur*[opposite(*dim*)]) **then**
    **begin** { constraint required here }
    add_constraint(*graph*,*source*,*sink*,spacing(*source*,*sink*,*dim*));
    { add the constraint to the graph }
    update_shadow(*shadowfront*,*source.ur*[*dim*],
            *source.ll*[opposite(*dim*)],*source.ur*[opposite(*dim*)]);
    { block appropriate part of shadow with source }
    **end**; { of adding a constraint }
  **end**; { of checking a cell against shadow }
**end**; { of shadow_cell }

**Figure 6.20** Algorithm for Constraint Generation by Shadowing

and also be above that edge in x. If the cell is in the shadow, the algorithm adds a constraint from it to *sink* and calls the *update_shadow* procedure to modify *shadowfront* so that one edge of the shadow coincides with the left edge of the cell that required the constraint. Figure 6.19 shows the progress of the algorithm as it traverses the layout − the intervals shown are the remaining subshadows that must be searched as the algorithm moves from left to right.

We can now see why consecutive x (or y) compactions do not improve the layout. The x constraints are determined by the y positions of the cells. Since y positions don't change during x compaction, the second step will solve the same constraints.

There are some difficulties in creating a robust implementation of shadowing. Some components may move through each other during compaction (two ends of a wire, for example), and the shadow must not be blocked until all cells that may enter the shadow have been considered.

The intervening groups method, introduced by Kingsley [Kin84], uses a simple heuristic to eliminate many, but not all, redundant constraints by remembering a path through the constraint graph that provides a lower bound

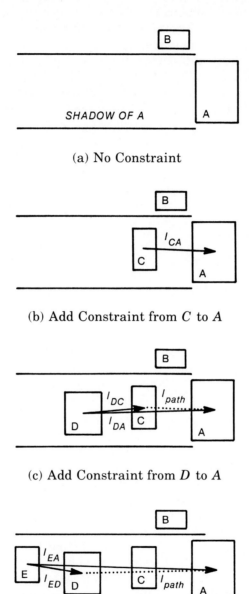

(a) No Constraint

(b) Add Constraint from $C$ to $A$

(c) Add Constraint from $D$ to $A$

(d) No Constraint

**Figure 6.21** Eliminating Redundant Constraints by Finding Intervening Groups

on the size of relevant constraints. To understand the algorithm, consider the sequence of events shown in Figure 6.21, where we are generating x-dimension constraints to $A$. The algorithm sweeps the layout backward from $A$ to generate constraint edges that terminate on $A$. In (a), no constraint is required between $A$ and $B$, because $B$ is not in $A$'s shadow. In (b), $C$ falls in $A$'s shadow; we get the required separation from the technology database and find that a constraint between them is in fact required. (If, for instance, $A$ were metal and $C$ were diffusion, no constraint would be needed even though $C$ is in $A$'s shadow.) We remember the distance $l_{path} = l_{CA}$ as the current bounding path. In (c), $l_{path} = l_{CA}$ helps to compute a lower bound on the length of a relevant constraint between $D$ and $A$ — if the length of the constraint required from $D$ to the head of the current path ($C$) plus $l_{path}$ is longer than $l_{DA}$, then the constraint from $D$ to $A$ is redundant. Assuming that $l_{DA} > l_{DC} + l_{path}$, the algorithm adds the constraint from $D$ to $A$ and updates the path, making $l_{path} = l_{DA}$ and the longest path's source $D$. In (d), the process is repeated; a constraint from $E$ to $A$ is added only if $l_{EA} > l_{ED} + l_{path}$.

The intervening groups algorithm is outlined in Figure 6.22 using an extension described by Hedges *et al.* [Hed85a]. The first function, *constrained*, returns **true** if the constraint graph includes a path of constraints between two cells that satisfies the spacing rule between them. It searches the graph for constraint paths to a depth *max_search_depth*, which is typically set to a value from 2 to 4. As the value of *max_search_depth* grows, the procedure eliminates more redundant constraints but takes longer to run. (If *max_search_depth* were set to ∞, the procedure would search the entire graph and never add a redundant constraint.)

The second procedure, *generate_constraint_graph*, considers each cell from left to right, generating constraints to the $i^{th}$ cell from all previously considered cells $1 \ldots i-1$. It adds a constraint between two cells only when no known path in the graph already satisfies the required spacing. It uses a Boolean array (which Hedges *et al.* call a bitmap) to keep track of what pairs of components are known to have constraint paths between them. If *cell[i].path_flags[j]* = **true**, there is a known constraint path from cell $j$ to cell $i$. Path existence is a transitive relation: if there is a path from $i$ to $j$ and from $j$ to $k$, there is a path from $i$ to $k$. Therefore, when a constraint from $j$ to $i$ is added, a simple **or** operation will add all the paths through component $j$ to the paths known to $i$. The speed with which the set of all constraint paths to a cell can be tested and updated allows the algorithm to eliminate redundant constraints very quickly.

The perpendicular-plane-sweep algorithm was developed by Doenhardt and Lengauer [Doe87]; its implementation in SPARCS extended the algorithm in several ways [Bur87]. The algorithm uses a scan line like that used in design rule checking algorithms but uses the scan line to perform a shadowing analysis. X-dimension constraints are generated by a vertical line that is swept from left to right across the layout. In design rule checking, because polygons

**function** constrained(*graph*: graph;
                     *source, sink, intermediate_cell*: node;
                     *distance, depth*: integer) : boolean;
{ Intermediate cell is a cell along a possible source → sink path. }
**begin**
*result* := **false**; { no constraint from source to sink yet }
**for** *constraint* **in** sink_edges(*intermediate_cell,graph*) **do**
  **if** *constraint.from_node* = *source* **then**
    { path ends here at source node − check total path length }
    *result* := *result* **or**
        *distance* + *constraint.weight* >= spacing(*source,sink*)
    **else if** *depth* > 0 **then** { if within search depth, keep looking }
    *result* := *result* **or**
        constrained(*graph,source,sink,constraint.source,*
          *distance*+*constraint.weight,depth*+1);
*constrained* := *result*; { return result of search }
**end**; { of constrained }


**procedure** generate_constraint_graph(*dim* : dimension);
**begin**
*sorted_cells* := sort(*cells*,low_to_high_sort_function(*dim*));
{ initialize path flags − every cell constrains itself }
**for** i **from** 1 **to** *n* **do** *sorted_cells[i].path,flags[i]* :=**true**;
  { all other entries are false }
{ generate constraints to i-th cell from cells below it }
**for** *i* **from** 1 **to** *n*, *tocell* := *sorted_cells[i]* **do**
  **for** *j* **from** *i-1* **downto** 1, *fromcell* := *sorted_cells[j]* **do**
    **if** in_shadow(*fromcell,tocell*,opposite(*dim*)) **and**
      { cells can pass through each other }
      **not** *tocell.path_flags[j]* **and**
      { have not yet seen constraint path from i to j }
      **not** constrained(*graph,fromcell,tocell,tocell*,0,*max_search_depth*)
      { can't find superceding constraint in graph }
      **then begin** { add constraint, update longest paths }
        add_constraint(*graph,fromcell,tocell,*
          spacing(*fromcell,tocell,dim*));
        *tocell.path_flags* := *fromcell.path_flags* **or** *tocell.path_flags*;
        **end**; { of adding constraint }
**end**; { of generate_constraint_graph }

**Figure 6.22**  Algorithm for Constraint Generation by the Intervening Groups
             Method

do not move, only polygons within a fixed distance of the scan line need be checked, but for compaction the scan line must collect arbitrarily distant elements. Constraints are generated from one element's right edge to another's left edge; an edge data structure holds the currently interesting edges to compute the shadows of all elements simultaneously. A right edge is added to the edge data structure and is checked to find left edges in the data structure against which it must be constrained. A left edge, because it blocks the shadow cast by a right edge, causes right edges to be removed from the data structure. The one-layer version of the perpendicular-plane-sweep algorithm has a time complexity of $O(n \log(n))$ where $n$ is the number of rectangles in the layout.

Once physical connectivity and separation constraints have been generated, the constraint set must be solved to find the longest path from $L$ to $U$. Depending on the layout model used and other information available, one of several special-purpose algorithms can be used to solve the constraint system.

If no physical connection has slop and user-defined constraints are not allowed as in CABBAGE and SLIM, the longest path algorithm for acyclic graphs shown in Figure 6.23 is sufficient. It solves the graph in time $O(v + e)$, where $v$ is the number of vertices and $e$ is the number of edges in the graph. If connections are allowed to have slop, the graph will contain cycles.

If no feasible solution is available, we can use an iterative algorithm [Wou83, Lia83, Coo84, Len84, Meh84]. An example graph and its solution are shown in Figure 6.24. An edge can be used to define an ordering on the vertices it connects — the edge's source is before its sink in the ordering. We can use this ordering principle to classify each edge as either *forward* or *backward*. All positive-weight edges are forward, as are any negative-weight edges that satisfy the partial ordering defined by the positive-weight edges; any other edges are backward. The subgraph of forward edges is acyclic, and we can use it to guide the solution. In an iteration we first apply *acyclic_longest_path* (without the initialization step) to find vertex positions that satisfy the forward edges, then we update that solution to also satisfy the backward edges. As the figure shows, a backward-edge update may move a node enough to violate one of the forward constraints, in which case we must repeat the procedure. The algorithm stops when a forward update/backward update iteration produces no change in the solution. This algorithm is of complexity $O(ve)$, but because constraint graphs have relatively few large strong components, its average behavior is much better.

Most compactors allow the designer to introduce extra constraints into the compaction problem. *User-defined constraints* allow the designer to, for example, separate two components during compaction so that two groups of cells can fall past each other. Most systems apply the user-defined constraints during all compaction steps, but it is also useful to be able to apply a constraint for only a specified compaction step. User-defined constraints may cause the compaction problem to be *infeasible*, that is, to have no solution. An infeasible set of

```
procedure acyclic_longest_path(graph: graph);
begin
{ initialize node solution and predecessor count }
for node in graph.nodes do
  with node do
    begin
    value := 0;
    predecessors_left := n_predecessors(node);
    end;
{ trace through the graph starting at the source }
enqueue(source_node(graph),queue); { start with L node }
while not empty(queue) do
  begin
  node := dequeue(queue);
  for edge in outward_edges(node) do
    begin { see if this edge pushes its sink outward }
    sink := edge.sink; { this edge's source is node }
    sink.value := max(sink.value,node.value + edge.weight);
    sink.predecessors_left := sink.predecessors_left - 1;
    if sink.predecessors_left = 0 then
       enqueue(sink,queue);
    end; { of evaluating an edge }
  end; { of testing all paths from node }
end; { of acyclic_longest_path }
```

**Figure 6.23** Algorithm for Longest Path through an Acyclic Graph

constraints appears in the graph as a cycle with a positive sum of weights. The graph solving algorithms can be adapted to find and report infeasible cycles to the designer. SPARCS [Bur86] uses an event-driven algorithm to identify illegal constraint cycles.

Constraints other than the simple separation and physical connectivity constraints may be useful during compaction. *Symmetry constraints*, which constrain one distance to be equal to another, can be useful in generating layouts with wires of matched length and therefore matched parasitic resistance and capacitance. (Note that an equal length constraint is not the same as an equal position constraint.) *Maximum-distance constraints*, which require that two cells can be no farther apart than a certain distance, are useful for placing an upper bound on the parasitic element values on a wire. The SPARCS system has explored techniques for solving both types of constraints.

**Virtual Grid Leaf Cell Compaction.**   The virtual grid model simplifies both constraint generation and solution. The virtual grid determines which components can interact. Virtual grid compaction systems determine spacing rules

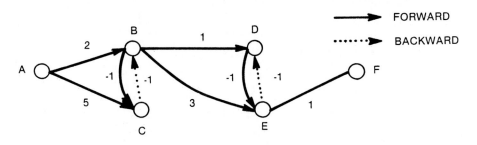

(a) Graph

| node | step 1 | | step 2 | | step 3 | |
|---|---|---|---|---|---|---|
| | *forward* | *backward* | *forward* | *backward* | *forward* | *backward* |
| A | 0 | 0 | 0 | 0 | 0 | 0 |
| B | 2 | 4 | 4 | 4 | 4 | 4 |
| C | 5 | 5 | 5 | 5 | 5 | 5 |
| D | 3 | 4 | 5 | 6 | 6 | 6 |
| E | 5 | 5 | 7 | 7 | 7 | 7 |
| F | 6 | 6 | 8 | 8 | 8 | 8 |

(b) Solution

**Figure 6.24**  Solution of a Constraint Graph with Cycles

and assign coordinates simultaneously, avoiding the need for an intermediate data structure to describe the constraints.

Compaction starts by assigning the leftmost virtual grid the mask position 0. The other virtual grid lines are assigned mask positions in order by looking at the components that have already been placed. The most commonly used method to find all the constraints that must be satisfied is the *most recent layers algorithm* [Boy83], shown in Figure 6.26. To understand the algorithm, consider the situation shown in Figure 6.25, where $vg_{xi}$ is being assigned a mask position and $vg_{x1}$ through $vg_{x(i-1)}$ have been assigned mask positions. The inner loop of the spacing algorithm determines the required spacing from $A$, a cell on $vg_{xi}$, and the cells to its left. The most recent layers table keeps track of the position of the last mask element seen on each layer. For example, the most recently seen diffusion element was on $vg_{x3}$; the required spacing is measured from the element's right edge, which is at $12\mu m$. That element hides an earlier diffusion element on $vg_{x1}$. The most recent layers table implements a simplified form of shadowing that takes advantage of the limitations placed on cell motion by the virtual grid.

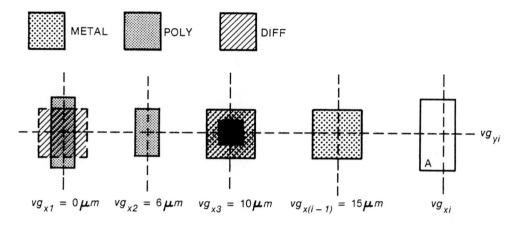

(a) Layout

| layer | responsible virtual grid | position |
|-------|--------------------------|----------|
| metal | $vg_{x(i-1)}$ | $17\mu m$ |
| poly  | $vg_{x2}$ | $9\mu m$ |
| diff  | $vg_{x3}$ | $12\mu m$ |

(b) Most Recent Layers Table

**Figure 6.25**  The Virtual Grid Spacing Operation

The design rule view of $A$ can easily be checked against the most recent layers table to find the lowest legal position of $A$. The virtual grid line is placed at the minimum position that satisfies all the spacing requirements. The virtual grid can be placed immediately because the constraints on it are part of an acyclic graph. Once $vg_{xi}$ has been placed, the most recent layers table can be updated and the algorithm's outer loop can place the next virtual grid.

After both x and y compaction, virtual grid compactors use an *xy compaction step* to reduce cell size. The one-dimensional compaction model produces a rectangular barrier around a component; as Figure 6.27 shows, this model is too pessimistic at the corners of the cell. The xy compaction step looks for adjacent components that can be moved closer together and adjusts the positions of their virtual grid lines. The xy compaction step is usually performed simultaneously with the second x or y compaction step since it must look at the same constraints. (The MACS constraint-graph compactor [Lo87] also uses an xy compaction step.)

```
procedure vg_compact(dim : dimension);
begin
vg[1].position := 0; { can position bottom vg independently }
{ initialize most recent layers from lowest vg }
for j from 1 to max_vg(opposite(dim)) do
   for rectangle in drc_view(cell_at(1,j)) do
      with most_recent[j,layer] do
         begin { set position and node number }
         location := rectangle.ur[dim] ; node := rectangle.node;
         end;
{ consider virtual grid lines 2..n }
for i from 2 to max_vg(dim) do
   begin
   position := empty;
   { find position of virtual grid line }
   for j from 1 to max_vg(opposite(dim)) do
      begin
      for rectangle in drc_view(cell_at(i,j)) do
         for layer in all_layers do
            position := max(position,
                             new_position(most_recent[j,layer],
                                          rectangle));
      end; { of finding position of i-th virtual grid line }
   vg[i].position := position;
   { update most recent layers }
   for j from 1 to max_vg(opposite(dim)) do
      for rect in drc_view(cell_at(i,j)) do
         with most_recent[j,rect.layer] do
            begin
            location := rect.layer.ur[dim] ; node := rect.node;
            end; { of updating j-th entry in most recent layers }
   end; { of positioning i-th virtual grid line }
end; { of vg_compact }
```

**Figure 6.26** Algorithm for Virtual Grid Leaf Cell Compaction

Because the virtual grid lines determine how components move, only one x and one y compaction step are necessary. Further x or y compactions will not reduce the cell size.

Boyer [Boy87b] introduced split-grid compaction which allows objects on a virtual grid line to move independently but does not allow connections with slop. Split-grid compaction produces layout closer in quality to constraint-graph compaction at the cost of more CPU time than required for simple virtual grid

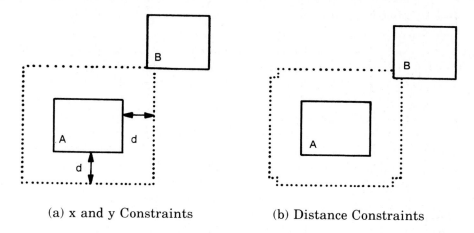

(a) x and y Constraints              (b) Distance Constraints

**Figure 6.27**  The xy Compaction Step

compaction. The virtual grid compactor of the NS system [Tan87] implements a number of techniques for improving layout quality. It was used to design the layout of a 400,000-transistor chip.

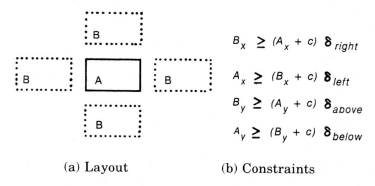

(a) Layout                    (b) Constraints

**Figure 6.28**  Two-Dimensional Compaction Constraints

**Two-Dimensional Leaf Cell Compaction.**  Two-dimensional compaction is an extension of one-dimensional constraint-graph compaction. As shown in Figure 6.28, two cells $A$ and $B$ can have four relative positions: $B$ can be above, below, left of, or right of $A$. In each case some spacing rule determines the minimum distance required between them. The minimum distance is expressed as a linear inequality, as in the 1-D constraint-graph model, but the inequality is multiplied by a *decision variable* that selects the constraints: when a decision variable is set to 1, the inequality constraint is included in the problem; when the decision variable is 0, the inequality is eliminated. Left-of/right-of placement constraints are expressed as functions of x-dimension variables and above/below constraints are written in terms of y-dimension variables.

This formulation is a special case of *mixed integer linear programming*, a well-known operations research problem which works on linear constraints and 0/1 decision variables. A simple insight helps us to discover algorithms for solving these problems — for any particular setting of the decision variables, the selected constraints form two 1-D constraint graphs, one for x and one for y. Solution algorithms choose settings for the decision variables (in effect selecting the relative locations of cells), solve the resulting 1-D compaction problems to get the cell positions, and evaluate the quality of the resulting layout.

The formulation of Schlag *et al.* [Sch83b] allows all four possible placements of component pairs. At worst, their algorithm requires $4n(n-1)$ separation constraints to describe the layout. (They demonstrated that layouts can have redundant constraints that can be eliminated but did not give an algorithm for generating a minimal or near-minimal constraint set.) This generality allows them to explore a wide variety of layouts. Their algorithm not only searches through all possible compactions of a given planar topology but also changes the planar topology — by moving a component from one side of a wire to the other, for instance. Although their problem is NP-complete, they gave some interesting proofs showing the amount of search required to produce a solution within some factor of the optimum.

Kedem and Watanabe [Ked84] used a slightly different problem formulation. They allow two cells only two of the four possible relative locations — $B$ may be either above or to the left of $A$, for example, but not below or to the right. In return for this restriction, any setting for the decision variables produces a legal layout. (In contrast, Schlag *et al.* search all the possible settings for the decision variables, which may require that one component be both below and above another.) Kedem and Watanabe's formulation is also NP-complete; they use branch-and-bound search to solve the 2-D constraint set. They also developed search heuristics that greatly speeded up solution time and algorithms for generating separation constraints from a symbolic layout.

No authors have attempted to extend the two-dimensional model to hierarchical compaction. One simple way to use these algorithms is to assemble 2-D-compacted leaf cells using the 1-D hierarchical algorithms of the next section.

**Constraint-Graph Hierarchical Compaction.**    Hierarchical compaction can handle both primitive and nonprimitive cells. There are two approaches to hierarchical constraint-graph compaction: we can treat subcells as fixed or allow them to stretch during compaction.

Fixed-cell hierarchical compaction was introduced by PYTHON and is also used in SPARCS. These programs' model of a subcell is the position of its ports and a *protection frame* used to determine cell-to-cell spacing. The protection frame includes material within the maximum design rule distance on each layer; this material determines the subcell's separation requirements. PYTHON and SPARCS create the required interface between subcells with routing.

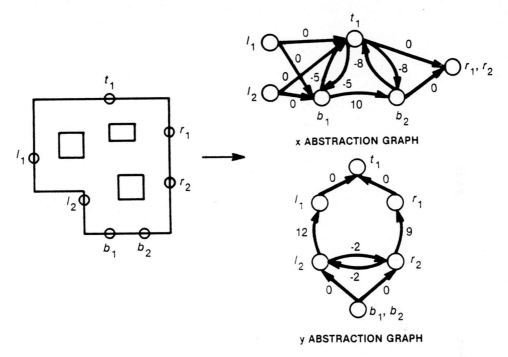

**Figure 6.29** The Port Abstraction

Pitchmatching — stretching cells so they can be connected by abutment — is simply recompacting the cells with additional constraints that require the connected ports to be at the same positions. Hierarchical compaction with stretchable cells requires the compactor to build a more complex model of a subcell than that required for fixed-cell assembly. The ideal model would contain information about how the cell's contents stretch during compaction — how the ports can move during compaction — and information to determine the spacing required between the cell and other cells. The most obvious model for a subcell is the cell itself, but this is equivalent to compacting a macroexpanded version of the layout hierarchy and is too expensive for large chips. The *port abstraction* model fully characterizes the stretching behavior of the cell with a much smaller constraint graph that includes only the ports as vertices and a flexible protection frame that describes the cell's boundary.

As shown in Figure 6.29, a port abstraction consists of two constraint graphs, one for x and one for y. The vertices in each graph represent the ports (with ports that move together, like $r_1$ and $r_2$ in the x graph, merged into a single vertex). The edges describe how the ports can move during compaction. The motion of the ports is determined by the components and wires in the cell's interior. If the cell contains $n$ components and wires, its x or y constraint graph can include at most $n(n-1)$ irredundant constraints. If it includes $p$ ports that move in one dimension, its abstraction graph can include at most $p(p-1)$ constraints. Because cells contain many fewer ports than components and wires, the port abstraction is a greatly simplified model of the cell.

The abstraction's constraints are determined by computing a subset of the transitive closure of the constraint graph. Because the abstraction contains only port vertices, only the longest paths between ports need to be computed. The simplest algorithm for calculating transitive closures is Floyd's algorithm [Flo62], which takes $v^3$ steps irrespective of the graph's sparseness. Fortunately, algorithms with better average behavior exist. The longest path algorithms described in Figure 6.23 can be applied once for each port to compute the partial transitive closure. These algorithms are of complexity $O(v^2 e)$ but perform much better for the relatively sparse graphs created during compaction. The partial closure contains redundant constraints; we need preserve only the negative-weight edges (which form connectivity constraints) and enough positive-weight edges to form a spanning tree of the nodes. Eliminating redundant constraints reduces both memory and solution time requirements. Experiments show that the number of irredundant constraints is slightly greater than linear in the number of ports [Wol85a, Eic86].

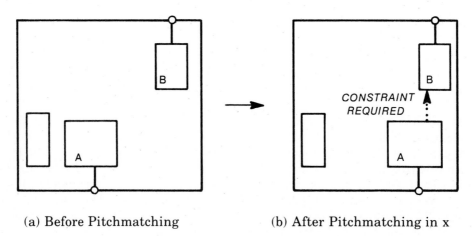

(a) Before Pitchmatching          (b) After Pitchmatching in x

**Figure 6.30** Interference Between x and y Compaction Steps

Port abstraction graphs generated directly from the x and y compaction graphs can produce illegal layouts because of unforseen interactions between the x and y movement of components. The *xy interlock* problem, illustrated in Figure 6.30, was identified by Eichenberger. The y port abstraction graph for the cell was computed from component positions before pitchmatching. If the subcell is pitchmatched first in x so that *A* moves as shown, the y abstraction graph will not reflect the constraint required between *A* and *B*. Two solutions to the problem which ensure that the port abstraction will produce only legal cell layouts have been proposed. The first [Per82] chooses a preferred dimension of compaction (usually the first compaction dimension specified by the cell's designer). The range of motion of components in the preferred dimension is computed by finding leftmost and rightmost solutions to the compaction graph. These ranges are used to build additional constraints in the other dimension — a constraint is added between two components if their ranges of motion overlap. This method is simple to compute, but it breaks down in some pathological cases (which have not been observed in practice), and it does not allow for more than one x and one y compaction of the subcell. The second method [Eic86] ensures that each pair of components is constrained by either an x or a y constraint. The longest paths between all pairs of components and wires in the cell are computed; if their ranges conflict in one dimension, then a constraint in the other dimension is introduced. This method is robust but costs much more to compute.

Finding the spacing required between two subcells during hierarchical compaction is much more complicated than finding the spacing required between primitive symbols because components in the subcell can move during compaction. Cell-to-cell spacing requires some knowledge of the components and wires near the subcells' boundaries. The port abstraction does not contain enough information to determine how cell-to-cell spacing changes as cells stretch, so most systems use worst-case spacing between subcells. Worst-case spacing may be acceptable for nMOS technologies but rarely for CMOS, where the ratio between smallest and largest spacing rules is much larger. The donut abstraction [Rei86] is an extension of the port abstraction in which vertices represent both ports and components near the cell boundary. The donut contains sufficient information to generate complete cell-to-cell spacing constraints, including those that allow cells to overlap. MACS [Lo87] compacts leaf cells along with primitive elements near the boundary of cells that will be abutted to them, then enforces pitchmatching constraints during assembly that ensure the abutted cells' boundaries will match.

**Virtual Grid Hierarchical Compaction.** Most virtual grid systems perform a simple two-level hierarchical compaction. First, all leaf cells are compacted, then all cells are assembled into the final design by a pitchmatching algorithm. During pitchmatching, the mask positions of virtual grid lines in the cells are adjusted to stretch the cells for abutment.

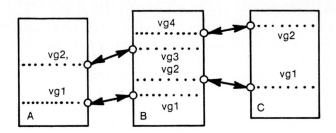

(a) Cells to Pitchmatch

| cell | virtual grid | location |
|------|-------------|----------|
| A | $vg\,1$ | $5\mu m$ |
|   | $vg\,2$ | $22\mu m$ |
|   |   |   |
| B | $vg\,1$ | $8\mu m$ |
|   | $vg\,2$ | $15\mu m$ |
|   | $vg\,3$ | $30\mu m$ |
|   | $vg\,4$ | $35\mu m$ |
|   |   |   |
| C | $vg\,1$ | $10\mu m$ |
|   | $vg\,2$ | $40\mu m$ |

(b) Original Virtual Grid Locations

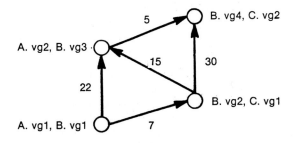

(c) Pitchmatching Constraints

**Figure 6.31**  Enforcing Cell-to-Cell Connections on the Virtual Grid

Stretching subcells during hierarchical virtual grid compaction is best formulated as a simple constraint problem. As shown in Figure 6.31, the virtual grid compactions of the subcells determine the minimum positions of their pins. The minimum distance of each pin from the base of its subcell can be written as a separation constraint, and a connection between pins can be written as an equality constraint. The resulting graph (shown in Figure 6.31 after merging nodes connected by equality constraints) is acyclic and can be solved with breadth-first search. Once the port positions have been found, we can update the positions of the other virtual grid lines for the subcells to maintain the required spacings.

The virtual grid model simplifies the determination of cell-to-cell spacing — the virtual grid allows the compactor to predict how components along the boundary will move. MULGA looks at the outermost virtual grid line on each side of a subcell to determine the spacing required to the cell. PANDA [Bul86] performs cell-to-cell spacing in two passes: first, using worst-case spacing, then pushing cells closer together based on a design rule analysis near the boundaries of abutted cells. COORDINATOR [Kol86] uses more efficient constraint-generation algorithms for leaf cell compaction, which make feasible the compaction of very large (10,000-transistor) cells.

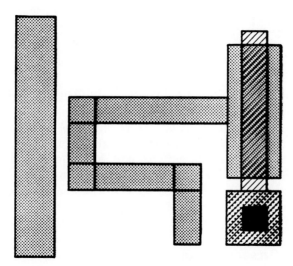

**Figure 6.32**  A Layout Compacted without Wire Length Minimization

**Wire Length Minimization.**   The layout of Figure 6.32 demonstrates the need for wire length minimization, particularly in constraint-graph compaction. The compaction algorithms described above have no criterion for placing components and wires that does not limit cell size. They generally put these cells at one end of their allowed range, creating unnecessarily long wires that add parasitic resistance and capacitance to the layout and slow down the circuit. A 1-D wire length minimization can be performed during a compaction step to clean up problems created by the compaction algorithm. The greater freedom of motion in constraint-graph compaction makes wire length minimization mandatory; however, it could also be used during hierarchical virtual grid compaction to find the positions of virtual grid lines without connected ports.

The first wire length minimization algorithm was described by Schiele [Sch83a]. Lava uses a linear programming formulation of the problem and solves it using network programming techniques; an alternative linear programming formulation is described in [Lin86].

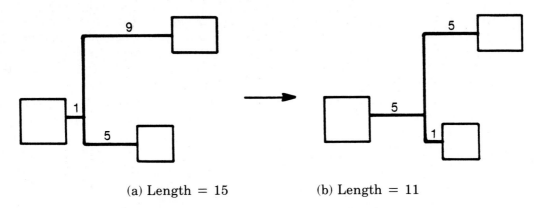

(a) Length = 15          (b) Length = 11

**Figure 6.33**  A Wire Length Minimization Step

Linear programming formulations of wire length minimization are based on the observation that if the ends of wires are not allowed to reverse order during minimization, the problem is convex and has one globally optimal value. The simplest algorithm for wire length minimization works directly on the compaction constraint graph. First, the compaction constraints are augmented with constraints that maintain the order of wire endpoints. The augmented graph is solved by repeatedly applying the minimization step illustrated in Figure 6.33. A vertex is chosen that, when moved, will reduce the weighted sum of wire lengths in the cell. Because the optimum solution to a linear program occurs at the simultaneous solution of two or more constraints, the wire length need be checked only at the ends of the vertex's range. When no vertex can be moved to reduce total weighted wire length, the cell's wire length is minimal.

Researchers have explored other methods for wire length minimization. Lakhani and Varadarajan [Lak87] developed an algorithm that performs grouping and shearing operations to minimize total wire length. Kingsley [Kin84] modified Schiele's algorithm to reassign the vertex-ordering constraints at each step to choose the ordering of vertices that gave the smallest wire length. Several experimenters [Rei85, Var86] have reported success with finding locally minimal wire lengths by not introducing the vertex-ordering constraints into the wire length minimization problem.

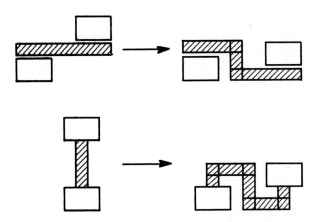

**Figure 6.34** Inserting Jogs during Compaction

**Incremental Cell Optimization.**   A number of authors have explored methods for incrementally modifying a cell during compaction. The most thoroughly understood method is *jog introduction* − adding a degree of freedom to a connection by changing a wire segment to a z-shaped wire. Figure 6.34 shows two prototypical situations for adding a jog. In both cases, allowing the ends of the wire to move separately allows the components to move closer together in y. Automatic jog introduction occurs at the start of a compaction step − wires are selected for jogging and jogs are added, then constraint generation and solving proceeds normally.

Early jog introduction algorithms did a limited amount of analysis to determine where to add jogs: CABBAGE used strictly local analysis and added all possible jogs; SLIM added jogs only to wires constraining the critical path but could add more jogs than were necessary. Unnecessary jogs can cause problems for later compaction steps by adding unwanted disorder into the cell. Supercompaction [Wol84, Wol88] introduced methods for analyzing the critical path to find a small set of jogs sufficient to reduce the cell size.

An example of critical path analysis is shown in Figure 6.35. The *critical path graph* is an undirected subset of the constraint graph that shows separation and connectivity constraints that determine the cell's size. Adding a jog to a wire removes a connectivity constraint; we can model this operation as the removal of the edge representing the connection from the critical path graph. The dotted edges represent opportunities to add jogs to break the critical path. To ensure that the layout will be made smaller in y, we must introduce jogs that break all critical paths from $L$ to $U$. In this case, we must add jogs to both wires to break all critical paths. Then the cell is guaranteed to shrink during the

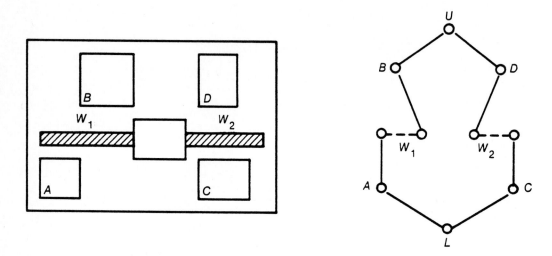

**Figure 6.35**  Analyzing the Critical Path for Optimizations

next compaction step. Finding a set of jogs that breaks all critical paths is equivalent to finding a cutset of the critical path graph that puts $L$ and $U$ in separate sets.

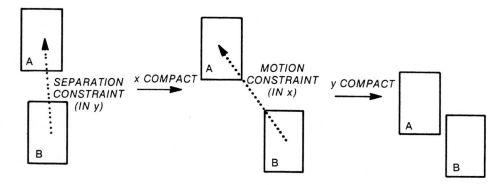

**Figure 6.36**  Breaking Separation Constraints

Critical path analysis also allows us to consider geometric reorganizations that break the critical path by moving cells to eliminate separation constraints, as shown in Figure 6.36. We can add jogs and do geometric reorganizations simultaneously to break critical paths that cannot be broken by jog introduction alone. Experiments show that doing both kinds of optimizations together produces smaller layouts than can be made by jog introduction alone.

The zone refining algorithm [Shi86] for compaction applies geometric reorganizations and jog introduction operations to sites both on and off the critical path. Preemptively applying optimizations off the critical path speeds the convergence to an optimal layout, though at the expense of extra CPU time and possibly unnecessarily increasing the cell size in the other dimension. Varadarajan [Var85] developed a jog introduction algorithm that adds jogs only to wires on the critical path. MACS [Lo87], rather than adding several jogs at a single step, adds them one at a time and uses fast algorithms to update the solution and evaluate the effects of the jog.

**Compaction Analysis Tools.**    Designers may want to analyze the results of a compaction to improve its results. The designer may change the cell's topology to remove cells from the critical path or change its geometry to change the separation constraints generated. He or she may also use a different compaction sequence or additional constraints during compaction to improve the final layout.

Beyond changing the stick diagram's topology, a few tools provided by the compactor to control its own behavior can let the designer make final optimizations to a layout while allowing the compactor to do most of the work. The simplest mechanism is the compaction sequence. The designer can change the number of compaction steps or reverse the order of x and y compaction to change the layout's size and aspect ratio.

But some changes may require more detailed control. The designer may wish to break separation constraints by moving components as shown in Figure 6.36. Geometric reorganizations can be done manually by adding a user-defined constraint to the compaction problem at a particular step. In the example, adding an x-dimension constraint during one step to push the components apart would allow them to move closer during the next y compaction. For virtual grid compactors without wire length minimization, a particularly useful user-defined constraint gives the minimum position of a virtual grid line, which the designer can use to push a lower virtual grid line closer to a higher one. The designer may occasionally want to add constraints that apply to all compaction steps in a given dimension. The compactor can supply a similar mechanism for jog introduction that allows the designer to add a jog to a particular wire, at a specified time during compaction.

Several types of information display help the designer decide where to apply these mechanisms or how to change the layout's topology. For leaf cells the most powerful information is the *critical path display*. The critical path allows the designer to quickly focus on the components and wires that are limiting the cell's size. After examining the critical path, the designer often needs to look at constraints between pairs of cells to determine how far something must be moved to change the critical path. A movie of the compaction process, made of one frame per compaction step, explains how the layout reached its current form. Once again, the critical path for each frame allows the designer to focus on key parts of the layout. For cell assembly, a display comparing the unstretched and stretched sizes of cells allows the designer to identify cells that grew much larger than intended. A hierarchical critical path display identifies the connections that forced cells to grow.

**Performance of Compaction Algorithms.**   There are two important questions to answer about compaction algorithms: how their results compare to optimal layouts and how the algorithms' results compare to each other. Researchers have performed a number of experiments to try to answer both questions.

Comparing compacted and optimal layouts is a difficult task that no one has adequately completed. Optimal layouts are much too expensive to generate for layouts of interesting size. Even a comparison with manually designed layouts is hard because of the time a person requires to create a good layout. A valid comparison would entail a large number of test cases. No one has yet been willing to spend the time required to produce high-quality, correct layouts both manually and with compaction. There has been enough experimentation, however, to draw some limited conclusions.

Williams [Wil77] justified the use of compaction based on the frequency of errors in hand-drawn layouts: he observed that 10 in 1000 rectangles drawn by a human designer were incorrect. Dunlop [Dun79] found that SLIM-compacted cells ranged from 15% smaller to 30% larger compared to hand-drawn cells and also verified the 1% error rate for humans. Cho *et al.* [Cho77] reported that FLOSS-compacted layouts averaged 30% smaller. Hsueh [Hsu79] reported results for one compacted cell that was 2% smaller than the hand-drawn design, though he modified the original cell's topology to achieve the smaller result. The most interesting experiment was conducted by Moscovitz [Mos86]. He organized a race at AT&T Bell Laboratories between experienced draftsmen using a mask-layout editor and experienced chip designers using the icon virtual-grid compactor. The groups designed the same cells and posted their best results for inspection by the other team. By using each other's design innovations, the two teams could produce cells of about the same size, though at any point in the race the virtual grid design was about 10% larger. Though none of these experiments is conclusive, they all show that a compactor can easily produce cells within 10% to 20% of a well-designed hand layout and that careful design of the symbolic layout can reduce the difference to zero.

The other question to answer is the relative performance of compaction algorithms. Virtual grid and constraint-graph compaction can be compared on three separate criteria: the amount of CPU time they take, the size of the compacted layout, and the predictability of the compaction process.

| example compactor | area ($\mu m^2$) | CPU time (sec)[*] | memory |
|---|---|---|---|
| **afa** | | | |
| MACS | 143 x 166 = 23738 | 9 | 1450K |
| SPARCS | 157 x 180 = 28260 | 11 | 356K |
| Symbolics | 160 x 189 = 30240 | 164K | |
| Zorro | 140.5 x 171 = 24025.5 | 430 | 647K |
| **afavg** | | | |
| MACS | 142 x 145 = 20590 | 5 | 1390K |
| SPARCS | 157 x 151 = 23707 | 8 | 372K |
| Symbolics | 154 x 154 = 23716 | 5 | 160K |
| Zorro | 128.5 x 151 = 19403.5 | 524 | 598K |
| **n28** | | | |
| SPARCS | 123 x 198 = 24354 | 8 | 356K |
| Zorro | 108 x 187 = 20196 | 378 | 704K |
| **C132** | | | |
| MACS | 627 x 354 = 221958 | 41 | 3000K |
| SPARCS | 685 x 339 = 232215 | 51 | 184K |
| Symbolics | 675 x 330 = 222750 | 57 | 1040K |
| Zorro | 660 x 322 = 212520 | 301 | 1413K |
| **mul2x2** | | | |
| MACS | 309 x 252 = 77868 | 16 | 2050K |
| SPARCS | 343 x 255 = 87465 | 47 | 215K |
| Symbolics | 370 x 270 = 99900 | 10 | 512K |
| Zorro | 312 x 252 = 78624 | 838 | 614K |
| **mul4x4** | | | |
| SPARCS | 649 x 601 = 390049 | 66 | 754K |
| Symbolics | 654 x 638 = 417252 | 54 | 840K |
| Zorro | 577 x 577.5 = 333217.5 | 1904 | 741K |
| **mul8x8** | | | |
| SPARCS | 1285 x 1285 = 1651225 | 89 | 2066K |
| Symbolics | 1276 x 1352 = 1725152 | 245 | 3200K |
| Zorro | 1138 x 1207.5 = 1374135 | 11738 | 7754K |
| **mul16x16** | | | |
| Symbolics | 2524 x 2780 = 7016720 | 1073 | 14400K |

[*]The Symbolics compactor runs on a Symbolics workstation. The CPU times were divided by a factor of 4 to scale them to an 8650.

**Figure 6.37** Compaction Times for Various Systems (from [Boy87a])

The most complete compilation of CPU times for compactors was compiled by Boyer [Boy87a], whose table is summarized in Figure 6.37. The results show that constraint-graph compactors can be comparable in speed to virtual grid compactors and produce smaller layouts. Eichenberger [Eic86] gives the results of a number of experiments on constraint-graph compaction algorithms, showing that constraint satisfaction and wire length minimization are linear in average performance and that port abstraction generation is slightly worse than linear.

Eichenberger also gives the total CPU time required to hierarchically compact two chips, including parsing and other overhead: a 1633-transistor chip required 951 CPU seconds, and a 3499-transistor chip required 12,777 CPU seconds on a VAX-11/780. Entenman and Daniel [Ent85] reported that the VIVID System generated a 1200-transistor chip layout in 360 seconds of real time on a VAX-11/780.

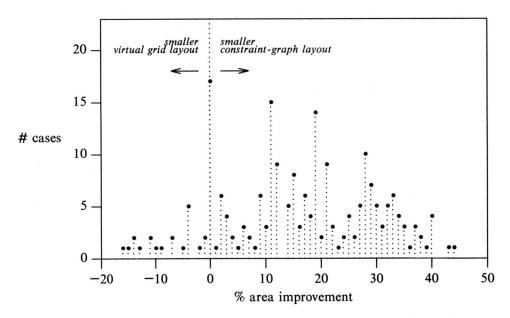

**Figure 6.38**  Area Improvement of Constraint-Graph over Virtual Grid Compaction (from [Wol85b])

Experiments by Wolf [Wol85b] compared the quality of layouts produced by constraint-graph and virtual grid compaction. One set of experiments compared the areas of layouts produced by the two algorithms. The graph of Figure 6.38 shows the result of an experiment that compared the area of cells compacted by constraint-graph and virtual grid algorithms. The average constraint-graph compaction result was 17% smaller than the virtual grid result. Other experiments showed that constraint-graph compaction is better at producing minimum-pitch layouts. Constraint-graph cell assembly is also noticeably better: experiments comparing MACS and MULGA [Lo87] showed that layouts assembled with constraint-graph algorithms were 16% to 31% smaller than when assembled with virtual grid techniques.

Another set of experiments by Wolf compared the *sensitivity* of the compaction process. A compactor that can create large differences in the compacted layout from small variations in the input layout is hard to use — the designer must edit the input several times to persuade the compactor to produce the desired layout. The experiments measured sensitivity by creating several different versions of a cell, each of which has small, random variations, and comparing the sizes of the compacted layouts. The results showed that constraint-graph compaction is significantly less sensitive than virtual grid compaction. Virtual grid compaction produced a wider spread of layout sizes for the different versions of a cell, and it was difficult to predict whether a change would make a cell larger or smaller; constraint-graph compaction produced more consistent results. Constraint-graph compaction is more predictable because components move independently; moving a component to a different virtual grid may change the spacing of all the components along the virtual grid line.

Developers are increasingly using constraint-graph techniques when building new compactors. Cho [Cho85] believes that the improved quality of layout generated by constraint-graph compaction more than pays for the increased CPU time required to generate it. The authors concur with this opinion. Compaction makes layout geometry design time such a small fraction of the chip design time that the difference in time required for constraint-graph and virtual grid compaction is negligible — CPU times are often determined as much by other programs in the design system or the software engineering skills of the implementor as by the compaction algorithms themselves. Because constraint-graph compaction gives smaller layouts and is less sensitive, fewer valuable designer-hours are consumed modifying stick diagrams to get a good compaction result. Also, in our experience constraint-graph compaction programs are not significantly more difficult to implement than virtual grid compactors. We believe that all these factors weigh the balance heavily in favor of constraint-graph compaction.

## Applications of Compaction

Freeing the designer from strict adherence to the design rules to concentrate on topological design of the layout has a number of advantages. It makes designs easier to update and reuse, it simplifies the task of the designer, and it simplifies the creation of programs that design layouts from abstract specifications. There are several different paradigms for the use of compaction in a layout design system.

One paradigm is compaction as a *design rule binding* operation. Design rule binding is the translation of a layout designed from a generic technology, such as nMOS or CMOS, to a particular fabrication technology's design rules. Although the designer must make the design decisions relevant to a generic technology, a great deal of the work involved in layout design comes from the adherence to the design rules. Symbolic layout and compaction can free the designer from a surprising variety of rules, such as tub insertion. Design rule binding allows a single design to be targeted to a number of different technologies, such as n-tub, p-tub, and twin tub CMOS from a single symbolic CMOS layout. It also allows a second-generation chip to be implemented in a new technology for cost reduction or higher performance. Technology independence is particularly important when the technology is not fully characterized before design starts, as often happens during very large chip design projects when chip and technology are developed in parallel.

A second paradigm is compaction as *assembly*. Hierarchical compaction allows the designer to specify the chip topology and connections to be enforced without having to design all cells to fit together. Using hierarchical compaction to assemble chips allows one to more easily build a library of reusable cells because the cells can be reshaped to fit different interconnection plans. It also makes changing the layout during design more practical — by adding an additional wiring track within a datapath, for example — because connectivity is automatically enforced.

The third paradigm is compaction as a *compilation target language*. Silicon compilers produce layouts from more abstract requirements; layout generation is much simpler if a symbolic layout to be compacted is the product. Chapter 7 describes several cell compilers that produce symbolic layouts that can be compacted. Hierarchical compaction allows these compiled layouts to be assembled into a chip with layouts designed with other methodologies.

Which of these models is most powerful depends on your view of the design process. For the cell designer, technology binding allows the reuse of cells designed for older technologies. For the chip designer, assembly greatly simplifies the management of changes to the design. For the logic designer or architect, a uniform compilation target language speeds design and allows the evaluation of different implementations of a circuit. But in every case the underlying advantage is design turnaround time. At each level of design, compaction allows the design to be generated and modified more quickly, encouraging experimentation with several different solutions to design problems. Compaction allows both layout and logic designers to explore more of the solution space to find the best possible design.

## 6.4 Summary

Symbolic layout and compaction are closely related topics. Symbolic layout captures more of a designer's intent than does pure layout and allows the layout to be updated to new technologies; compaction frees the designer from the tedium of adhering to the design rules during layout design. The two techniques together form a powerful tool for automating the design process. Most importantly, symbolic layout and compaction are general-purpose design tools − they impose very few restrictions on the type of circuit to be designed or the methodology of design. The universal utility of symbolic layout and compaction make them especially attractive building blocks for a VLSI computer-aided design system.

## Acknowledgements

We would like to thank Dwight Hill, Kurt Keutzer, Chi-Yuan Lo, Mark Reichelt, Ravi Varadarajan, and our co-authors of this book for their enlightening comments and criticism of this chapter. Discussions with Bill Crocker and Dwight Hill helped us clarify some ideas on technology databases, and discussions with Jeff Burns helped us to understand primitive-independent technology databases. One of us (Wolf) learned a lot about hierarchical compaction while working with Mark Reichelt. Kurt Keutzer was an important source of knowledge on graph algorithms. Dave Boyer provided the results of the International Conference on Computer Design compactor benchmark tests. We would like to thank a number of people for discussions on symbolic layout and compaction over the years: Sheldon Akers, Dave Boyer, Jeff Burns, Bill Crocker, Peter Eichenberger, Herman Gummel, Dwight Hill, Min-Yu Hsueh, Kurt Keutzer, Chris Kingsley, Chi-Yuan Lo, Rob Mathews, John Newkirk, Dan Perkins, Mark Reichelt, Tim Saxe, Dan Schweikert, Don Sharfetter, Dan Siewiorek, and Ravi Varadarajan. We would also like to thank our management at AT&T Bell Laboratories for their support of this project.

## Exercises

**6.1**    With $L=0$, find two solutions for the graph of Figure 6.39: with slack nodes assigned their lowest possible values and assigned their highest values. Identify the forward and backward edges. Identify the critical path from $L$ to $U$ through the graph.

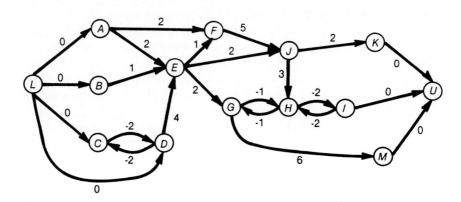

**Figure 6.39**  A Sample Compaction Graph

**6.2**   Figure 6.40 illustrates the shear-line compaction algorithm.  A *shear line* cuts the layout into disjoint sections.  Excess space is removed along the shear line, making the layout smaller.  Shear-line compaction alternately finds shear lines through the cell, then removes space along the shear line to reduce the cell size.

Compact by hand the layout of Figure 6.41 as follows:

**a)**   Virtual grid − x followed by y.

**b)**   Virtual grid − y followed by x.

**c)**   Shear line − x followed by y.

**d)**   Shear line − y followed by x.

**e)**   Constraint-graph − x followed by y.

**f)**   Constraint-graph − y followed by x.

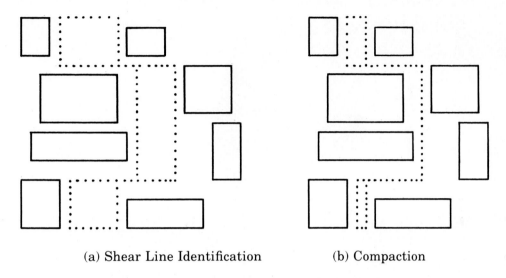

(a) Shear Line Identification       (b) Compaction

**Figure 6.40** A Shear-Line Compaction Step

**6.3** In the symbolic layout language of your choice, write a description of the layout of Figure 6.41 in three styles:

**a)** Using only numbers for positions (no symbolic constants).

**b)** Using symbolic constants for positions, but without using arithmetic expressions.

**c)** Using arithmetic expressions written with symbolic constants.

Compare these three descriptions on ease of specification, ease of modification, and complexity of the language translator required.

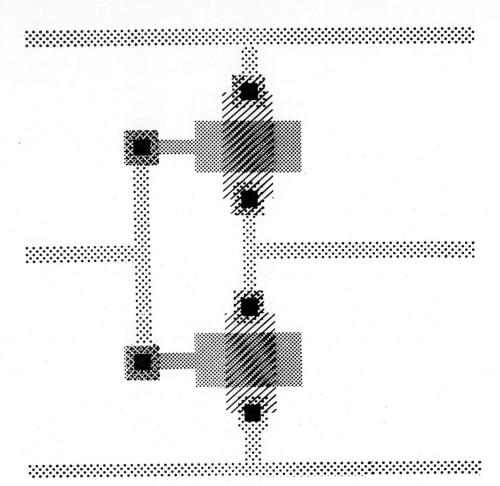

**Figure 6.41** A Sample Layout

**6.4**  Create implementation views for a technology database of primitive symbols whose shape changes during compaction:

**a)**  A wire that stretches in x, whose endpoints are represented by two x variables and one y variable.

**b)**  A stretchable rectangle, whose endpoints are represented by two x and two y variables.

**6.5†**  Shear-line compaction can be modeled using constraint graphs; this model shows the algorithm to be a simple form of constraint solving by breadth-first search.

**a)**  Show how to find a shear line through a cell using the cell's constraint graph.

**b)**  Show how to use the shear-line-finding algorithm to solve the constraint graph to minimize the distance $U - L$.

**c)**  Compare the complexity of the shear-line algorithm to that of *acyclic_longest_path*.

**6.6†**  It is often useful to have a simple program to eliminate some redundant rectangles in a mask layout description, if only to reduce the amount of storage required before processing with edge-based programs. Develop an algorithm that takes as input a set of rectangles on a mask layer and produces a new set of rectangles, no larger in size than the input set, that covers *exactly* the same region. This algorithm need not produce the smallest possible covering set, but it should produce a good solution. The algorithms should have good asymptotic bounds since it is easy to generate a mask with 100,000 or more rectangles.

**6.7†**  Many nodes in a constraint graph are required to be a fixed distance apart. These nodes can be merged into a single equivalent node to reduce the amount of time required for constraint solving. Develop data structures and an algorithm for merging fixed-distance nodes. The algorithm should take as input a constraint graph and return a new, reduced constraint graph and a table mapping nodes in the given graph into the new graph's nodes. The table must show which new node represents a given node and the offset from the given node to the new node's position. (Remember that as nodes are merged, constraint weights must also be updated.)

**6.8†**  Develop an algorithm to produce a legal true-size layout from a unit-size symbolic layout. The algorithm must add wires to keep all wiring paths Manhattan as the components change to their true size.

**6.9†**  Extend the two-dimensional compaction model to include jog introduction. Your model should consider wires with $n$ or fewer jogs per wire, so that search can generate layouts that are 1-jog optimal, 2-jog optimal, *etc.* The insertion of a jog should be controlled by a decision variable.

**6.10‡** Symbolic layout systems that solve constraints and compaction systems that allow the user to add constraints to a compaction problem can both produce constraint graphs that are infeasible. The infeasible constraints in the simplest case form a loop, or they may form the union of several loops. Develop an algorithm to determine all the infeasible constraint loops in a constraint graph. The algorithm should produce sets of constraints, each of which contains only the constraints that form an infeasible problem.

**6.11‡** Ideally, a symbolic layout system should automatically insert tub ties into a CMOS layout. The general problem is to insert tub ties into a layout, minimizing total area, such that the resistance from any point in the tub to the power supply is less than some value. (Excess resistance in the tub connections to $V_{dd}$ and $V_{ss}$ encourages latch-up.) This optimization problem is made difficult by the cost of the resistance calculation and obstacles in the layout that restrict tub tie placement.

Design an algorithm to insert tub ties into a layout. The algorithm takes as input a compacted layout, from which it estimates initial resistances. The new layout with tub ties inserted can be compacted to produce a final design.

**6.12‡** Due to the limitations of the machines used to make masks, mask data must be supplied to specified tolerances. For instance, many mask-making machines can only create polygons whose vertices are on a $1/4\mu m$ grid. Develop an algorithm that takes as input a constraint graph and its solution and produces a new solution that satisfies all the constraints and whose values are all multiples of $\varepsilon$, the mask-making resolution. The algorithm should change the given solution as little as possible, to minimize changes to both the cell size and the results of wire length minimization.

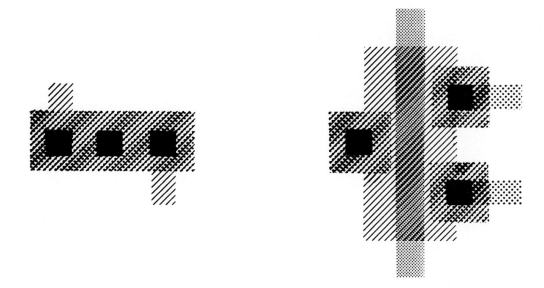

**Figure 6.42** Hard Cases for Symbolic Layout Extraction

**6.13‡** A program to reverse-engineer mask layouts into symbolic layouts is a very useful tool. Unfortunately, the problem is harder than it may first appear, thanks to the numerous special cases one encounters. Most of these special cases are manifestations of *implied connectivity* − mask layout polygons that touch to make an electrical connection but whose symbolic equivalents would not touch. Two simple examples are shown in Figure 6.42: the first case is a transistor source and via connected without a wire; the second case is two wires whose edges touch but whose centerlines do not.

Investigate the problems of reverse-engineering mask layouts:

**a)** Enumerate as many cases of implied connectivity as you can.

**b)** Assume that transistors and vias have been found using layout verification algorithms and eliminated from the layout. Only polygons that form wires should be left. Develop an algorithm to find the centerlines of the symbolic wires formed by these polygons. (Hint: think about shrinking the polygons to zero width.)

**c)** In what cases will your algorithm leave wires unconnected? What methods can be used to make all required connections?

# References

[Aho86] Aho, A. V., R. Sethi, and J. D. Ullman, *Compilers: Principles, Techniques, and Tools*, Addison-Wesley, Reading, MA, 1986.

[Aho74] Aho, A. V., J. E. Hopcroft, and J. D. Ullman, *The Design and Analysis of Computer Algorithms*, Addison-Wesley, Reading, MA, 1974.

[Ake70] Akers, S. B., J. M. Geyer, and D. L. Roberts, "IC mask layout with a single conductor layer," *Proc., 7th Design Automation Workshop*, pp. 7-16, 1970.

[Ayr83] Ayres, R. F., *VLSI: Silicon Compilation and the Art of Automatic Microchip Design*, Prentice-Hall, Inc., Englewood Cliffs, NJ, 1983.

[Bal82] Bales, M. W., *Layout Rule Spacing of Symbolic Integrated Circuit Artwork*, Master of Science thesis, University of California, Berkeley, May 4, 1982.

[Bar75] Barnes, D. E., and T. L. Davis, "Graphics layout systems for faster tooling from faster designs," *Wescon Professional Program*, Wescon Electronic Show and Convention, September 1975.

[Baz77] Bazaraa, M. S., and J. J. Jarvis, *Linear Programming and Network Flows*, John Wiley and Sons, New York, NY, 1977.

[Boy83] Boyer, D. G., and N. Weste, "Virtual grid compaction using the most recent layers algorithm," *Digest Intl. Conf. on Computer-Aided Design*, pp. 92-93, September 1983.

[Boy87a] Boyer, D. G., "Symbolic layout compaction benchmarks − results," *Digest Intl. Conf. on Computer-Aided Design*, pp. 209-217, October 1987.

[Boy87b] Boyer, D. G., "Split grid compaction for a virtual grid symbolic design system," *Digest Intl. Conf. on Computer-Aided Design*, pp. 134-137, November 1987.

[Bul86] Bullman, W. R., L. A. Davieu, H. S. Moscovitz, and G. D. O'Donnell, PANDA − a module assembler for the IDA system, private communication, September 1986.

[Bur83] Buric, M. R., C. Christensen, and T. G. Matheson, "The plex project: VLSI layouts of microcomputers generated by a computer program," *Digest Intl. Conf. on Computer-Aided Design*, pp. 49-50, ACM/IEEE, September 1983.

[Bur86] Burns, J. L., and A. R. Newton, "SPARCS: a new constraint-based IC symbolic layout spacer," *Trans., IEEE Custom Integrated Circuits Conf.*, pp. 534-539, IEEE, May 1986.

[Bur87]  Burns, J. L., and A. R. Newton, "Efficient constraint generation for hierarchical compaction," *Proc., Intl. Conf. on Computer Design*, pp. 197-200, IEEE Computer Society, October 1987.

[Cho85]  Cho, Y. E., "A subjective review of compaction," *Proc., 22nd Design Automation Conf.*, pp. 396-404, ACM/IEEE, June, 1985.

[Cho77]  Cho, Y. E., A. J. Korenjak, and D. E. Stockton, "FLOSS: an approach to automated layout for high-volume designs," *Fourteenth Annual Design Automation Conf.*, pp. 138-141, ACM/IEEE, 1977.

[Com85]  EDIF Technical Committee, *EDIF Electronic Design Interchange Format*, EDIF Technical Committee, November 1985.

[Coo84]  Cook, P. W., "Constraint solver for generalized IC layout," *IBM Journal of Research and Development*, pp. 581-589, September 1984.

[Cro87]  Crocker, W. H., R. Varadarajan, and C.-Y. Lo, "MACS: a module assembly and compaction system," *Proc., Intl. Conf. on Computer Design*, pp. 205-208, IEEE Computer Society, October 1987.

[Dan63]  Dantzig, G. B., *Linear Programming and Extensions*, Princeton University Press, Princeton, NJ, 1963.

[Dav83]  Davis, T., and J. Clark, *SILT: A VLSI Design Language*, Technical Report, Stanford University, June 1983.

[Doe87]  Doenhardt, J., and T. Lengauer, "Algorithmic aspects of one-dimensional layout," *IEEE Transactions on Computer-Aided Design*, vol. CAD-6, no. 5, pp. 863-878, IEEE, 1987.

[Dre69]  Dreyfus, S. E., "An appraisal of some shortest-path algorithms," *Operations Research*, vol. 17, pp. 395-412, 1969.

[Dun79]  Dunlop, A. E., *Integrated Circuit Mask Compaction*, Ph.D. thesis, Carnegie-Mellon University, October 1979.

[Dun80]  Dunlop, A. E., "SLIM: the translation of symbolic layouts into mask data," *Proc., 17th Design Automation Conf.*, pp. 595-602, ACM/IEEE, 1980.

[Dun81]  Dunlop, A. E., "SLIM: the translation of symbolic layouts into mask data," *Journal of Digital Systems*, vol. V, no. 4, pp. 429-451, 1981.

[Dun78]  Dunlop, A. E., "SLIP: symbolic layout of integrated circuits with compaction," *Computer Aided Design*, vol. 10, no. 6, pp. 387-391, November 1978.

[Eic86]  Eichenberger, P., *Fast Symbolic Layout Translation for Custom VLSI Integrated Circuits*, Ph.D. thesis, Stanford University, April 1986.

[Ent85]    Entenman, G., and S. W. Daniel, "A fully automatic hierarchical compactor," *Proc., 22nd Design Automation Conf.*, pp. 69-75, ACM/IEEE, June 1985.

[Fie84]    Fiebrich, R.-D., Y.-Z. Liao, G. Koppelman, and E. Adams, "PSI: a symbolic layout system," IBM *Journal of Research and Development*, vol. 28, no. 5, pp. 572-580, September 1984.

[Fis85]    Fishburn , J. P., and A. E. Dunlop, "TILOS: a posynomial programming approach to transistor sizing," *Digest Intl. Conf. on Computer-Aided Design*, pp. 326-328, IEEE Computer Society, November 1985.

[Flo62]    Floyd, R., "Algorithm 97, shortest path," *Communications of the ACM*, vol. 5, no. 6,  p. 345, June 1962.

[Fol82]    Foley, J. D., and A. Van Dam, *Fundamentals of Interactive Computer Graphics*, Addison-Wesley, Reading, MA, 1982.

[Fow69]    Fowler, B. R., "XYMASK," *Bell Laboratories Record*, vol. 47, no. 6, pp. 204-209, July 1969.

[Gal86]    Gallo, G., and S. Pallottino, "Shortest path methods: a unifying approach," *Mathematical Programming Study*, vol. 26, pp. 38-64, North Holland, Amsterdam, the Netherlands, 1986.

[Gib76]    Gibson, D., and S. Nance, "SLIC — symbolic layout of integrated circuits," *Proc., 13th Design Automation Conf.*, pp. 434-441, June 1976.

[Hed85a]  Hedges, T., W. Dawson, and Y. E. Cho, "Bitmap graph build algorithm for compaction," *Digest Intl. Conf. on Computer-Aided Design*, pp. 340-342, September 1985.

[Hed85b]  Hedlund, K. S., "Electrical optimization of PLAs," *Proc., 22nd Design Automation Conf.*, pp. 681-687, June 1985.

[Hil84]    Hill, D. D., "ICON: a tool for design at schematic, virtual-grid and layout levels," *IEEE Design & Test*, vol. 1, no. 4, pp. 53-61, 1984.

[Hil86]    Hill, D. D., "Stretchable routing in a symbolic, constraint-based environment," *Proc., Intl. Conf. on Computer Design*, pp. 543-546, October 1986.

[Hil87]    Hill, D. D., K. Keutzer, and W. Wolf, "Overview of the IDA system: a toolset for VLSI layout synthesis," *VLSI CAD Tools and Applications*, pp. 233-263, Kluwer Academic Publishers, Norwell, MA, 1987.

[Hil80]    Hillier, F. S., and G. J. Lieberman, *Introduction to Operations Research* (third edition), Holden-Day, Inc., San Francisco, 1980.

[Hsu79]    Hsueh, M.-Y., *Symbolic Layout and Compaction of Integrated Circuits*, Ph.D. thesis, University of California, Berkeley, December 1979.

[Inf78]  Infante, B., D. Bracken, B. McCalla, S. Yamakoshi, and E. Cohen, "An interactive graphics system for the design of integrated circuits," *Proc., 15th Design Automation Conf.*, pp. 182-187, June 1978.

[Int85]  Intermetrics, Inc., *VHDL Language Reference Manual, version 7.2*, Intermetrics, Inc., Bethesda, MD, 1985.

[Jen74]  Jensen, K., and N. Wirth, *Pascal User Manual and Report*, Springer-Verlag, New York, 1974.

[Joh81]  Johnson, S. C., The ic layout language, Bell Laboratories Internal Memorandum, 1981.

[Joh80]  Johnson, S. C., and S. A. Browning, The LSI design language i, Bell Laboratories Internal Memorandum, 1980.

[Ked84]  Kedem, G., and H. Watanabe, "Graph-optimization techniques for s-2IC layout and compaction," *IEEE Trans. on Computer-Aided Design of Integrated Circuits and Systems*, vol. CAD-3, no. 1, pp. 12-20, 1984.

[Ker78]  Kernighan, B. W., and D. M. Ritchie, *The C Programming Language*, Prentice-Hall, Englewood Cliffs, NJ, 1978.

[Kin82]  Kingsley, C., *Earl: An Integrated Circuit Design Language*, Master of Science Thesis, California Institute of Technology, June, 1982.

[Kin84]  Kingsley, C., "A hierarchical, error-tolerant compactor," *Proc., 21st Design Automation Conf.*, pp. 126-132, June 1984.

[Kol86]  Kollaritsch, P., and B. Ackland, "COORDINATOR: a complete design-rule enforced layout methodology," *Proc., Intl. Conf. on Computer Design*, pp. 302-307, October 1986.

[Lak87]  Lakhani, G., and R. Varadarajan, "A wire-length minimization algorithm for circuit layout compaction," *Proc., ISCAS-87*, pp. 276-279, May 1987.

[Lar71]  Larsen, R. P., "Computer-aided preliminary layout design of customized MOS arrays," *IEEE Trans. on Computers*, vol. C-20, no. 5, pp. 512-523, May 1971.

[Lar73]  Larsen, R. P., "Interactive computer-aided editing of custom MOS device layout designs," *Proc. of the 1973 Intl. Microelectronics Symposium*, pp. 2A-7-1-2A-7-10, IEEE, October 1973.

[Lar78]  Larsen, R. P., "Versatile mask generation techniques for custom microelectronic devices," *Proc., 15th Design Automation Conf.*, pp. 193-198, June 1978.

[Len84]  Lengauer, T., "On the solution of inequality systems relevant to IC-layout," *Journal of Algorithms*, vol. 5, pp. 408-421, 1984.

[Lia83]    Liao, Y.-Z., and C. K. Wong, "An algorithm to compact a VLSI symbolic layout with mixed constraints," *IEEE Trans. on Computer-Aided Design of Integrated Circuits and Systems*, vol. CAD-2, no. 2, pp. 62-69, 1983.

[Lin86]    Lin, S. L., and J. Allen, "Minplex — a compactor that minimizes the bounding rectangle and individual rectangles in a layout," *Proc., 23rd Design Automation Conf.*, pp. 123-130, June 1986.

[Lip83]    Lipton, R. J., S. C. North, J. Valdes, G. Vijiyan, and R. Sedgewick, "VLSI layout as programming," *ACM Trans. on Programming Languages and Systems*, vol. 5, no. 3, pp. 405-421, 1983.

[Lo87]    Lo, C.-Y., R. Varadarajan, and W. H. Crocker, "Compaction with performance optimization," *Proc., Intl. Conf. on Circuits and Systems*, pp. 514-517, 1987.

[Mal85]    Maley, F. M., "Compaction with automatic jog introduction," pp. 261-283 in *1985 Chapel Hill Conf. on VLSI*, ed. Henry Fuchs, Computer Science Press, Rockville, MD, May 1985.

[Mar86]    Marple, D., and A. El Gamal, *Area-delay Optimization of Programmable Logic Arrays,* Technical Report, Stanford University, September 1986.

[Mat85a]    da Mata, J. M., "ALLENDE: a procedural language for the hierarchical specification of VLSI layout," *Proc., 22nd Design Automation Conf.*, pp. 183-189, June 1985.

[Mat82]    Mathews, R., J. Newkirk, and P. Eichenberger, "A target language for silicon compilers," *Compcon Proc.*, pp. 349-353, Spring 1982.

[Mat85b]    Matson, M. D., "Optimization of digital MOS VLSI circuits," pp. 109-126 in *1985 Chapel Hill Conf. on VLSI*, ed. Henry Fuchs, Computer Science Press, Rockville, MD, May 1985.

[May86]    Mayo, R. N., "Mocha chip: a system for the graphical design of VLSI module generators," *Digest Intl. Conf. on Computer-Aided Design*, pp. 74-77, November 1986.

[McW78]    McWilliams, T. M., and L. C. Widdoes, Jr., "SCALD: structured computer-aided logic design," *Proc., 15th Design Automation Conf.*, pp. 271-277, 1978.

[Mea80]    Mead, C., and L. Conway, *Introduction to VLSI Systems*, Addison-Wesley, Reading, MA, 1980.

[Meh84]    Mehlhorn, K., *Data Structures and Algorithms 2: Graph Algorithms and NP-Completeness,* Springer-Verlag, Berlin, 1984.

[Mos86]    Moscovitz, H., personal communication. October 1986.

[Mos87]    Mosteller, R. C., A. H. Frey, and R. Suaya, "2-D compaction: a Monte Carlo method," *Proc., Conference on Advanced Research in VLSI*, pp. 173-197, MIT Press, 1987.

[New79]   Newman, W. M., and R. F. Sproull, *Principles of Interactive Computer Graphics*, McGraw-Hill, New York, 1979.

[Ng84]   Ng, C. H., "A symbolic-interconnect router for custom IC design," *Proc., 21st Design Automation Conf.*, pp. 52-58, June 1984.

[Nye83]   Nye, W. T., *Delight, An Interactive System for Optimization-Based Engineering Design*, Ph.D. Thesis, EECS Dept., University of California, Berkeley, 1983.

[Ost84]   Osterhout, J. K., G. T. Hamachi, R. N. Mayo, W. S. Scott, and G. S. Taylor, "Magic: a VLSI layout system," *Proc., 21st Design Automation Conf.*, pp. 152-159, June 1984.

[Per82]   Perkins, D., private communication. 1982.

[Pos86]   Posluszny, S. D., "SLS: an advanced symbolic layout system for bipolar and FET design," *IEEE Trans. on Computer-Aided Design of Integrated Circuits and Systems*, vol. CAD-5, no. 4, pp. 450-458, October 1986.

[Pre85]   Preparata, F. P., and M. Ian Shamos, *Computational Geometry: An Introduction*, Springer-Verlag, New York, 1985.

[Rab86]   Rabbie, H., and J. Jacobsson, "Gridless channel routing and compaction for cell-based custom IC layout," *Proc., IEEE Custom Integrated Circuits Conf.*, pp. 297-299, May 1986.

[Rei86]   Reichelt, M., and W. Wolf, "An improved cell model for hierarchical constraint-graph compaction," *Digest Intl. Conf. on Computer-Aided Design*, pp. 482-485, September 1986.

[Rei85]   Reichelt, M., private communication. October 1985.

[Rog85]   Rogers, C. D., J. B. Rosenberg, and S. W. Daniel, "MCNC's vertically integrated symbolic design system," *Proc., 22nd Design Automation Conf.*, pp. 62-68, June 1985.

[Row78]   Rowson, J., and D. Fairbarn, "Icarus: an interactive integrated circuit layout program," *Proc., 15th Design Automation Conf.*, pp. 188-192, June 1978.

[Rub83]   Rubinstein, J., P. Penfield, Jr., and M. A. Horowitz, "Signal delay in RC tree networks," *IEEE Trans. on Computer-Aided Design of Integrated Circuits and Systems*, vol. CAD-2, no. 3, pp. 202-211, July 1983.

[Rue77]   Ruehli, A. E., P. K. Wolff, Sr., and G. Goertzel, "Analytical power/timing optimization technique for digital system," *Proc., 14th Design Automation Conf.*, pp. 142-146, June, 1977.

[Sap79]   Sapiro, S., "SIDS — an interactive color graphics system for the symbolic layout of integrated circuits," *Proc. of Euro-Graphics Conf.*, pp. 317-328, October 1979.

[Sas82]   Sastry, S., and A. Parker, "The complexity of two-dimensional compaction of VLSI layouts," *Proc., Intl. Conf. on Circuits and Computers*, pp. 402-406, September 1982.

[Sax83]   Saxe, T., *CLL — A Chip Layout Language (Version 4)*, Technical Report, Stanford University, September 1983.

[Sch83a]  Schiele, W. L., "Improved compaction by minimized length of wires," *Proc., 20th Design Automation Conf.*, pp. 121-127, 1983.

[Sch83b]  Schlag, M., Y. Z. Liao, and C. K. Wong, "An algorithm for optimal two-dimensional compaction of VLSI layouts," *Integration*, vol. 1, no. 2, 3, pp. 179-209, September 1983.

[Sco84]   Scott, W. S., and J. K. Osterhout, "Plowing: interactive stretching and compaction in Magic," *Proc., 21st Design Automation Conf.*, pp. 166-172, June 1984.

[Shi86]   Shin, H., A. L. Sangiovanni-Vincentelli, and C. H. Sequin, "Two-dimensional compaction by 'zone refining'," *Proc., 23rd Design Automation Conf.*, pp. 115-122, June 1986.

[Smi85]   Smith, P., and S. Daniel, "The VIVID system approach to technology independence: the master technology file system," *Proc., 22nd Design Automation Conf.*, pp. 76-81, June 1985.

[Sza84]   Szabo, K. S. B., and M. I. Elmasry, "SYMPLE: a process independent symbolic layout tool for bipolar VLSI," *Proc., Intl. Conf. on Computer Design*, pp. 474-479, October 1984.

[Tan87]   Tan, D., and N. Weste, "Virtual grid symbolic layout 1987," *Proc., Intl. Conf. on Computer Design*, pp. 192-196, October 1987.

[Tar83]   Tarjan, R. E., *Data Structures and Network Algorithms*, Society for Industrial and Applied Mathematics, Philadelphia, PA, 1983.

[Tri81]   Trimberger, S., "Combining graphics and a layout language in a single interactive system," *Proc., 18th Design Automation Conf.*, pp. 234-239, June 1981.

[Ull84]   Ullman, J. D., *Computational Aspects of VLSI*, Computer Science Press, Rockville, MD, 1984.

[van77]   vanCleemput, W. M., "An hierarchical language for the structural description of digital systems," *Proc., 14th Design Automation Conf.*, pp. 377-385, June 1977.

[Var85]   Varadarajan, R., *Algorithms for Circuit Layout Compaction of Building Blocks*, Master of Science thesis, Texas Tech University, December 1985.

[Var86]   Varadarajan, R., private communication. August 1986.

[Wes81]   Weste, N., "Virtual grid symbolic layout," *Proc., 18th Design Automation Conf.*, pp. 225-233, June 1981.

[Wil78]   Williams, J., "STICKS — a graphical compiler for high level LSI design," *AFIPS Conf. Proc.*, pp. 289-295, 1978.

[Wil77]   Williams, J., *STICKS — A New Approach to LSI Design*, SM thesis, Massachusetts Institute of Technology, 1977.

[Wol88]   Wolf, W. H., R. G. Mathews, J. A. Newkirk, and R. W. Dutton, "Algorithms for optimizing, two-dimensional symbolic layout compaction," *IEEE Trans. on Computer-Aided Design of Integrated Circuits and Systems*, vol. CAD-7, no. 4, pp. 451-466, April 1988.

[Wol85b]  Wolf, W., "An experimental comparison of 1-D compaction algorithms," pp. 165-180 in *1985 Chapel Hill Conf. on VLSI*, ed. Henry Fuchs, Computer Science Press, Rockville, MD, May 1985.

[Wol85a]  Wolf, W., Memory requirements for cell assembly, AT&T Bell Laboratories internal memorandum, July 1985.

[Wol84]   Wolf, W., *Two-dimensional Compaction Strategies*, Ph.D. thesis, Stanford University, March 1984.

[Wol63]   Wolfe, P., and L. Cutler, "Experiments in linear programming," pp. 177-200 in *Recent Advances in Mathematical Programming*, ed. R. L. Graves and P. Wolfe, McGraw-Hill, New York 1963.

[Wou83]   van der Woude, M., and X. Timermans, "Compaction of hierarchical cells with minimum and maximum compaction constraints," *Proc., Intl. Symposium on Circuits and Systems*, pp. 1018-1021, 1983.

# Chapter 7

# Module Generation
# and Silicon Compilation

**Daniel D. Gajski**

*University of California*
*Irvine, California*

**Y-L. Steve Lin**

*Tsing Hua University*
*Hsinchu, Taiwan*

## 7.1 Introduction

The term *silicon compilation* was introduced by Dave Johannsen [Joh79] to describe the concept of assembling parameterized pieces of layout. Recently, this term has gained popularity throughout the IC CAD community, where it has been used in a variety of different contexts.

In a narrow sense, silicon compilation is an extension of the standard cell approach, in which standard cells are replaced by parameterized cell compilers [Nan83]. These compilers allow users to customize cell functionality, as well as electrical and layout parameters. In the case of simple cells, such as NAND and NOR gates, the user can specify the number of inputs, choose among several drive buffers, and select the position of some I/O ports on the boundary of the cell. More recently, compilers for microarchitectural components such as ROMs, RAMs, PLAs, and ALUs were added to the basic SSI set. Since the number of options has increased in complexity, special forms (or menus) have been provided for specification of compiler parameters.

In a much broader sense, silicon compilation can be defined as the translation process from a higher-level description into layout. (Here, a higher-level description is defined as a description that hides some level of detail from the user, not just a textual equivalent of the layout.) This translation process can be broken into several steps, and each step can be considered a compiler for the next lower level. For example, a *logic compiler* translates a description into a set of logic gates and flip-flops, and a *microarchitecture compiler* translates an instruction-set description into a set of registers, buses and ALUs.

In order to represent different approaches to silicon compilation, this chapter uses the tripartite representation of design, called a *Y-chart* [Gaj83]. (See Figure 7.1a.) Multiple levels of abstraction (or detail) are represented along each of the three axes. The axes represent three different domains of description: behavioral, structural and geometric. The levels of abstraction increase as one moves away from the vertex.

In the behavioral or *functional domain*, the designer is interested in what a design does and not how it is built. The design is treated as a black box with a specified set of inputs, a set of outputs, and a set of functions describing the behavior of each output as a function of inputs and time. For example, the Boolean expression $x = a'b + ab'$ *after* $30ns$ (see Figure 7.1b) indicates only the function of the cell whose inputs are $a$ and $b$ and whose output $x$ will become *one 30 ns* after $a$ and $b$ become different, or *zero* $30ns$ after $a$ and $b$ become the same. Note that on the more abstract levels of description, such as finite-state-machine and register-transfer description, time is replaced by the concept of state. On the algorithmic level, however, time is replaced by the concept of statement sequence, which prescribes only the order of execution.

Several levels of abstraction can be identified in the behavioral domain [Wal85]. Differential equations are used on the circuit level and Boolean expressions on the logic level. On the microarchitecture level, a register transfer (or finite-state machine) description is used which specifies for each control the condition to be tested and all register transfers to be executed. The description also designates the next control state to be entered. An algorithmic description defines all the data structures and a sequence of transformations that manipulate these structures. At an algorithmic level, variables or data structures are not bound to registers or memories, and operations are not bound to any functional unit or control state. The system or *architectural description* defines gross operational characteristics using performance specifications without being concerned about how data is manipulated or what algorithm is used.

A *structural representation* is the bridge between the functional and geometric representations. It is a one-to-many mapping of a functional representation onto a set of components and connections (under constraints such as cost, area, and time). If, for example, the Boolean expression from Figure 7.1b is mapped onto a set of components consisting only of two-input NAND gates, with max delay of $10ns$, then one of the structural representations will consist of four NAND gates, as shown in Figure 7.1c. The structural representation does not specify any physical parameters, such as the position of the four NAND gates on a printed circuit board or a silicon chip.

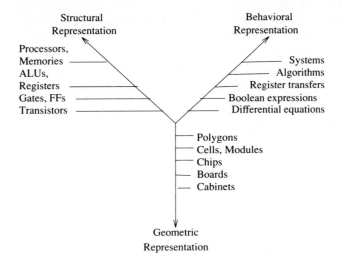

(a) Y-Chart

$$x = \bar{a}b + a\bar{b} \ after \ 30ns$$

(b) Behavioral Representation

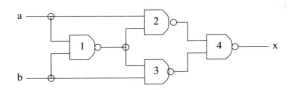

(c) Structural Representation

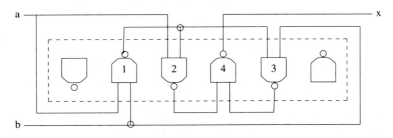

(d) Geometric Representation

**Figure 7.1**  Design Representation for Silicon Compilation

Most commonly used levels of structural representation can be identified by the basic structural elements used. On the circuit level, the basic elements are transistors, resistors and capacitors. On the logic level, the basic elements are gates and flip-flops. ALUs, registers, RAMs and ROMs can be used to represent register-transfer as well as algorithmic structures. However, at the algorithmic level, microarchitectural components tend to be grouped into datapaths, control units and data storage. The concern is more with synchronization and communication among the components than with their implementation. Processors, memories and switches are used on the system level.

The geometric or *physical representation* ignores, as much as possible, what the design is supposed to do; it binds its structure in space (physical design) or to silicon (geometric design). For example, if a gate array consisting of two-input NAND gates arranged in cells of six gates each is used, then a possible binding of the structure from Figure 7.1b is shown in Figure 7.1d. Each NAND gate and each connection in the structure are assigned a physical location. The structure-to-geometry mapping can be defined as a two-step process. The first step, usually called symbolic or topological layout, determines relative or approximate positions for all structural elements. In the second step, the absolute positions are determined after substituting layout pieces for symbols and compacting the composed layout. Although symbolic layout can be viewed as an independent design representation, the discussion includes it in geometric representation in order to simplify the representation space. The detailed exposition of symbolic layout and compaction can be found in Chapter 6.

The most commonly used levels in geometric representation are polygons, cells, modules, chips, boards and cabinets. Note again that one geometric level may cover several different levels in two other representational domains. On the system level, for example, there may be physical partitioning, placement and routing on the chip, board or cabinet levels, all depending on the size of the system and technology used.

The Y-chart can be used to represent different design methodologies and explain pictorially differences among design systems based on silicon compilers.

The methodologies used currently are mostly *bottom-up* and only partially hierarchical. They require a designer to start with basic components (such as transistors or gates) and build a structure (such as a cell), define its function, and use it hierarchically to build higher-level structures. Once the design is finished, it is flattened into the structure of basic components for simulation, placement, and routing (Figure 7.2a). This methodology does not efficiently exploit the hierarchical nature of the design since simulation, placement and routing are performed on the lowest level of abstraction where these operations are most expensive. In order to simplify the design process, frequently used pieces of design, called components, are stored in a library. Adding new components and maintaining this library over different tools, machines, and technologies is a time-consuming and expensive task. Since library components are designed without any application in mind, they are usually overdesigned with many options that are not needed in every instantiation. Furthermore, fixed functionality as well as fixed electrical and layout properties of the library components that do not match the external environment lead to an inefficient layout.

Silicon compilation tries to overcome these deficiencies by providing flexible components that can be fine-tuned to match the design environment. Furthermore, the components are specified on a higher level of abstraction such as ALUs, PLAs, RAMs, datapaths, controllers, and processors. The designer does not have to be concerned with electrical and layout details. Silicon compilers are programs that translate a component description into a layout description (geometric model), a corresponding simplified structural description for simulation (simulation model), and a timing analysis description (timing model). The functionality and the electrical and layout properties for each basic component are passed as parameters to the corresponding silicon compiler. Hierarchical methodology is supported by allowing silicon compilers to call other silicon compilers. According to the level of the component description, today's silicon compilers can be divided into cell, module, processor and system compilers.

An *ideal silicon compiler* is shown in Figure 7.2b. A translation from behavioral to structural representation is usually called synthesis. A translation from structure into geometry (usually called layout or physical design) generally requires partitioning, placement, routing and compaction. The ideal silicon compiler would start from an algorithm or program and produce the design that "best" or "optimally" executes this algorithm or program.

Unfortunately, the ideal compiler is not available today. Module compilers (sometimes called silicon compilers) generate a module frame or tiling template in which each tile is hand-designed or generated by a cell compiler. Module compilers are usually packaged together with physical design systems into *silicon compilation-based system*s called *structural silicon compiler*s. If they include processor compilers and allow a processor to be specified by a finite-state machine-like algorithmic description or by an instruction set, they are called functional or *behavioral silicon compiler*s. Thus, today's designers must specify the design using a behavioral or structural description as shown in Figure 7.3. In the case of a behavioral description, five different tasks must be performed before the design is ready for fabrication:

- The behavioral description is synthesized into a structural description of well-defined components whose compilers are in the compiler library (from $A$ to $B$ in Figure 7.3).

- The layout of each structural component is instantiated by a cell or module compiler (from $C$ to $F$ and from $D$ to $E$ in Figure 7.3).

- All compiled layouts are placed on silicon and routed (from $B$ to $G$ in Figure 7.3).

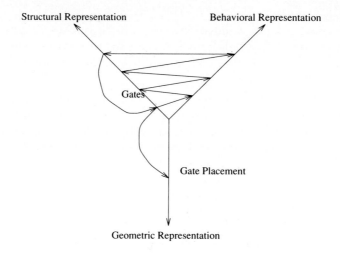

(a) Bottom-Up Methodology

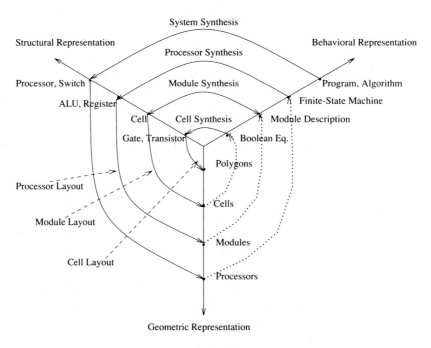

(b) Ideal Silicon Compiler

**Figure 7.2**  Design Methodologies

- Packaging is selected.

- A test vector set is generated.

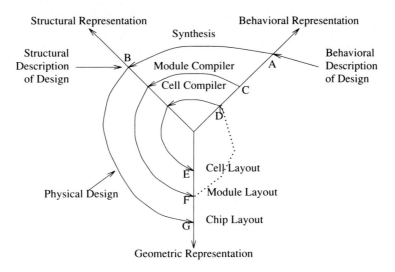

**Figure 7.3** Present Silicon-Compilation Methodology

In the case of a structural description, the synthesis task is not needed since the designer specifies the design structure.

The rest of this chapter describes several techniques for translating structural or behavioral descriptions into layout. Different levels of design call for different types of description. Section 7.2 describes the translation from symbolic description into layout. Sections 7.3 and 7.4 describe module generation for random and regular structures. Section 7.5 deals with chip and processor compilation from structural and behavioral descriptions. Finally, Section 7.6 describes the concept of intelligent compilation.

## 7.2 Layout Languages

Layout (or IC) languages can be graphic, textual, or both. Since graphic representation is much easier for designers to work with, some layout design systems connect text to graphics and *vice versa*. Usually the layout is captured in symbolic form and converted into text. This textual representation can then be converted into a compiler or procedural layout generator through parameterization. Parameterization uses procedural constructs to instantiate different layout versions for different sets of parameters. Thus, a procedural generator does not have a graphic representation.

In general, IC languages are based on three types of layout architectures: fixed, virtual and relative grid [New87].

In *fixed-grid architecture*, the layout area is divided into a uniformly spaced grid in both directions. The grid size must satisfy the "worst case spacing" rule for a particular process. A symbol is defined for each combination of mask layers that may exist at a grid location. Each symbol may be thought of as a uniquely colored tile that the designer uses to obtain a specific floor pattern. Figure 7.4a shows an example from [Wes85] of a fixed-grid symbolic language [Lar78], while Figure 7.4b shows the corresponding geometry.

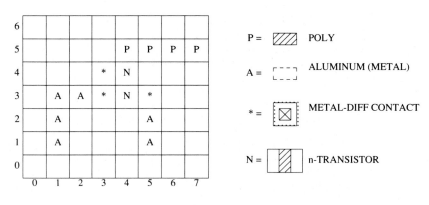

(a) Language

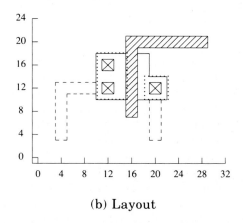

(b) Layout

**Figure 7.4** Fixed-Grid Architecture

The compilation process consists of selecting the proper tile and replacing each symbol with a corresponding layout. For example, two different versions are used for the *P* symbol in Figure 7.4a. The proper version is determined by looking at the neighboring symbols. Although fixed-grid layout languages simplify layout specification and compilation, they lack silicon efficiency because worst case spacing is used for grid spacing.

In *virtual grid architecture*, transistors, contacts and wires are placed on a grid for easy design capture and simplified interface to other tools. The final geometric spacing between neighboring grid lines is determined by the "worst case spacing" rule between the layout component on those grid lines. ICDL [Wes85] and ABCD [Ros82] are both virtual grid languages. For example, the description of an inverter and its layout are shown in Figures 7.5a and 7.5b [Wes85]. It consists of five parts: *size*, *device*, *wire*, *contact* and *pin* specifications. Pins are used for interfacing to other cells. Note that symbolic binding allows the specification of points by name rather than by absolute value. Thus, the metal-to-diffusion contact at *(2, 2)* is denoted by *nt.d*, that is, the drain point of the *nt* device.

The compilation process consists of compacting the virtual grids and possibly stretching the virtual grid for pitchmatching with the neighboring cells.

The *relative grid architecture* uses a grid only to indicate relative placement of transistors, contacts, and wires and to determine electrical connectivity of the circuit. The relative grid allows layouts to be designed at a conceptual level in which size and position of the layout components do not need to be specified. Such a layout language (a) makes the layout task more like programming then editing, (b) eliminates the need for design rule checking after the layout is generated, (c) permits the creation of easy to use cell libraries, and (d) allows a hierarchical description in which most of the detail at one level is truly hidden from all higher levels. Examples of relative grid languages are DPL [Bat80], ALI [Lip82] and L [Bur88].

Figure 7.6 shows a simple example of the relative grid language ALI [Lip82]. An ALI user specifies a layout by declaring the necessary rectangles (also called boxes) and stating the relationships between them. ALI then generates a minimum area layout that satisfies all the specified relationships between boxes. The program in Figure 7.6 consists of a declarative part and an executable part. To declare a box, the designer specifies its name (horizontal or vertical in this example) and its type (metal, for instance). The standard box types correspond to the layers of the physical layout. As the example also shows, the ALI user can define structured objects such as arrays.

The relationships between rectangles are specified through calls to a set of primitive operations in the executable part of the program. These operations take as arguments boxes and possibly values of standard types (integers in this example). In the **begin** statement, the primitive *above* specifies that its first argument must appear above the second one in the final layout. The primitive

```
          BB 0 0 6 8
nt:    device n (3,2) or=east
pt:    device p (nt,6) or=east w=2
       wire alum (0,0) (6,0)
       wire alum (0,8) (6,8)
       wire poly (3,2) (3,6)
       wire alum vss nt.s
       wire alum pt.s vdd
       wire alum nt.d pt.d
       contact md nt.d
       contact md nt.s
       contact md pt.d
       contact md pt.d
       contact md pt.s
       contact vss (nt.d,0)
       contact vdd (nt.d,8)
vss:   pin alum (0,0) vss
vdd:   pin alum (0,8) vdd
       pin poly (3,4) a
       pin poly (4,4) z
```

## (a) Language

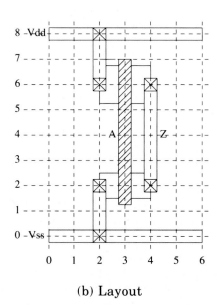

## (b) Layout

**Figure 7.5**  Virtual Grid Architecture

```
chip   simple;
    const
        hnumber = 10;
        length = 20;
        width = 6;
    boxtype
        htype : array   [1..hnumber] of   metal;
    var
        i : integer;
    box
        horizontal : htype;
        vertical : metal;
    begin
        for   i := 1   to   hnumber-1   do begin
            above (horizontal[i], horizontal[i+1]);
            glueright (horizontal[i], vertical);
            xmore (horizontal[i], length)
        end;
        glueright (horizontal[hnumber], vertical);
        xmore (horizontal[hnumber], length);
        xmore (vertical, width)
    end.
```

(a)  Language

(b)  Layout

**Figure 7.6**  Relative Grid Architecture

*glueright* extends its first argument to the right to intersect its second argument. The primitive *xmore* makes the size of its first argument along the *x* axis at least as large as the value of the second argument. In this example, ALI accesses a table of design rules to determine the minimum separation between the horizontal elements, as well as the minimum size of boxes not specified by *xmore* (such as the height of the horizontal metal lines).

When an ALI program is executed, it generates two kinds of information. First, it produces a set of linear inequalities involving box corner coordinates in the layout as variables. These inequalities, which embody the relationships between layout rectangles, are then solved to generate the positions and sizes of the layout elements. Second, the program produces connectivity information about the rectangles. After the program produces a layout, this information is used by a switch level simulator to predict the behavior of the circuit.

An in-depth survey of IC languages and their features can be found in Chapter 6.

## 7.3  Random Logic Module Generation

A *random logic module* is defined as an irregular structure of basic components such as transistors, gates and flip-flops. It is difficult to create a library of such modules because they vary widely in complexity and type. Thus, they must be designed from scratch. Compared to other regular modules in the same chip, random logic modules occupy less silicon area but take most of the design time.

Random module generators can be classified according to the architecture style, dimensionality and type of input description. Layout architecture style can be constrained or flexible. In a *constrained architecture*, basic topological relationships are known ahead of time. These topological relationships define orientation of devices and order and layer of power lines and connections. A constrained architecture may be thought of as a partially completed module. Examples of constrained architectures are Weinberger arrays, gate matrices, standard cells, and PLAs. In a *flexible architecture*, there are no constraints on placement of cells, devices, contacts or wires.

Module dimensionality refers to the number of dimensions a module may grow in relation to the increase in the number of components. In a *linear array* the module grows in one dimension with respect to the number of components (Weinberger array) or with the number of nets (gate matrix). In a *quadratic architecture*, modules grow in both directions. Flexible architectures are always quadratic.

The input description can be a circuit or transistor schematic, a logic schematic, a mixed schematic, or a set of Boolean equations. This section will present an example of each.

### Constrained Linear Architecture

In a linear array style of layout, transistors or gates are placed in a one-dimensional array and connected by a set of parallel wires. This section describes two popular linear array layout approaches: Weinberger array and gate matrix.

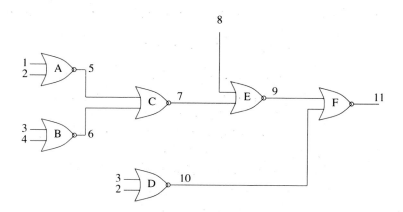

**Figure 7.7** An Example of NOR Logic

**Weinberger Array.** The Weinberger array [Wei67] was the first attempt to stylize the layout implementation of multilevel Boolean functions. A circuit consisting only of NOR gates (see Figure 7.7) is converted into a one-dimensional array of nMOS gates, one gate per column. (See Figure 7.8.) Each column consists of two vertical metal wires. One wire is connected to the pull-up transistor and serves as the gate output port, while the other wire is extended to the ground power line. (In a real application, two neighboring gates share a common ground wire.) The input signals to the gates are derived from a set of parallel polysilicon rows, laid out horizontally across all gates. A transistor is formed by the intersection of a diffusion segment between the output and ground lines, and an extension from a horizontal polysilicon row. For example, two transistors, $a$ and $b$, are formed by connecting rows 1 and 2 to diffusion segments in gate $A$. The output of gate $A$ is connected to row 5, which drives gate $C$.

The size of a Weinberger array layout is proportional to the product of the number of gate columns and connecting rows. In a plain layout, each gate occupies a single column, and each signal (input, output and internal) occupies a single row. This approach is simple but it results in a very sparse array. In practice, both column and row numbers can be reduced; however, these two reductions are not independent.

In a plain layout, each signal line occupies a whole horizontal row, while only an interval of that row may be sufficient for the required connection. For example, net *5* occupies only the interval from gate $A$ to gate $D$, while net *10* uses only the interval from gate $D$ to gate $F$. To compact the layout, connections can be placed in the same row if their intervals do not overlap. Thus, nets *5* and *10* could share the same row. This "row-assignment" problem is defined

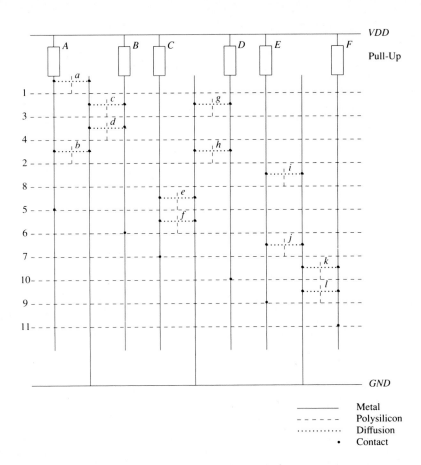

**Figure 7.8** A Weinberger Array Layout of the Circuit of Figure 7.7

as the assignment of all intervals using as few rows as possible. The left-edge algorithm [Has71], described in Chapter 5, is usually used to obtain an optimal solution of the row assignment problem.

Figure 7.9 shows the Weinberger array of Figure 7.8 after the left-edge algorithm has been used for row assignment. Note that 11 signals share 6 rows in the new layout.

Although row assignment may be simple, different gate placements result in different interval sets and, therefore, different row assignments. For a circuit of $N$ gates, there are $N!$ possible gate placements. It is impractical to explore exhaustively all placements for a large circuit. Many heuristics have been developed to search for a good, but not necessarily optimal solution in a

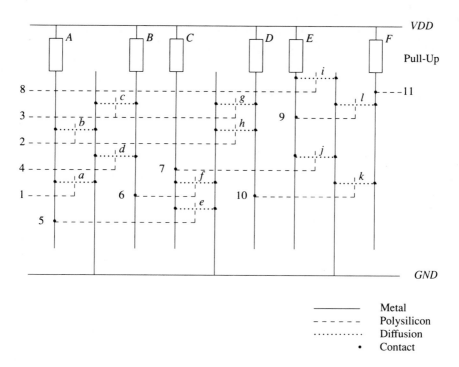

**Figure 7.9** Row Assignment Using Left-edge Algorithm

reasonable amount of time [Asa82]. Note that it is not known how good a place-ment is until row assignment has been performed. Usually, the estimated total wire length is used to determine the quality of the placement.

The natural way to reduce the number of columns used is to let two gates share a single column. This technique is called folding. A pair of gates can be folded if they use the same signal line and their signal lines do not interleave with each other. For example, gates *A* and *C* of Figure 7.8 can be folded. How-ever, the folding of one pair of gates may prevent the folding of another pair, since folding imposes restrictions on the permutation of the signal lines. For example, nets *1*, *2* and *5* used by gate *A* must be above nets *5*, *6* and *7* used by gate *C*. (See Figure 7.10.)

**Gate Matrix Layout.** *Gate matrix layout* was first introduced by Lopez and Law of Bell Laboratories [Lop80] as a systematic approach to large scale layout prob-lems. It is an extension of the cell layout in which a complex CMOS gate (con-sisting of series and parallel-connected transistors) is laid out in a single row of *p-devices* and a single row of *n-devices*, one row above the other. In such a lay-out, nets connecting the source and drain of two neighboring transistors can be

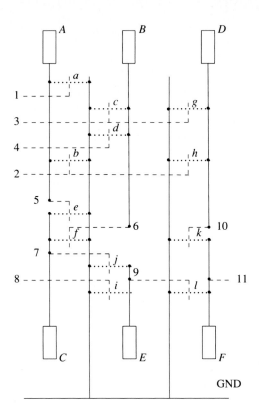

**Figure 7.10**  A Folded Weinberger Array

implemented in the diffusion layer.  It has been shown [Ueh81] that such a complex gate can be laid out in a single line of diffusion if, and only if, there are two Euler paths with the same labeling through each *p*- and *n*-part of the gate. This idea has been used in several working compilers, such as Sc2 [Hil85].  Figure 7.11a shows a circuit example, while Figure 7.11b shows one of its possible gate matrix layout implementations.

In a gate matrix layout, a vertical polysilicon wire corresponding to an input, internal, or output net is placed in every column (lines *A*, *B*, *C*, *D*, and *Z* in Figure 7.11b).  All transistors using the same gate signal are constructed along the same column (transistors *1*, *3*, *7* and *10* in column *A* of Figure 7.11b). Connections among transistors are made by horizontal metal lines.  Connections to *Vdd*/*Vss* are in second layer metal (shown as an arrow in Figure 7.11b).  We will define a net as a series of metal lines and transistors.  Net $N_1$ in row *a* of

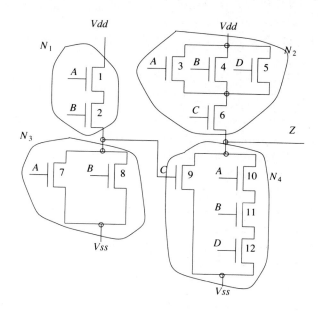

(a)  Transistor Circuit

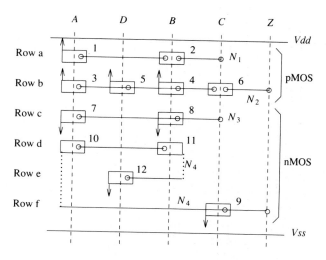

(b)  Gate Matrix Layout

**Figure 7.11**  An Example of Gate Matrix Layout

Figure 7.11b spans from column $A$ to $C$ and includes two transistors (1 and 2) and two metal lines. If a net is assigned to more than one row, a short vertical diffusion line is used. (See net $N_4$ which occupies rows $d$, $e$ and $f$ in Figure 7.11b.)

The size of a gate matrix layout is proportional to the product of the number of columns and rows. To minimize a gate matrix layout area, the number of rows used must be reduced, since the number of columns is fixed to the number of nets in the circuit schematic. Because a row can be shared by more than one net, the number of rows depends heavily on both the column ordering and the net assignment to rows. Figure 7.12 shows how column ordering affects the size of the layout area.

To simplify the problem, only the nMOS part of the circuit is considered, since the column ordering of the pMOS part will be similar to the nMOS part.

Wing solved the gate matrix layout problem using a graph-theoretic technique [Win82, Win85]. In his solution, a gate matrix layout, $L$, is represented as an interval graph, $I(L) = \{V, E\}$, where $V$ is the set of all nets and $E$ is the set of edges connecting vertices whose nets overlap in one or more columns. The number of rows used by a layout is equal to the size of the largest dominant clique of $I(L)$. (A clique of a graph, $G$, is a fully connected (complete) subgraph of $G$.) The size of a clique is equal to the number of its vertices. A dominant clique of $G$ is a clique which is not a proper subgraph of any other clique of $G$. Wing transformed the gate matrix optimization problem into the problem of finding an interval graph in which the size of the largest dominant clique is minimized. From an interval graph, the optimal column ordering can be obtained by applying the algorithm of Ohtsuki *et al.* [Oht79]. The left-edge algorithm of Hashimoto and Stevens [Has71] can be used to assign nets to rows.

A layout is not *realizable* if any vertical diffusion lines must cross over transistors or contacts in the same column. An unrealizable layout can be turned into a realizable layout in one of two ways: by permuting rows and thus removing obstacles (see Figure 7.13a) or by increasing the space between two columns (see Figure 7.13b).

If the binding of serial connected transistors is committed too early, *i.e.*, before the gate assignment, then it is possible that an optimal layout will never be achieved. Hwang *et al.* [Hwa86b] proposed a *dynamic net list* representation scheme to deal with this binding problem. In their scheme, a group of serial connected transistors can be permuted in arbitrary order without changing their logic functionality. By using the dynamic representation scheme, Hwang *et al.* delayed the binding process until the column ordering was determined. The results have been shown to be better than previous *fixed net list* schemes. Leong [Leo86] then applied a simulated annealing technique to Hwang's dynamic representation scheme. He considered gate assignment and transistor binding simultaneously during the annealing process.

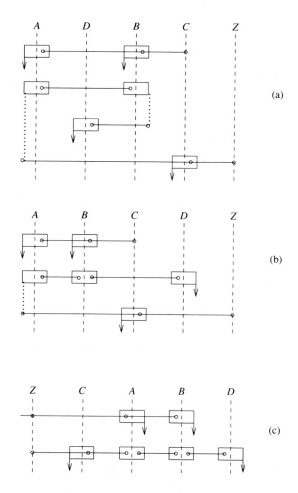

**Figure 7.12** Column permutation affects row usage.

All the solutions mentioned previously concentrated only on either the pMOS or nMOS part of the circuit, based on the assumption that the permutations of both the pMOS and nMOS sides are the same. Practically, this is not true. Nakatani *et al.* [Nak86] proposed a heuristic that considers both pMOS and nMOS simultaneously, thereby improving the previous results.

The advantages of the gate matrix layout are that it is a systematic layout style that can be formally described and that symbolic layout can be easily

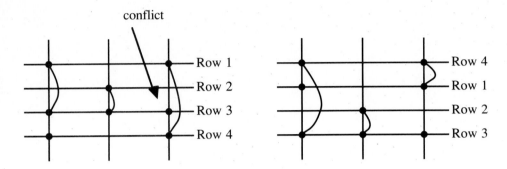

(a) Realizable Layout by Row Permutation

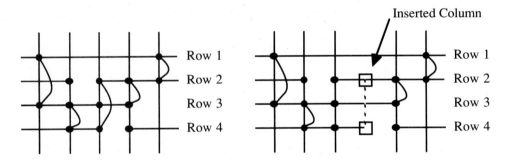

(b) Realizable Layout by Column Spacing

**Figure 7.13** Realizable Layout

updated for new design rules. Its disadvantages are its inflexibility of pin assignment, its long average wire length for large circuits, and its requirement for 2-metal layers.

### Constrained Quadratic Architectures

Standard cell architecture divides the layout area into a number of parallel rows separated by routing channels, as shown in Figure 7.14.

A standard cell layout system uses a placement and routing approach. It is so named because it divides the layout problem into two independent subtasks: placement and routing.

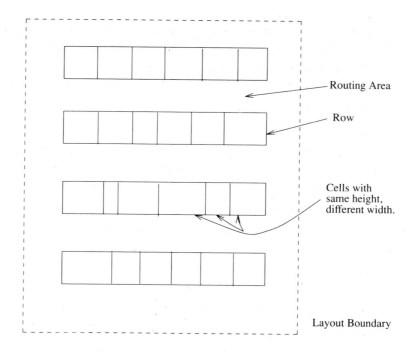

**Figure 7.14** Standard Cell Architecture

The design system maintains a library of predesigned cells representing the layout of all possible basic units in the circuit schematic. All library cells have the same fixed height but may have different widths. Signal ports are on the top and bottom boundaries of the standard cell. Power pins are on the left and right boundaries.

Module generation starts by selecting from the cell library the standard cells that correspond to units in the input schematic. A placement subtask is invoked next to decide appropriate locations and orientations for every standard cell. Finally, the routing subtask is invoked to connect placed standard cells according to the specification in the input description.

The objective of the cell placement task is to make the routing task easier. The quality of placement can be approximated by total wire length, which is the sum of all net lengths. A net length is the length of the wires used to connect a signal net. Exact net length is not available until detailed routing has been performed. A net's length can be approximated by taking the sum of the height and width of its enclosing rectangle. The enclosing rectangle is the smallest rectangle that encloses all the cells in the net.

The objective of the routing task is to realize all necessary cell interconnect using as little space as possible. The routing task is divided into two subtasks: global routing and detailed routing. A global router divides the routing area into channels and assigns each net to one or more channels. The global router does not perform the actual layout. Its objective is to find the shortest path while preventing congestion in the channels. Many different solutions exist for the channel routing problem as explained in Chapter 5.

The advantage of a standard cell system is in the simplicity of the compilation process. The fact that it requires only placement and routing tasks made it a popular system, leading to the development of many placement and routing algorithms. Another advantage of standard cell systems is their reliability. Since all cells are characterized before being stored in the library, the design is error free with predictable performance. The disadvantage is in the low area utilization and the maintenance of the cell library. Because of the fixed sizes, pin positions, and functionality of standard cells, 60%-70% of the standard cell layout area is commonly used by routing. Furthermore, any change in the fabrication process requires a redesign of the library since standard cells are usually designed for a particular process and are not easily scalable.

Although the standard cell concept is simple, it has been proved that both placement and routing problems are NP-complete. Many heuristic algorithms have been developed to tackle this difficulty. Chapters 4 and 5 discuss placement and routing problems and survey techniques for solving them.

**Programmable Logic Array.** A *programmable logic array* (PLA) maps a set of Boolean functions in canonical, two-level sum-of-product form into a geometrical structure [Fle75]. A PLA consists of an AND-plane and an OR-plane. (See Figure 7.15.) For every input variable in the Boolean equations, there is an input signal to the AND-plane. The AND-plane produces a set of product terms by performing an AND operation. The OR-plane generates output signals by performing an OR operation on the product terms fed by the AND-plane.

A PLA layout follows a fixed architecture in which connections and power lines are deployed in advance, regardless of the functions to be implemented. Possible transistor locations are in a matrix pattern. A PLA layout generator determines the presence of transistors in the matrix according to the input specifications. For example, the PLA of Figure 7.15 realizes three functions ($F_1$, $F_2$, and $F_3$) out of four input variables ($X_1$, $X_2$, $X_3$, and $X_4$).

The area of a PLA is proportional to the product of the number of columns and rows used, where the number of rows equals the number of product terms, and the number of columns equals the sum of the number of output signals and twice the number of input signals. Reducing either of these two numbers results in a more compact PLA. Two techniques have been developed: *logic minimization* for reducing the number of rows and PLA *folding* for reducing the number of columns.

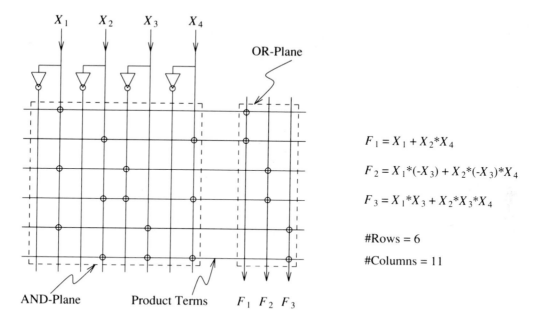

$$F_1 = X_1 + X_2 * X_4$$

$$F_2 = X_1 * (-X_3) + X_2 * (-X_3) * X_4$$

$$F_3 = X_1 * X_3 + X_2 * X_3 * X_4$$

#Rows = 6

#Columns = 11

**Figure 7.15**  A Standard PLA Stick Diagram

Using the logic minimization technique can possibly reduce the number of product terms while still realizing the same set of Boolean functions. For example, Figure 7.16 shows a PLA implementing the same functions as those of Figure 7.15, but only 4 rows are used instead of 6. This technique is the same as finding the minimal number of prime implicants for a set of Boolean equations. It can be found in most books dealing with logic design [Bra84].

PLA folding allows two or more signals to share a single row or column and thus reduces the total number of rows or columns. Figure 7.17 shows the result of folding the PLA in Figure 7.16 by allowing signals $X_1$ and $X_2$ to share a single column. This column folding reduces the total number of columns from $11$ to $9$, while keeping the number of rows unchanged. Furthermore, the pair $F_2$ and $F_3$, as well as the pair $X_1$ and $X_4$, can also be placed into a single column. Row folding and segmented folding can also be used to reduce PLA layout area.

The folding of one signal pair may block another. The object of PLA folding is to find the maximum number of pairs that can be folded simultaneously. Many algorithms and heuristics have been developed to solve this problem. The simplest one is the branch-and-bound search [Lew82]. Although it is simple and able to find the optimal solution, the branch-and-bound search is only useful for small circuits because its computation time grows exponentially with circuit

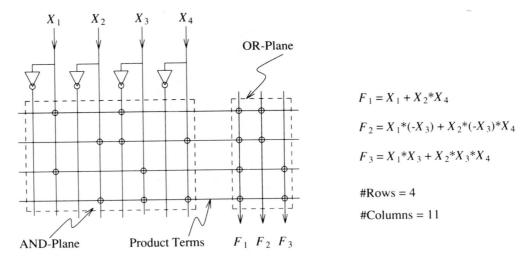

$$F_1 = X_1 + X_2 * X_4$$

$$F_2 = X_1 * (-X_3) + X_2 * (-X_3) * X_4$$

$$F_3 = X_1 * X_3 + X_2 * X_3 * X_4$$

#Rows = 4

#Columns = 11

**Figure 7.16** A Programmable Logic Array with Logic Minimization

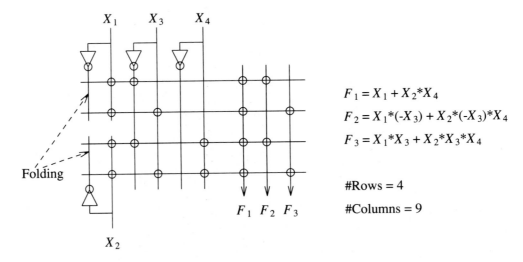

$$F_1 = X_1 + X_2 * X_4$$

$$F_2 = X_1 * (-X_3) + X_2 * (-X_3) * X_4$$

$$F_3 = X_1 * X_3 + X_2 * X_3 * X_4$$

#Rows = 4

#Columns = 9

**Figure 7.17** A Folded Programmable Logic Array

size. Therefore, many heuristics have been developed to find good, but not optimal solutions in a reasonably short period of time. Hwang *et al.* [Hwa86a] used a best-first search algorithm to find a near-optimal result. Ullman [Ull84] used a graphic algorithm to find a feasible solution in a time complexity no worse than $O(wc^2)$, where $w$ is proportional to the number of rows and $c$ is the number of columns. Hachtel *et al.* [Hac82] proposed an algorithm for both row and column foldings. Wong *et al.* [Won86] applied the simulated annealing technique to folding problems. Several authors proposed multiply-folded architectures [Chu82, DeM83].

By adding storage elements and simple feedback connections, a PLA can very easily be used to implement a sequential circuit. A circuit of this kind acts exactly like a finite-state machine. Therefore, one of the popular applications of PLA is the design of microcontrollers.

A *storage/logic array (SLA)* [Leu84, Smi82] is an extension of the PLA concept. Unlike a PLA, where AND and OR operations are performed on different planes, an SLA mixes both operations on a single plane. (See Figure 7.18.) Furthermore, an SLA allows higher level constructs such as flip-flops and inverters to be placed at grid points, while a PLA allows only transistors to be placed at grid points.

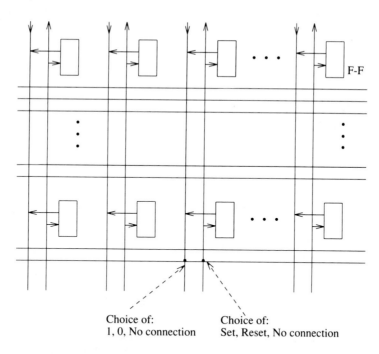

**Figure 7.18**  A Storage/Logic Array Stick Diagram

The main advantage of a PLA approach to cell compilation is that its regular grid pattern makes the generation task very simple. Design time is short and cost is low. However, there are several disadvantages with the PLA module generator. First, it may be very inefficient for some types of circuits. For example, a PLA may implement an n-input function by using as many as $O(2^n)$ product terms, which is not practical compared to other layout approaches. Second, long signal lines increase circuit parasitics, which in turn slows down circuit performance. Hence, PLAs are not convenient for large circuits. Finally, the designer cannot control PLAs I/O pin positions and aspect ratio.

### Flexible Architectures

Since flexible architecture does not impose any constraints on the layout, compilation becomes a complex task. The compilation process proceeds from conversion of the logic schematic into a transistor schematic, to placement of transistors, to routing, and finally to compaction.

Dumbo [Wol83] was one of the early cell compilers. It used five phases to translate mixed transistor/logic schematics into stick layouts. In the first phase, placement of each component was determined with a force-directed placement algorithm. The orienter determined orientation of each component, while the expander converted a description of compound components into a description containing only transistors. The wirer broke each net down into a set of pairwise connections (branches) with assigned layers and introduced vias where necessary. The router translated the branches into a Manhattan stick diagram. The output from Dumbo was a stick diagram that needed a stick compactor for translation to layout.

Complexity of module compilation with flexible architecture provides a very good ground for application of AI techniques. Recently, researchers have started to attack layout problems using a knowledge-based approach [Kim86]. The rest of this section describes two systems that use this approach. The first system, TOPOLOGIZER, uses a transistor schematic as the input, while the second one, LES, uses a logic schematic. A complete survey of expert systems in physical design automation can be found in Chapter 9.

**TOPOLOGIZER.** TOPOLOGIZER [Kol85] is a rule-based, CMOS layout generator. It uses a layout style in which all *p*-type transistors are put in rows parallel to the *Vdd* line, while all *n*-type transistors are put in rows parallel to the *Vss* line.

TOPOLOGIZER's inputs are a transistor net list and an *environment constraint*. The transistor net list specifies transistor types (either *p* or *n*) and their connections with each other. The *environment constraints* include layout size and pin constraints. Layout size can be specified by the maximum height and width, or by an aspect ratio. (The aspect ratio is defined by the number of rows and columns available for transistor placement. For example, an aspect ratio of 6/4 means that a maximum of 6 transistors should be placed in each of 4 rows.) Pin constraints are made up of side, layer, location and loading. The pin location constraint can be described absolutely by a set of coordinates or relatively with respect to other pins.

TOPOLOGIZER's output is a symbolic file of CMOS layout, making it free of technology-dependent design rules. Using a symbolic layout system such as MULGA [Wes81], TOPOLOGIZER's output can be translated into a mask layout.

Figure 7.19 shows a TOPOLOGIZER input description for a carry generator. Part (a) of the figure describes circuit connectivity. Circuit connectivity is defined by listing all transistors in the circuit schematic. The first line (with the keyword **PART**) indicates that the circuit named **CARRY** has three input signals ($a$, $b$, and $c$) and one output ($O$) signal. Then, every entry in the list describes a type of device and the nets to which that device is connected. For example, the second line defines a $p$-type transistor which has its source connected to net $o$, its gate connected to net $c$, and its drain connected to net $w$.

Part (b) of the figure specifies the environmental constraints imposed on TOPOLOGIZER. The first line,

$$ASPECT\_RATIO = 5/2$$

means that the transistors should be placed in two rows of five transistors each. Each of the succeeding lines defines a pin. For example, line 6 indicates that pin $c$ must be connected to the bottom side of the layout with a 1-unit-wide polysilicon wire, and its position must be somewhere between pins $vss$ and $vss.r$. Similarly, the last line indicates that pin $vdd.r$ must be connected to the top-right corner of the layout with a 2-unit-wide metal wire. Note that pin position can also be specified with an absolute value as in the case of pin $a$ (line 3).

Like other module generators, TOPOLOGIZER synthesizes cell layout using two expert systems for transistor placement and routing. The placement expert's objective is to simplify the routing task by reducing the number of wires needed to connect transistors. The number of wires is reduced by applying two strategies:

- Placing connected transistors next to each other in the same row so that the connection can be realized through the diffusion layer.

- Placing transistors with the same gate connection in the same column so that the gates can be connected through the polysilicon layer.

First, the placement expert randomly places transistors to satisfy the aspect ratio specification. Then, it applies its knowledge to improve the transistor placement. There are three ways that the placement expert can modify the transistor placement:

- Moving a transistor from one location to another.

- Exchanging two transistors.

- Rotating a transistor, *i.e.*, exchanging its drain and source positions.

Figure 7.20 shows how TOPOLOGIZER improves placement by applying the transistor exchange rules. Two pins from two different transistors are considered "matched" if they are connected to a common net and placed next to

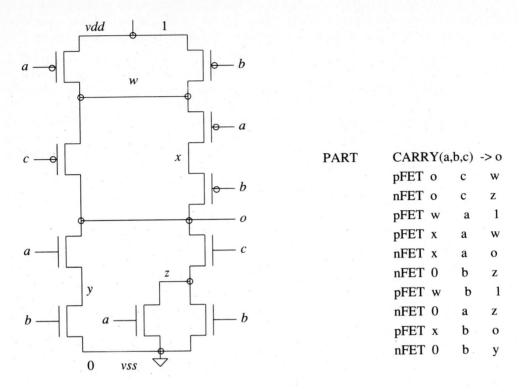

| PART | CARRY(a,b,c) | | -> o |
|------|------|------|------|
| pFET | o | c | w |
| nFET | o | c | z |
| pFET | w | a | 1 |
| pFET | x | a | w |
| nFET | x | a | o |
| nFET | 0 | b | z |
| pFET | w | b | 1 |
| nFET | 0 | a | z |
| pFET | x | b | o |
| nFET | 0 | b | y |

(a) Transistor Connectivity Description

ASPECT,RATIO=5/2

TOPLEFT vdd layers=(ALUM) width=2

LEFT a layers=(ALUM) width=1 position=4

LEFT b layers=(ALUM) width=1 position,between=(vss vdd)

BOTLEFT vss layers=(ALUM) width=2

BOT c layers=(POLY) width=1 position,between=(vss vss.r)

BOTRIGHT vss.r layers=(ALUM) width=2

TOP o layers=(PDIFF) width=1 position,between=(vdd vdd.r)

TOPRIGHT vdd.r layers=(ALUM) width=2

(b)  Environment Description

**Figure 7.19**  An Example of TOPOLOGIZER's Input Description

each other. In Figure 7.20a, the placement expert detected that the number of matches will increase by 2 if it swaps transistors $A$ and $B$. The result is shown in Figure 7.20b. Secondly, if transistor $C$ is rotated after swapping $C$ and $D$, a gain of 1 will be achieved. (See Figure 7.20c.) Finally, a gain of 2 can be obtained by swapping transistors $D$ and $A$ while rotating $D$. The resulting optimal placement is shown in Figure 7.20d.

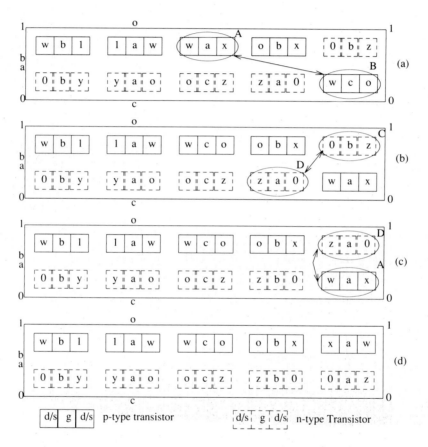

**Figure 7.20** TOPOLOGIZER's Placement Example

After transistor positions are decided upon, TOPOLOGIZER invokes its routing expert to connect transistors. The routing expert's objective is to complete all connections using as little silicon area and the fewest number of jogs as possible. TOPOLOGIZER's routing expert consists of a pre-router and a refinement router. First, the pre-router produces "rough routing" by assigning a unique track to each pair of terminals to be connected. (See Figure 7.21.) Although the layout is technically correct, its density is not high.

The refinement router then improves the rough routing by applying a set of rules. Figure 7.22 shows two rules used by the routing expert. In (a), the U-turn elimination rule is used to straighten the U-turn, saving one row, two jogs, and four contacts. In (b), nets $b$ and $c$ are allowed to share the same row, saving one row. Figure 7.23 shows the final result.

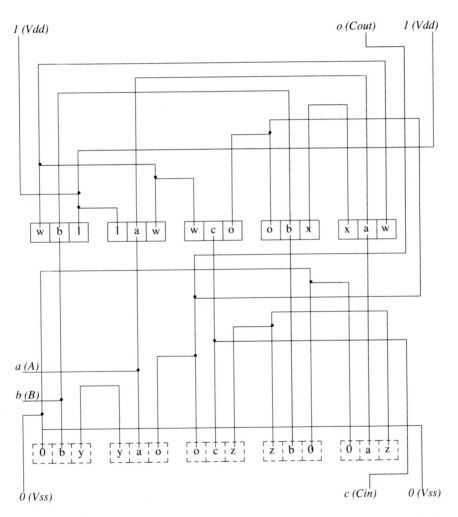

**Figure 7.21** TOPOLOGIZER's Rough Routing (One Row per Connection)

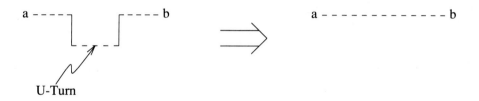

(a)  U-Turn Elimination

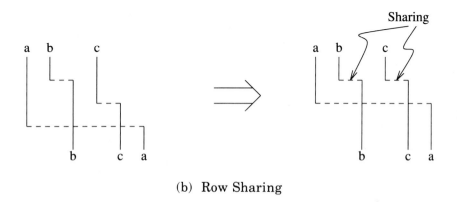

(b)  Row Sharing

**Figure 7.22**  Examples of TOPOLOGIZER's Routing Rules

Advantages of TOPOLOGIZER's approach are that it is simple and it generates high quality layout for small circuits. The designer can interrupt the expert system after each task, modify the result, and continue. Another advantage is that it is easy to improve TOPOLOGIZER's performance by modifying its rule base. TOPOLOGIZER's weakness is its inability to deal efficiently with large circuits.

**LES.** *LES* [Lin87] is a rule-based CMOS cell compiler that attempts to remove the weakness of TOPOLOGIZER. As with TOPOLOGIZER, the input to LES is a set of boundary constraints and a list of components. The boundary constraints include layout area (height/width or aspect ratio) and I/O pin positions. The component list specifies the functionalities and connectivities, and its basic units are pass-transistors, gates, and complex gates.

The output from LES is a symbolic layout file in which only topological relationships among transistors, contacts and wires are specified, and not their absolute geometry. A compactor is used to convert the symbolic file into a mask artwork. (See Chapter 6 for more information on compaction.) Figure 7.24a

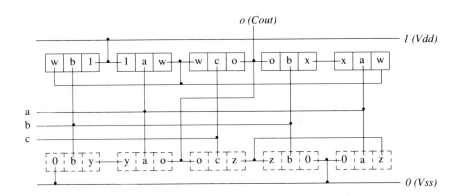

**Figure 7.23**  TOPOLOGIZER's Improved Layout

shows the logic schematic of a D-type flip-flop and Figure 7.24b shows the resulting mask layout. LES's target technology is single-metal single-poly CMOS.

Like other cell compilers, LES places and routes devices from the schematic. However, LES has two features that distinguish it from other placement and routing tools: it does not create routing channels, and it does not use a cell library. Instead of using routing channels, LES makes connections among leaf cells with *through-the-cell routing*. The cell library is replaced by *FLEX*, a flexible leaf cell layout expert.

FLEX is a rule-based system. Its input is a completely characterized cell with size, functionality, pin positions, and through-the-cell routing. Figure 7.25 shows the input specification for a typical NAND gate. FLEX can perform three tasks: estimation, consultation, and layout. The estimation task provides LES with the approximate cell size. The consultation task supplies LES with information about possible pin positions. The layout task generates the actual symbolic layout.

LES uses a two-dimensional layout style. A layout block is sliced into a number of parallel strips. Each strip has a pair of *Vdd/Vss* lines along its borders. Two neighboring strips share a common power line. Leaf cells within a strip are placed in one dimension. A *p*-well is formed close to the *Vss* line.

LES consists of seven experts, which are organized in a blackboard architecture [Nii86]: analysis, architecture, placement, characterization, flexible layout (FLEX), evaluation, and optimization experts.

The Analysis Expert (*AE*) extracts useful information, such as shape and pin distribution, from boundary constraints. AE also finds a set of seed paths,

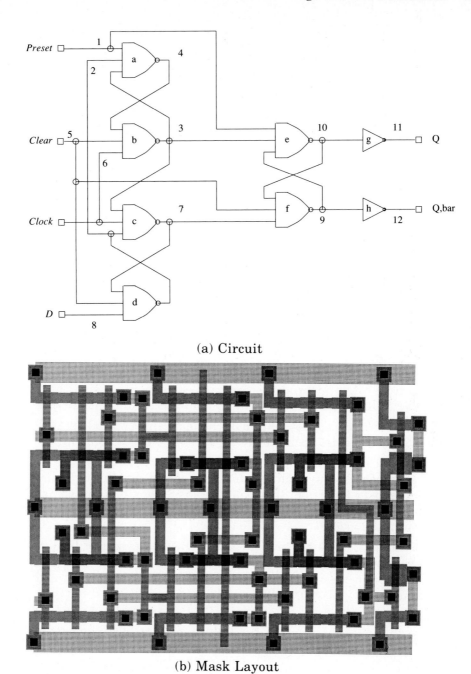

(a) Circuit

(b) Mask Layout

**Figure 7.24** A D-type Flip-Flop Circuit and its Layout Generated by LES

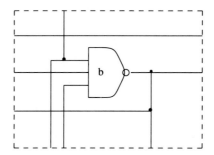

**Figure 7.25** A Typical Characterization Cell

which will be used to help the architecture expert. Examples of the criteria for seed path selection are path criticality, overlap and orientation.

LES then invokes the Architecture Expert (*ARE*) to determine the layout style according to the information obtained from the AE. A layout area consists of several parallel strips. AE determines three parameters: the number and orientation of the strips and the power line distribution.

The Placement Expert (*PE*) determines the relative positions of components, *i.e.*, the "strip id" and the order of components inside the strip. The objectives of PE are: (a) simplified routing, and (b) area minimization.

To obtain an accurate estimation of the leaf cells, LES calls FLEX in the estimation mode. PE uses the estimates to adjust the initial placement.

The Characterization Expert (CE) determines routing inside the module. Since LES does not use any routing channels, wires run through the leaf cells. For every required connection, there may be more than one route. When selecting the route, CE tries to minimize wire length, number of jogs, and congestion in the center of the module.

The Flexible Layout Expert (FLEX) generates a symbolic layout according to the cell specification. It returns either a layout description or a failure report. A layout description consists of a symbolic file, plus pin bindings which are then propagated to the neighboring cells.

If FLEX fails to produce a satisfactory layout, it reports the reasons for failure. To eliminate the cause of the failure, LES modifies the current cell and in some situations neighboring cells.

After all leaf cells have been successfully generated, the Evaluation Expert (EE) checks the results against the input specification. If the specification is not satisfied, or if the layout can be improved, LES repeats the layout process. This time, more information is provided for AE and ARE, allowing them to produce an "optimal" design.

After the symbolic layout is generated, the Optimization Expert (OE) tries to improve it by applying its optimization knowledge. The objectives of optimization are: (a) to improve the circuit's performance, and (b) to reduce the layout area. The performance is improved by eliminating unnecessary jogs and replacing polysilicon wires with metal wires wherever possible. To reduce the layout area, OE compacts the layout by rearranging components and connections.

To control optimization, OE uses *goal-directed reasoning*, where optimization is achieved by issuing a goal for layout change. OE chooses the candidate with maximum gain and minimum cost and obstacles. Obstacles to this change are removed by issuing subgoals for each obstacle. When all obstacles have been removed, the OE performs the original optimization. The remaining candidates are then updated, and the optimization cycle is repeated until no optimization rule is applicable.

Experiments have shown that LES is superior to standard-cell systems because it uses 25%-90% of standard-cell area. In comparison with a human expert, LES generated an area only 6% larger.

## 7.4  Regular Module Generation

This section describes methodologies for regular module generation. A *module* can roughly be defined as a microarchitectural entity that performs specific functions and consists of arrays of cells or tiles of a specific type. Examples of modules are: PLAs, ROMs, RAMs, register stacks, multipliers, ALUs, counters and datapaths.

Each module is defined by a template or frame and a set of cells that populate the template. For example, a generic PLA template [Wes85] is shown in Figure 7.26a. It consists of an AND array of type *A* cells, an OR array of type *O* cells, and boundary cells including input drivers (IN), output drivers, (OUT), the AND-plane pull-ups (LA), and the OR-plane pull-ups (TO). An example of the cell types for a dynamic CMOS PLA with a two-phase clock [Wes85] is shown in Figure 7.26b. A self-timed clock for precharging the OR-plane is implemented using the additional cells TL, TA, and TM.

A module compiler describes the template, tiles, interfaces, template personalization, and models to be used by other tools.

The template is described by a procedural language or an embedded language with some array constructs for the template assembly. Since procedural languages tend to obscure geometrical relationships, some module compilers use layout editors for the tile specification and for the template's textual description. More recent tools for module design use menus [Law85] or graphics [May86] to specify the template.

| TL | TA | TA | TA | TA | TM | TO | TO | TO | TR |
|----|----|----|----|----|----|----|----|----|----|
| LA | A | A | A | A | AO | O | O | O | RO |
| LA | A | A | A | A | AO | O | O | O | RO |
| LA | A | A | A | A | AO | O | O | O | RO |
| LA | A | A | A | A | AO | O | O | O | RO |
| LA | A | A | A | A | AO | O | O | O | RO |
| BL | IN | IN | IN | IN | BM | OUT | OUT | OUT | BR |

| | |
|-----|----------------------------|
| A | AND-plane programming cell |
| O | OR-plane programming cell |
| AO | AND-OR communication cell |
| IN | AND-plane input cell |
| OUT | OR-plane output cell |
| LA | Left AND-plane cell |
| RO | Right OR plane cell |
| BL | Bottom left cell |
| BM | Bottom middle cell |
| BR | Bottom right cell |
| TL | Top left cell |
| TA | Top AND cell |
| TM | Top middle cell |
| TO | Top OR cell |
| TR | Top right cell |

**Figure 7.26a**  PLA Generic Floor Plan

The interfaces between cells are usually by abutment, where the designer specifies the bounding box and alignment for each cell. For all possible combinations of cells, the designer must also verify that the design rules are not violated and the module is functionally correct. This task can be simplified by inserting some of this knowledge into the compiler. Instead of specifying several versions for each cell, only one cell needs to be specified. The compiler performs pitchmatching. Pitchmatching is performed by stretching cells so that their I/O pins match on the common boundary. When *pitchmatching* does not work, a compiler may use river routing or some other form of switchbox routing. (See Chapter 5 for more information on river and switchbox routing.) Furthermore, a compiler may perform compaction by overlapping cells so that they share buses and power lines or by spacing them apart to avoid geometrical design rule violations.

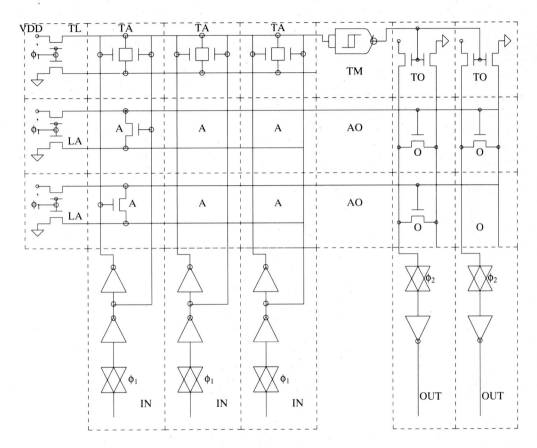

**Figure 7.26b**  Dynamic CMOS 2φ PLA Circuits

*Template personalization* is achieved by specifying a personalization matrix and defining a function that maps the matrix's symbols into a set of tiles. Some tools allow module optimization by selected replacement of tiles. As an example, consider the PLA in Figure 7.26b. The type A tiles in rows 2 and 3 and columns 3 and 4 can be simplified by removing the bottom horizontal wire passing through the cell.

In addition to producing layouts, module generators must generate models to be used with other tools, such as the logic model used with a logic simulator or the interface model used with a floorplanner.

Many different module generators have been developed, and they are available commercially from silicon compiler vendors. The most popular are those for producing RAMs, ROMs, PLAs, multipliers, datapaths, controllers, and microprocessor compilers.

The next two paragraphs briefly describe two sample generators: a PLA generator and a multiplier generator.

APSS [Ste82] is an automatic, *technology-independent* PLA generator. It accepts a set of Boolean equations as input and produces a PLA layout. Its tasks include logic translation, Boolean minimization, PLA folding, PLA topology generation, and interfacing with placement and routing tools. An excellent survey of PLA issues and tools is found in [DeM85]. The logic translation task converts the input Boolean equations into *sum-of-product* form and sets that into a personality matrix. The minimization task reduces the number of product terms and literals. The folding task tries to reduce the number of columns and rows used. The topology generation task maps the personality matrix into a complete PLA topology. The tiles used guarantee proper interface.

Chu and Sharma of Bell Laboratories developed a multiplier generator which implements a modified Booth's algorithm [Chu84]. A designer may specify the design rules, power dissipation, static or dynamic circuit type, *2*'s complement or sign magnitude representation, sizes of the multiplicand and the multiplier, rounding scheme, number of pipelined stages, ripple or carry look ahead addition, and positions of inputs and outputs.

### Generator-Generators

A *module generator-generator* allows a designer to create his or her own module generators, just like a compiler-compiler enables easy construction of a new programming language compiler. The flexibility of this approach allows a module generator to be tailored to fit the specific design environment instead of adjusting it to fit available generators. This section describes two generator-generators with two different approaches: the structure compiler program and Mocha Chips.

**Structure Compiler.** Law and Mosby developed a *structure compiler program* [Law85] which allows designers to build module generators without modifying the program. A similar compiler has been described by Bamji [Bam85]. With this program, the designer supplies the compiler with two types of information: a parameterized block diagram called the array-structure template and a library of tiles or master cells to be used. The compiler then produces a customized module generator according to the specifications.

The *array-structure template* carries the technology-independent floorplan information. Each template is composed of one or more rectangular blocks. Each block will contain a single array of tiles from the user-provided library. The placement of master cells within a block is defined in the personality matrix input to the module generator. Block sizes are defined either explicitly in the block properties list or implicitly by the dimensions of the personality matrix. The interface between blocks in a module is specified by matching corners of neighboring blocks.

Each block has a list of parameters, including: the part of the personality matrix to be used, array type (homogeneous or heterogeneous), the mapping of personality matrix symbols into the symbols representing tiles in the library, the number of columns and rows (optional), cell orientation, and an optional procedure to further optimize the structure.

The designer enters the tiles using a layout editor. The only limitation on the contents of cells is that they must have an enclosing rectangle. The enclosing rectangle provides the size of the tile to the structure compiler, allowing it to assemble arrays that have different spacing for different columns and rows.

**Mocha Chip.** *Mocha Chip* [May86] uses an *assembly diagram* to specify the structure of the module generator to be produced. The designer draws the diagram using a graphic editor. Assembly diagrams show the relative placement and orientation of the module subcells. Each subcell is either another diagram or a part of a primitive.

Subcell orientation is specified by a flag. Subcell placement is specified by a number of cutlines. A cutline defines the relative position of a subcell (or a group of subcells) with regard to another subcell (or a group of subcells). A vertical cutline defines an above/below relationship, while a horizontal cutline defines a left/right relationship.

In addition to placement and orientation of subcells, assembly diagrams can take parameters, compute local variables, and pass values to subcells at lower levels. Conversely, subcells may access parameters and variables defined at higher levels. A parameter with a default value can be accessed if it is not defined when the subcell is invoked.

Two special subcells make use of parameters: the *Array* subcell and the *Case* subcell. The *Array* subcell provides a form of iteration, and the *Case* subcell provides a form of conditional selection. The *Array* and *Case* cells are analogous to *loop* and *if-then-else* constructs in programming languages.

To specify a two-dimensional, homogeneous array-block, an *Array* subcell needs three parameters: number of columns, number of rows, and type of tile. The designer can define a more complicated structure by using a combination of *Array* and *Case* subcells. For example, a heterogeneous array-block can be defined by replacing the device type with a *Case* subcell.

## 7.5  Silicon Compilation

A VLSI chip can be described by a structure of components in which each component is either designed manually, a copy of a cell or module in the library, or instantiated by a module or cell compiler. The description is then converted to a layout by a program called a *structural silicon compiler*. A structural silicon compiler calls cell or module compilers, and then places, routes and compacts generated component layouts. On the other hand, a chip can be thought of as a black box with well-described interfaces and behavior. A *behavioral silicon compiler* synthesizes a structural description from behavior and calls a structural silicon compiler to complete the task. An *intelligent silicon compiler* is a constraint-driven behavioral silicon compiler. The difference between behavioral and intelligent silicon compilers is in the amount of interaction required by the designer. With a behavioral silicon compiler, the designer must rewrite the input description to change the generated design. With an intelligent silicon compiler the designer achieves the same effect by specifying lower and/or upper bounds on time, cost, area, power, testability and design style, or by partially specifying the design.

A typical silicon compiler system is shown in Figure 7.27. The designer specifies the design structure through a schematic capture subsystem. The behavioral description of each component in the structure is captured by a menu/form package. The behavioral description usually consists of a set of functional, electrical, and geometrical parameters that are passed to corresponding module compilers. A *technology file* contains all process-relevant design rules used for generating the geometric model (*layout*). In an interactive environment, layout or schematic editors may be used to alter compiler outputs. In this case, however, the *correct-by-construction* property of silicon compilers is lost. In addition to geometric models, timing and logic models are generated for each component in the structure. These models are linked together and passed to a timing analyzer, simulator, or some other analysis tool. Geometric models are then linked together with placement and routing tools to form a chip composite. For interactive placement and routing, a composition editor can be used. Next, a package editor provides packaging information to the foundry. Lastly, the simulator can provide a test set for testing the assembled IC.

A *behavioral silicon compiler* is a structural silicon compiler with a *synthesizer*, a translator from the behavioral into a structural description. An *intelligent silicon compiler* is a *behavioral silicon compiler* that selects different design styles, components, or optimization goals in order to satisfy design constraints. The compiler then evaluates the generated design and feeds the design quality measures back to the synthesizer.

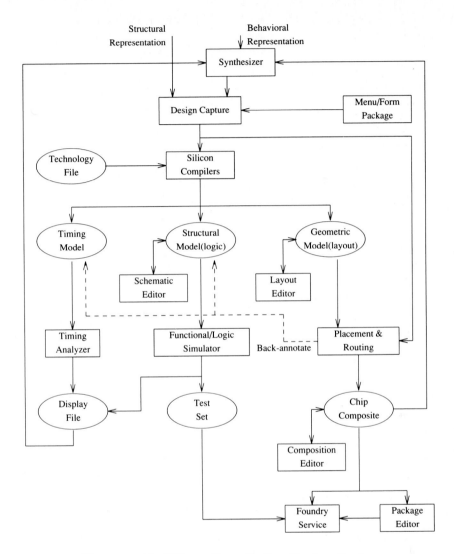

**Figure 7.27** Silicon Compilation Design System

## Structural Silicon Compilers

The basic issues in structural silicon compilation are the designer and technology interface, controllability of the design process, level of integration, richness of the component generator set, and performance.

The *interface*s define how the designer will specify the initial structure and the technology-dependent design rules for the compiler. One alternative to using interfaces is to create a block-diagram using a graphic terminal, with menus for each component. This is a natural way for designers who are used to manipulating two-dimensional schematics. The other alternative is to use a procedural language. A procedural language has easy mechanisms for passing parameters and recursive calls to already created modules. It does not, however, allow visualization of interconnectivity or the floorplan, as does a schematic. Therefore, it is useful for preliminary designs to have a schematic capture capability associated with a language. The schematic can be converted later into a procedural description.

Silicon compilers usually produce output in CIF or GDSII format. In any case, the compiler must interface with the design rules imposed by the particular fabrication technology. Some compilers are designed through tiling schemes in which each tile is generated with a layout editor. The layout is usually defined in terms of *lambda* rules. Different *lambda*s are used for different technologies, resulting in linear change of the layout. (See Figure 7.28a.) When a procedural language is used for describing the layout, each spacing is defined as a variable. The variable values are obtained from a technology file at compilation time. Using a procedural language allows a non-linear, dynamic compaction (Figure 7.28b), that is, a different scaling of each design rule.

*Design controllability* refers to the level of human interaction allowed during the design generation. There is no reason why a design system should not allow layout or schematic editing after the design has been generated. In this case, a silicon compiler generates an initial design, which a designer may then modify. The *correct-by-construction* property of silicon compilation is exchanged for the gain in flexibility. The problem with allowing design modification is in propagating designer-introduced changes to other design modules used by available tools. A possible alternative is to allow the designer to pre-specify the design style, as with pre-placement and pre-routing during physical design.

*Integratability* of a silicon compiler refers to the level of coupling of all tools needed for design. Presently silicon compilers are one of two types:

- Complete systems, with their own set of proprietary tools (such as simulators, time verifiers, and routers) for complete IC design.

- A set of cell and module compilers integrated into a standard CAE workstation, with most of the support tools provided by the host workstation environment.

The advantage of the first approach is the possibility of a more efficient database and better interfaces between different tools. Furthermore, a well-established design practice, such as clocking strategy, can be easily enforced in a tightly integrated system. The advantage of the second approach is the easy upgradability from present manual methodology and the reuse of already designed parts.

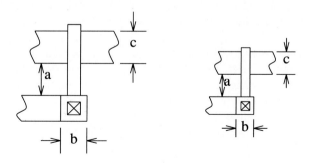

(a) Linear Scaling:  a=60%, b=60%, c=60%

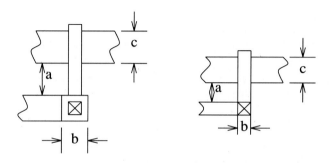

(b) Dynamic Scaling:  a=60%, b=50%, c=80%

**Figure 7.28**  Design Rule Scaling

The *richness of a silicon compiler* refers to the availability of cell and module compilers.  Module types can be divided into basic computational and storage modules whose operations take one control state (such as SSI logic, registers, ALUs, PLAs, RAMs, ROMs and datapaths) and sequential modules whose descriptions encompass more than one control state (such as counters, finite-state machines, controllers, and processors).

*Performance* can be characterized by different metrics.  Three types of quality measures can be defined to evaluate silicon compilers: transistor density in square mils/transistor, compiler performance in transistors/hour, and design time in transistors/person-hour.  The quality measure used depends on the complexity and regularity of the design.  Until standard benchmarks are established, these measures only indicate capabilities of silicon compilation and should not be used to compare different systems.

In addition to design functionality, transistor density also depends on the fabrication process. The transistor densities currently reported by commercial silicon compilers range from 2.0 square mils/transistor in 3-micron CMOS, to 1.5 square mils/transistor in 2-micron CMOS. Seattle Silicon recently reported 0.36 square mils/transistor on a 12000-transistor DSP chip, which is better than some existing microprocessors such as the Intel 80386. The density measure is similar to the computer performance measure of instructions/second. It is not very accurate but is a very popular approximation of performance. A better measure of density would use the design functionality as a normalizing factor instead of transistor count, since the existence of each transistor may be difficult to justify.

The compiler performance defines compiler runtime without plotting and should be normalized to the same machine. Current compilation speeds vary from 5K-25K transistors/hour for different chips.

The design time depends heavily on design complexity, as well as the abstraction level of the compiler. Typical design times range from 6-25 transistors/hour for various designs. Better design times (300-400 transistors/hour) are achieved by using core compilers, which take most of the chip area, or functional silicon compilers, such as McPitts [Sou83]. These numbers indicate an obvious trade-off between design time, compiler complexity, and the level of input specification.

## Behavioral Silicon Compilers

Behavioral silicon compilers start with the behavioral description of a black box with well-defined interfaces. Any programming language can be used to create the behavioral description. However, programming languages assume a very simplistic implementational model in which there is no notion of time, delay, performance or connectivity. There is only an execution order imposed by the language's control constructs and sequencing of statements. Each variable or array represents a storage container accessible without delay. Each operator is a single-function unit that is available at any time and in any quantity and that takes no time to execute. The movement and transformation of data also takes no time.

In order to simplify the compilation process, most behavioral silicon compilers assume a well-defined design model or *target architecture*. Such a model is shown in Figure 7.29. The entire design consists of communicating processing elements (PEs). (These PEs correspond to the notion of communicating processes.) Each PE consists of a Control Unit (CU) and a Datapath (DP). There may be several types of CUs; however, each CU must contain a state register for storing the "state id" and a control logic for controlling the DP and communicating with other PEs. A DP consists of storage elements (such as registers, counters, register banks, and memories) and functional units (such as ALUs, shifters and multipliers) connected through sets of buses.

Access to registers, units or I/O ports is controlled by the CU. If several buses are used as sources to a storage or only functional unit, a selector controlled by the CU must be added at the input. Some DP models use only point-to-point connection with selectors only and no buses. PEs communicate through DP ports to the CU or DP (nets *a* and *b* in Figure 7.29) or through CU ports to the CU or DP (nets *c* and *d*). Each PE may run under a different clock, and each clock may have several phases.

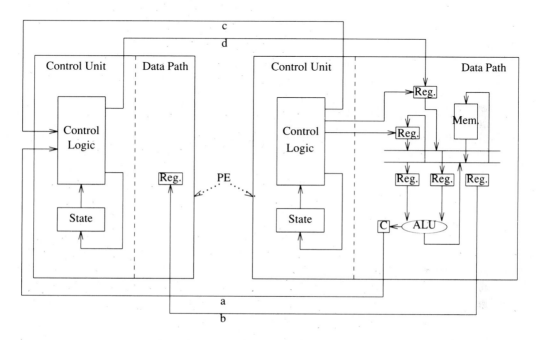

**Figure 7.29**  Design Model

Note that in the above model an adder may be represented as a PE with no CU but with a DP (having one output port, two input ports, and no storage elements). Similarly, a flip-flop can be modeled as a DP with no functional units or as a CU with no DP and no control logic. Thus, this model is complete in the sense that it can model any digital system.

Behavioral silicon compilation consists of binding language constructs to structural components in the design model. In other words, every variable must be allocated to a storage element and every operator to a functional module. When a variable is modified, a connection path or bus must be allocated to send the present value to the functional unit, then back to the storage element. Furthermore, a loop statement must be converted into a sequential control unit operation, and each operation must be given a different time interval or machine state. This transformation is usually achieved by converting the language description into a *control/data flow graph* in which all unnecessary data and/or control dependencies have been removed [Sno78, Cam85, Ora86]. The graph is then partitioned (or sliced) into states, with an operation or series of operations in each state taking no more than the pre-defined time. This process is sometimes called *scheduling* since it is similar to the task of job scheduling.

Scheduling can be performed using *as-soon-as-possible* (ASAP) or *as-late-as-possible* (ALAP) strategies that produce schedules with a minimum number of machine states [Lan80]. Figure 7.30b and 7.30c show an ASAP and an ALAP schedule for executing the program shown in Figure 7.30a. Both of these strategies generate maximally parallel implementations of the datapath and use suboptimally large numbers of components. For example, DAA [Kow85] and Emerald [Tse84] use ASAP strategy. The number of functional units, registers, and connections − the cost of the design − is minimized using node mobilities. Node mobilities are defined as the difference in state assignment between ASAP and ALAP schedules. They can be used to minimize the number of same type functional units by balancing their usage as evenly as possible across all states [Pau86]. Such an optimized assignment requires only one unit of each type in all states, except state number 1 as shown in Figure 7.30c (by coincidence, this figure is an ALAP assignment). Note that an ASAP assignment in this example requires two shifters and two min/max units in addition to the two absolute-value units.

In general, two types of scheduling algorithms exist. The first type tries to minimize the cost of the given design [Pau86, Nes86, Par86]. The second type tries to minimize time − the number of control states − given the design cost in terms of functional units, storage elements, and connections [Gir85, Pan86]. For example, Figure 7.30d shows a time minimal schedule used when the design is limited to one functional unit of each type.

Once operations are assigned to states, registers must be assigned to variables used in more than one state. Some systems have a one-to-one correspondence of language variables to registers, while others optimize the number of registers by sharing. Register sharing allows a register to store two or more variables that have disjoint lifetimes. Figure 7.31a shows a graph with four machine states. For each variable (an output of a node in the graph), the lifetime is indicated by the letter $L$ in the lifetime table (Figure 7.31b). The variable lifetimes are used to create a connectivity graph. (See Figure 7.31c.) Two variables are connected in the connectivity graph if their lifetimes do not overlap.

Variables are merged into registers by partitioning the connectivity graph into maximal cliques. A maximal clique is a fully connected subgraph that is not contained in any other fully connected subgraph. Four possible options are indicated in Figure 7.31c. For a given set of functional units, each option has a

$$x = \max(|a|, |b|)$$
$$x = \min(|a|, |b|)$$
$$result = \max(0.875x + 0.5y, x)$$

(a) Program

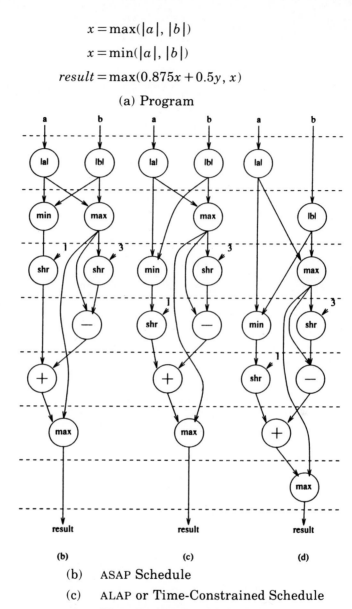

(b)    ASAP Schedule

(c)    ALAP or Time-Constrained Schedule

(d)    Unit-Constrained Schedule

**Figure 7.30** State Binding

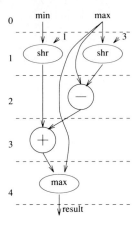

(a) Dataflow Graph

|   | min | max | max/8 | min/2 | 0.875max | sum | result |
|---|-----|-----|-------|-------|----------|-----|--------|
| 0 | $L$ | $L$ |       |       |          |     |        |
| 1 |     | $L$ | $L$   | $L$   |          |     |        |
| 2 |     | $L$ |       | $L$   | $L$      |     |        |
| 3 |     | $L$ |       |       |          | $L$ |        |
| 4 |     |     |       |       |          |     | $L$    |

(b) Lifetime Table

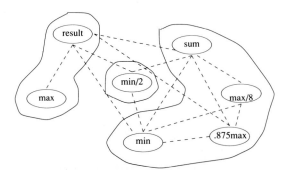

(c) Connectivity Graph

**Figure 7.31** Register Sharing through Variable Merging

Option 1:  $R_1 = \{max, result\}$
$R_2 = \{min, max/8, 0.875max, sum\}$
$R_3 = \{min/2\}$

Option 2:  $R_1 = \{max, result\}$
$R_2 = \{min, max/8, 0.875max\}$
$R_3 = \{min/2, sum\}$

Option 3:  $R_1 = \{max, result\}$
$R_2 = \{max/8, 0.875max, sum\}$
$R_3 = \{min, min/2\}$

Option 4:  $R_1 = \{max, result\}$
$R_2 = \{max/8, 0.875max\}$
$R_3 = \{min, min/2, sum\}$

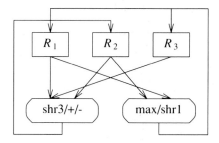

(d) Option 1 & 4 Implementation

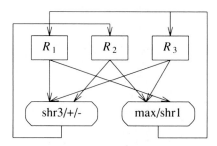

(e) Option 2 & 3 Implementation

**Figure 7.31** Register Sharing through Variable Merging (Cont.)

different connectivity cost. Options 2 and 3 require one more connection than options 1 and 4, as shown in Figures 7.31d and 7.31e.

Once variables are assigned to registers, registers can be merged into *register files*, or *register banks*, if their read and write times do not overlap. Using master-slave semantics allows the reading and writing of a register in the same clock. A read/write table is shown in Figure 7.32a. Since the read and write times of registers $R_1$ and $R_3$ do not overlap in any control state, the registers can be combined. This register combining saves one point-to-point connection, as shown in Figure 7.32b.

After operations have been assigned to states, and variables have been assigned to registers, functional units can be created by merging operators that are not used in the same state. Merging operators should lower the design cost since the cost of two combined operators should be smaller than the sum of single-operator costs. Only compatible operators are merged. Operators are incompatible if there is an increase in delay or if there is no cost gain when they are combined. Merging is achieved by constructing a connectivity graph and finding maximal cliques that minimize connectivity cost. For example, consider the partial design shown in Figure 7.33a. Its connectivity graph is shown in Figure 7.33b. Assuming that max operator is not compatible with "+" and "−" operators, two options (shown in Figure 7.33c and 7.33d) are possible. Option 1 has smaller connectivity cost.

Connection bindings represent the allocation of connections between hardware components to create the necessary information paths. Connections are either point-to-point or bussed. The bus model allows greater interconnection sharing by creating a larger number of paths for the same number of connections.

Once operators and variables have been assigned, connections between registers and units can be derived from the dataflow graph. These connections can be merged into buses by combining connections used in different states. The same clique-partitioning method can be used. For example, the lifetime table for the design in Figure 7.33c is shown in Figure 7.34a. The resulting bus structure is shown in Figure 7.34b. The method using datapath allocation based on clique partitioning was introduced by Tseng and Siewiorek [Tse84].

In addition to partitioning, other approaches have been used for datapath synthesis. For example, DAA [Kow85] uses a two-pass expert system. The first pass assigns components to nodes in the flow graph. The second pass merges those virtual components to form the actual components. MAHA [Par86] determines the critical path through the flow graph, then schedules the operation along the critical path and synthesizes the datapath for the critical path. The operation on non-critical paths is allocated to existing components whenever possible. EMUCS' [Tho83] binding is based on a heuristic that minimizes the change of the previously allocated datapath. On each iteration of the min-max algorithm, the cost for binding each unbound graph node to each available

| | $R_1$ | | $R_2$ | | $R_3$ | |
|---|---|---|---|---|---|---|
| 0 | max | W | min | W | | |
| 1 | max | R | max/8 | R W | min/2 | W |
| 2 | max | R | .875max | R W | min/2 | |
| 3 | max | | sum | R W | | R |
| 4 | result | R W | | R | | |

Combine

(a) Lifetime Table

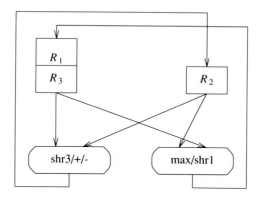

(b) Implementation

**Figure 7.32**  Register Merging

functional unit is computed. The node with the largest difference in the two least costly bindings is bound to the functional unit with the lowest cost. Finally, SPLICER [Pan86] uses a branch-and-bound algorithm to synthesize the datapath from the flow graphs one state at a time with several states look ahead.

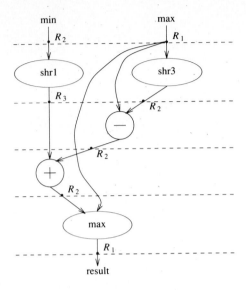

(a) Dataflow Graph with State and Register Assignment

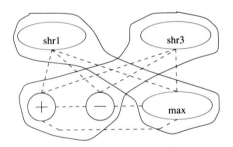

(b) Connectivity Graph

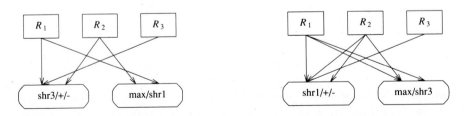

(c) Option 1 Implementation    (d) Option 2 Implementation

**Figure 7.33** Unit Merging

|   | A | B | C | D | E | F | G | H |
|---|---|---|---|---|---|---|---|---|
| 1 | L |   |   | L |   | L |   | L |
| 2 | L |   | L |   |   | L |   |   |
| 3 |   |   | L |   | L | L |   |   |
| 4 |   | L |   | L |   |   | L |   |

Buses = {A, B, E} {C, D} {F} {G, H}

(a) Lifetime Table ·

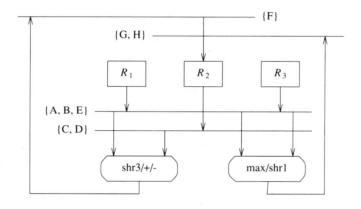

(b) Implementation

**Figure 7.34** Connection Merging

## Design-Process Model

The previous section showed how to construct a design from its behavioral description by merging variables, operators and connections. Merging was accomplished by partitioning connectivity graphs into maximal cliques. However, maximal cliques do not necessarily yield an optimal design. So a cost function was introduced, which was equivalent to the number of point-to-point connections. Yet even with this simple function it was difficult to select the best possible design. The difficulty came from two requirements: first, for variable merging the designer needed to know functional units used. Second, for operator merging the designer needed to know register assignments in order to compute the cost function. In reality, more complex heuristic cost functions are used [Tse84]. The other alternative would be to perform all bindings at the same time, while minimizing area, time, power and other quality measures. Since this method creates a large search space, the translation of behavior to structure — the bindings of language constructs to the components in the design model — is performed in several steps.

Thus, the design process can be defined as a sequence of bindings with their associated algorithms and optimization goals. There are two approaches to basic binding: constructive and iterative. The constructive method creates an initial design (seed design) by merging variables, operators and connections. It removes redundancy from the input description by minimizing the number of components without making any trade-offs. In other words, the constructive method always produces the same design, regardless of the imposed constraints, since the cost functions are built into the algorithms.

The iterative method modifies the design to satisfy imposed constraints. It may produce a partial (local) redesign or a complete new design in a new design style. This modification is usually a trade-off in favor of the most important constraint.

Since the binding problem is so difficult, most behavioral silicon compilers use the constructive approach or divide binding responsibilities between the designer and the compiler. In general, a high level description language requires more work from the compiler, while a lower level description requires more work from the designer. Language bindings allow explicit specification of part of the design in the language. An example of using language binding is specifying operations to be performed during each state. This specification is accomplished by allowing serial or parallel operators, which enable explicit specification of two or more operations in the same state. The advantages of explicit specification are that the designer has tighter control over the design process, and it makes the compiler simpler. The disadvantages are that the compiler is restricted from performing optimizations and trade-offs on the user-bound areas, and the designer has to do some bindings manually. Because the compiler is restricted from performing trade-offs, the designer must rewrite the description to explore another design in the design space.

Typically, language bindings in current systems are at the state-binding level. Compilers like McPitts [Sou83], SILC [Bla85], and SYCO [Jer86] allow the designer to specify the operations performed during each state, eliminating the need for the compiler to perform state bindings. Language variables store data between operations. If the machine states are specified in the language description, the variables correspond to hardware storage elements, such as registers. McPitts, SILC, and SYCO are examples of systems that assign language variables directly to registers in the resulting machine. This eliminates the need for compiler variable-to-register binding.

Other compilers, such as the Yorktown silicon compiler [Bra88] and CATHEDRAL [Rab88], allow the designer to assign functional units to some or all nodes in the flow graph. Since a functional unit cannot perform two operations in the same machine state, this binding forces the compiler to slice the graph into machine states. Nodes that are not assigned by the designer are assigned to functional units by a set of optimization rules. Similarly, register and connection allocations are performed by a set of heuristics.

Presently, no compilers allow the designer to allocate connections or buses. With the growing importance of connectivity in VLSI design, assignment of bussing structure is becoming more important. In the next generation of silicon compilers, a designer should be able to partially allocate states, storages, functional units, and interconnections, with the compiler completing the allocation.

## 7.6 Future Trends

Present-day silicon compilers translate an input description into layout in a unique way. If a designer working with a behavioral silicon compiler wishes to consider alternatives, the behavioral description must be rewritten. In other words, he or she must understand the translation process built into the compiler and modify the input description in a way that forces the compiler to produce the desired results. When working with structural compilers, the designer must be able to evaluate a design and choose a different style or a component. An intelligent silicon compiler, on the other hand, incorporates knowledge about the design process and uses this knowledge to guide the transformation of the input specification through several design iterations until the pre-specified set of constraints is met.

In summary, there are four basic tasks on each level of design: *style selection*, *refinement*, *optimization*, and *strategy formulation*. The selection of styles depends on the goals assigned to a particular implementation. To automate refinement, the process of translating overall design constraints into different design styles must be solved and captured in one form or another. Refinement (sometimes called partitioning or synthesis) involves translating a behavioral description into a structure of pre-defined components from the next lower level of the design. As mentioned before, this translation is not unique; usually several different styles can be selected. For example, the next-state function in the control unit can be implemented as a one-way or two-way branch; multiple inputs to a functional unit can be implemented with a bus or a multiplexer; an adder can be designed as a ripple-carry adder or a carry-look-ahead adder; a chip may use a one-phase or a two-phase clocking scheme; I/O ports on a cell can be on different sides; a cell layout may have different aspect ratios; or a vertical or horizontal power distribution can be used in each cell. Most structural silicon compilers offer different output buffer sizes, aspect ratios for RAMs and ROMs, latching positions in datapaths, direct, tristate, and precharged outputs, low and high speed PLAs, LSSD and scanpath testability, *etc.*

The optimization step improves utilization of allocated resources such as the silicon area in PLA folding or layout compaction, the number of tracks in channel routing, or the number of functional units in a microarchitecture. An optimization can usually be defined by an algorithm. Compaction, placement, and routing optimizations are typical, while clock and power sizing and PLA folding are offered by some silicon compilers. The McPitts and SILC compilers perform sharing or merging of functional units and buses on the microarchitectural level. An optimization can be performed with or after the refinement.

A *strategy* is a sequence of different style selection, refinement, and optimization steps. The type and order of steps in the sequence characterize a strategy. Strategies are better understood at levels closer to physical design than on the higher, more abstract levels. It is natural, for example, that symbolic layout is followed by geometric layout, which is followed by compaction. However, the order of register, unit, and bus allocation and their optimization on the register transfer or microarchitectural level is not clear. Presently, all existing silicon compilers allow only one fixed strategy. The whole design process may be thought of as a set of refinements and optimizations for each style at each level of design.

In order to formulate a proper design strategy, three more mechanisms must be added. The *constraint propagation* process partitions a design module's constraints into constraints for each of its components. This task is presently performed by the designer. The second mechanism deals with *evaluating* the final design of each component, estimating how well constraints have been satisfied, and finding the cause of failure in cases where they were not satisfied. The evaluation process is supported by providing time analysis, cost, and power reports. The third task is performing trade-offs in cases where one or more constraints were not satisfied. The trade-offs are made by using different design styles, such as choosing a ripple-carry adder over a carry-look-ahead adder if the smaller area goal is more important than the higher speed goal.

An intelligent silicon compiler model must be supported by two other features. First, it needs an input language with minimal explicit binding so the binding decisions can be postponed until compile time. Second, it needs an internal representation flexible enough to allow trade-off analysis and transformations based on user constraints. An example is shown in Figure 7.35, where the contents of two registers are incremented and exchanged. Depending on the constraints at compile time, the compiler may perform the exchange in one state if two adders are available (Figure 7.35a), in two states if only one adder and three buses are available (Figure 7.35b), or in three states if only one adder and two buses are available (Figure 7.35c).

Presently, goal definition, style selection, constraint propagation, design evaluation, and trade-offs are made by a designer with the help of CAD tools such as simulators and timing analyzers. New generation silicon compilers will bring automation of the remaining tasks.

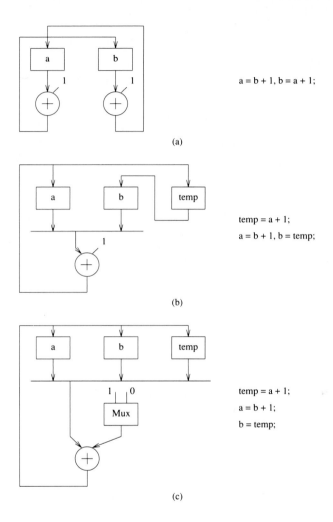

$a = b + 1, b = a + 1;$

(a)

$temp = a + 1;$
$a = b + 1, b = temp;$

(b)

$temp = a + 1;$
$a = b + 1;$
$b = temp;$

(c)

**Figure 7.35** Design Trade-offs of Execution States, Buses and Adders

## Acknowledgements

We would like to thank Scott Baeder, Alfred Dunlop, Michael Lorenzetti, and Wayne Wolf for the many suggestions that improved the quality of this chapter.

## Exercises

**7.1**   Explain the difference between

   **a)**   Behavioral, structural, and geometrical representations.

   **b)**   Synthesis and physical design.

**7.2†**   Define a language for laying out a CMOS NAND gate in fixed-grid architecture.

**7.3†**   Lay out the circuit of Figure 7.7 in gate matrix style.

**7.4†**   Describe a 2-input NAND gate using the language shown in Figure 7.5.

**7.5†**   Write a program to perform the left-edge algorithm for Weinberger array layout.

**7.6‡**   Develop an algorithm/heuristic for folding Weinberger arrays.

**7.7†**   Specify several functions for evaluating the quality of standard cell routing.

**7.8†**   Obtain a minimal, folded PLA for the following set of Boolean equations:

   **a)**   $f_1 = x_1(\bar{x}_3 x_6 + x_3 \bar{x}_5) + x_2 x_5$

   **b)**   $f_2 = \bar{x}_1 \bar{x}_6 + x_1 x_3 \bar{x}_5$

   **c)**   $f_3 = \bar{x}_2 \bar{x}_4 + x_3 x_4 x_5$

**7.9†**   Use TOPOLOGIZER's rules to generate the optimal placement of Figure 7.19 with 4 rows of 3 transistors.

**7.10‡** Write a program to do TOPOLOGIZER's routing refinement.

**7.11†** Develop a ROM template similar to the PLA template shown in Figure 7.26.

**7.12‡** Develop a PLA generator for CMOS technology.

**7.13†** Define several quality measurements for chip layout.

**7.14‡** Develop an algorithm for unit-constraint scheduling with each unit having equal delay time.

**7.15‡** Write a program for register-sharing using the number of connections as a quality measure.

**7.16**   Explain a design process model.

**7.17**   Define several trade-offs for:

   **a)**   Layout.

   **b)**   Logic.

   **c)**   Microarchitecture design levels.

# References

[Asa82]   Asano, T., "An optimum gate placement algorithm for MOS one-dimensional arrays," *Journal of Digital Systems*, vol. 5, no. 1, pp. 1-27, 1982.

[Bam85]   Bamji, C. S., C. E. Hauck, and J. Allen, "A design by example regular structure generator," in *Proc. of 22nd Design Automation Conf.*, pp.16-22, 1985.

[Bat80]   Batali, J., and A. Hartheimer, "The design procedure language manual," *A. I. Memo 598, MIT*, Artificial Intelligence Laboratory, 1980.

[Bla85]   Blackman, T., J. Fox, and C. Rosebrugh, "The SILC silicon compiler: language and features," in *Proc. of 22nd Design Automation Conf.*, pp. 232-237, June 1985.

[Bra84]   Brayton, R. K., C. McMullen, G. D. Hachtel, and A. Sangiovanni-Vincentelli, *Logic Minimization Algorithms for VLSI Synthesis*. Kluwer Academic Publishers, Boston, MA, 1984.

[Bra88]   Brayton, R. K., R. Camposano, G. De Michelli, R. H. J. M. Otten, and J. van Eijndhoven, "The Yorktown silicon compiler," in *Silicon Compilation*, Addison-Wesley, Reading, MA, Chapter 6, 1988.

[Bur88]   Burich, M., "Design of module generators and silicon compilers," in *Silicon Compilation*, Addison-Wesley, Reading, MA, Chapter 2, 1988.

[Cam85]   Camposano, R., and W. Rosentiel, "A design environment for the synthesis of integrated circuits," in *Proc. 11th EUROMICRO*, September 1985.

[Chu82]   Chuquillanqui, S., and T. Perez Segovia, "PAOLA: A tool for topological optimization of large PLAs," in *Proc. of 19th Design Automation Conf.*, pp. 300-306, 1982.

[Chu84]   Chu, K-C., and R. Sharma, "A technology independent MOS multiplier generator," in *Proc. of 21st Design Automation Conf.*, pp. 90-97, 1984.

[DeM83]   De Micheli, G., and A. M. Sangiovanni-Vincentelli, "Multiple constrained folding of programmable logic arrays, theory and applications," *IEEE Trans. Computer-Aided Design*, vol. CAD-2, pp. 151-167, 1983.

[DeM85]   De Micheli, G., M. Hoffmann, A. R. Newton, and A. M. Sangiovanni-Vincentelli, "A design system for PLA-based digital circuits," *Advances in Computer-Aided Engineering Design, Volume 1*, Jai Press, Greenwich, CT, 1985.

[Fle75]   Fleisher, H., and L. I. Maissel, "An introduction to array logic," *IBM Journal of Research and Development*, vol. 19, no. 2, pp. 98-109, March 1975.

[Gaj83]    Gajski, D., and R. H. Kuhn, "New VLSI tools," *IEEE Computer*, pp. 11-14, December 1983.

[Gir85]    Girczyk, E. F., R. J. Buhr, and J. P. Knight, "Applicability of a subset of ada as an algorithmic hardware description language for graph-based hardware compilation," *IEEE Trans. on Computer-Aided Design*, vol. CAD-4, no. 2, pp. 134-142, April 1985.

[Hac82]    Hachtel, G. D., A. R. Newton, and A. L. Sangiovanni-Vincentelli, "Techniques for programmable logic array folding," in *Proc. of 19th Design Automation Conf.*, pp. 147-155, 1982.

[Has71]    Hashimoto, A., and J. Stevens, "Wire routing by optimizing channel assignment within large apertures," in *Proc. of 8th Design Automation Conf.*, pp. 155-169, 1971.

[Hil85]    Hill, D. D., "Sc2: a hybrid automatic layout system," *Digest, Intl. Conf. on Computer-Aided Design*, pp. 172-174, 1985.

[Hwa86a]   Hwang, D. K., W. K. Fuchs, and S. M. Kang, "An efficient approach to gate matrix layout," in *Digest, Intl. Conf. on Computer-Aided Design*, pp. 312-315, November 1986.

[Hwa86b]   Hwang, S. Y., R. W. Dutton, and T. Blank, "A best-first search algorithm for optimal PLA folding," *IEEE Trans. on CAD*, vol. CAD-5, no. 3, pp. 433-442, July 1986.

[Jer86]    Jerraya, A. A., P. Varinot, R. Janier, and B. Courtois, "Principles of the SYCO Compiler," in *Proc. of 23rd Design Automation Conf.*, pp. 715-721, June 1986.

[Joh79]    Johannsen, D., "Bristle blocks: a silicon compiler," in *Proc. of 16th Design Automation Conf.*, pp. 310-313, 1979.

[Kim86]    Kim, J., and J. McDermott, "Computer aids for IC design," *IEEE Software*, pp. 38-47, March 1986.

[Kol85]    Kollaritsch, P. W., and N. H. E. Weste, "TOPOLOGIZER: an expert system translator of transistor connectivity to symbolic cell layout," *IEEE Journal of Solid-State Circuits*, pp. 799-804, June 1985.

[Kow85]    Kowalski, T. J., D. J. Geiger, W. Wolf, and W. Fichtner, "The VLSI design automation assistant: from algorithms to silicon," *IEEE Design and Test*, pp. 33-43, August 1985.

[Lan80]    Landskov, D., S. Davidson, B. Shriver, and P. Mallet, "Local microcode compaction techniques," *ACM Computing Surveys*, vol. 12, no. 3, September 1980.

[Lar78]    Larson, R. P., "Versatile mask generation techniques for custom microelectronic devices," in *Proc. of 15th Design Automation Conf.*, pp. 193-198, June 1978.

[Law85]    Law, H-F. S., and J. D. Mosby, "An intelligent composition tool for regular and semi-regular VLSI structures," in *Digest, Intl. Conf. on Computer-Aided Design*, pp. 169-171, 1985.

[Leo86]    Leong, H. W., "A new algorithm for gate matrix layout," in *Digest, Intl. Conf. on Computer-Aided Design*, pp. 316-319, November 1986.

[Leu84]    Leung, C. K. C., S. S. Patil, and H. Ravindra, "The Storage/Logic Array (SLA) approach to IC design," *VLSI Design*, pp. 54-61, January 1984.

[Lew82]    Lewandowski, J. L., and C. L. Liu, "A branch and bound algorithm for optimal PLA folding," in *Proc. of the 21st Design Automation Conf.*, pp. 426-433, 1984.

[Lin87]    Lin, Y-L. S., and D. D. Gajski, "LES: a layout expert system," in *Proc. of the 24th Design Automation Conf.*, pp. 672-678, June 1987.

[Lip82]    Lipton, R. J., S. C. North, R. Sedgewick, J. Valdes, and G. Vijayan, "ALI: a procedural language to describe VLSI layouts," in *Proc. of 19th Design Automation Conf.*, pp. 467-474, June 1982.

[Lop80]    Lopez, A., and H. Law, "A dense gate matrix layout method for MOS VLSI," *IEEE Trans. on Electronic Devices*, vol. ED-27, no. 8, pp. 1671-1675, August 1980.

[May86]    Mayo, R. N., "Mocha Chip: a system for the graphical design of VLSI module generators," in *Digest, Intl. Conf. on Computer-Aided Design*, pp. 74-77, 1986.

[Nan83]    Nance, S., C. Starr, B. Duyn, and M. Kliment, "Cell-layout compilers simplify custom IC design," *EDN*, September 15, 1983.

[Nak86]    Nakatani, K., T. Fujii, T. Kikuno, and N. Yoshita, "A heuristic algorithm for gate matrix layout," in *Digest Intl. Conf. on Computer-Aided Design*, pp. 324-327, November 1986.

[Nes86]    Nestor, J. A., and D. E. Thomas, "Behavioral synthesis with interfaces," in *Digest Intl. Conf. on Computer-Aided Design*, pp. 112-115, November 1986.

[New87]    Newton, A. R., "Symbolic layout and procedural design," in *Design Systems for VLSI Circuits: Logic Synthesis and Silicon Compilation*, edited by G. De Micheli, Martinus Nighoff Publishers, Norwell, MA, Chapter 4, 1987.

[Nii86]    Nii, H. P., "Blackboard systems: the blackboard model of problem solving and the evolution of blackboard architectures," *The AI Magazine*, vol. 7, no. 2, pp. 38-53, Summer 1986.

[Oht79]   Ohtsuki, T., H. Mori, E. S. Kuh, T. Kashiwabara, and T. Fujisawa, "One-dimensional logic gate assignment and interval graphs," *IEEE Trans. on Circuits and Systems*, pp.675-684, September 1979.

[Ora86]   Obrailoglu, A., and D. Gajski, "Flow graph representation," in *Proc. of 23rd Design Automation Conf.*, pp. 503-509, June 1986.

[Pan86]   Pangrle, B. P., and D. Gajski, "State synthesis and connectivity binding for microarchitecture compilation," in *Digest, Intl. Conf. on Computer-Aided Design*, pp. 210-213, November 1986.

[Par86]   Parker, A. C., J. Pizarro, and M. Milnar, "MAHA: a program for datapath synthesis," in *Proc. of 23rd Design Automation Conf.*, pp. 461-466, June 1986.

[Pau86]   Paulin, P. G., J. P. Knight, and E. F. Girczyk, "HAL: a multi-paradigm approach to automatic datapath synthesis," in *Proc. of 23rd Design Automation Conf.*, pp. 263-270, June 1986.

[Rab88]   Rabaey, J., H. DeMan, S. Vanhoof, G. Goossens, and F. Catthoor, "CATHEDRAL," in *Silicon Compilation*, Addison-Wesley, Reading, MA, Chapter 8, 1988.

[Ros82]   Rosenberg, J., and N. Weste, "ABCD — a better circuit description," *Microelectronics Center of North Carolina Tech Report*, 82-01, 1982.

[Smi82]   Smith, K. F., T. M. Carter, and C. E. Hunt, "Structured logic design of integrated circuits using the storage/logic array (SLA)," *IEEE Trans. on Electron Devices*, pp. 765-776, April 1982.

[Sno78]   Snow, E. A., "Automation of Module Set Independent Register-Transfer Level Design," Ph.D. Dissertation, Carnegie Mellon University, April 1978.

[Sou83]   Southard, J. R., "McPitts: an approach to silicon compilation," *Computer*, vol. 16, no. 12, pp. 74-82, December 1983.

[Ste82]   Stebnisky, M. W., M. J. McGinnis, J. C. Werbickas, R. N. Putatunda, and A. Feller,, "A fully automatic, technology-independent PLA macrocell generator," in *Proc. ICCC*, pp. 156-159, 1982.

[Tho83]   Thomas, D. E., C. Hitchcock, T. Kowalski, J. Rajan, and R. Walker, "Automatic datapath synthesis," *Computer*, vol. 16, no. 12, pp. 59-70, December 1983.

[Tse84]   Tseng, C., and D. P. Siewiorek, "Emerald: a bus style designer," in *Proc. 21st Design Automation Conf.*, pp. 315-321, June 1984.

[Ueh81]   Uehara, T., and W. M. vanCleemput, "Optimal layout of CMOS functional arrays," *IEEE Trans. on Computers*, vol. C-30, no. 5, pp. 305-312, May 1981.

[Ull84]   Ullman, J. D., *Computational Aspects of VLSI*. Computer Science Press, Rockville, MD, 1984.

[Wal85]   Walker, R. A., and D. E. Thomas, "A model for design representation and synthesis," in *Proc. 22nd Design Automation Conf.*, pp. 453-459, June 1985.

[Wei67]   Weinberger, A., "Large scale integration of MOS complex logic: a layout method," *IEEE Journal of Solid-State Circuits*, vol. SC-2, no. 4, pp. 182-190, December 1967.

[Wes81]   Weste, N., "MULGA — An interactive symbolic layout system for the design of integrated circuits," *The Bell System Technical Journal*, vol. 60, no. 6, pp. 823-857, July-August 1981.

[Wes85]   Weste, N., and K. Eshraghian, *Principles of CMOS VLSI Design: A Systems Perspective*. Addison-Wesley, Reading, MA 1985.

[Win82]   Wing, O., "Automated gate matrix layout," in *Intl. Symposium on Circuits and Systems*, pp.681-685, 1982.

[Win85]   Wing, O., S. Huang, and R. Wang, "Gate matrix layout," *IEEE Trans. on CAD*, vol. CAD-4, no. 3, pp. 220-231, July 1985.

[Wol83]   Wolf, W., J. Newkirk, R. Mathews, and R. Dutton, "Dumbo, a schematic-to-layout compiler," in *Proc. of 3rd Caltech Conf. on VLSI*, pp. 379-393, 1983.

[Won86]   Wong, D. F., H. W. Leong, and C. L. Liu, "Multiple PLA folding by the method of simulated annealing," in *Proc. Custom Integrated Circuits Conf.*, pp. 351-355, May 1986.

# Chapter 8

# Layout Analysis and Verification

**Thomas G. Szymanski**

**Christopher J. Van Wyk**

*AT&T Bell Laboratories*
*Murray Hill, New Jersey*

## 8.1 Introduction

The focus of most chapters of this book is on circuit *synthesis* — creation or improvement of the layout of an integrated circuit. In contrast, the present chapter is concerned with layout *analysis* — recovering properties that the chip would have if it were manufactured as specified by a given layout. The continually increasing size of chips (measured in either area or number of transistors), and the waste involved in fabricating chips that do not work, make layout analysis an important part of physical design automation.

Imagine writing a program in a language that is subject to frequent changes, whose compilation lasts several weeks, costs tens of thousands of dollars, and produces object code that is very difficult to debug. The sensible approach in such a situation would be to do as much as possible to ensure the program's correctness before submitting it to compilation. This is precisely the role of layout analysis in chip design. The "language" is the rules of the technology in which the chip will be manufactured; even if they are stable during the design of a single chip, the rules are likely to be different for chips designed several months apart. "Compilation" is the process of chip fabrication. The "object code" is the chip itself: once testing shows that it works, the designer's job is done; but if it doesn't work, only a limited number of ways are available to find out what is wrong.

We discuss three kinds of layout analysis tools in this chapter:

*   Design-rule checkers, which detect violations of rules that govern the technology in which the chip is to be made.

*   Netlist extractors and comparators, which tell the designer what circuit is expressed by a layout, for comparison to another representation of the circuit.

*   Parameter extractors, which provide information about electrical characteristics that can be used for full-scale simulation.

Simulation can be performed at a variety of levels; common possibilities, in order of increasing amount of detail, include functional, gate-level, transistor-level, timing, and analog waveform or circuit simulations. True circuit simulation is usually so costly that it is used only on small pieces of a chip layout whose characteristics are thought to be critical to overall performance. Moreover, no level of simulation has much practical use until a design is approved by the various layout analysis tools discussed here. We will not discuss simulation any further, since it is such a broad topic that it deserves at least one book of its own.

This chapter consists of nine sections. Section 8.2 sets the stage for a discussion of layout analysis with a discussion of the role of computational geometry. Section 8.3 describes several methods that are used to represent the geometric features of a chip. Section 8.4 presents circuit extraction problems and solutions. Section 8.5 presents design-rule checking problems and solutions. These four sections are written assuming that the input to layout analysis is the raw geometric features that will be used to fabricate the masks; any other information supplied by the user, such as hierarchical relationships among cells on a chip, is simply ignored in these sections.

In the remainder of the chapter, this viewpoint is relaxed. Section 8.6 discusses the use that layout analysis can make of the hierarchy that exists in a layout. Section 8.7 discusses the possibilities of incremental analysis that can be performed while a chip is being designed. Section 8.8 describes algorithms for comparing the extracted circuit information with an abstract representation of the desired circuit. We offer some concluding remarks and observations in Section 8.9.

## 8.2  The Role of Computational Geometry in Layout Analysis

Integrated circuit artwork depicts a chip layout as a collection of polygons of various colors, with different colors corresponding to different mask levels in the fabrication process. Layout analysis obtains information about the chip from computation on the polygons. Analysis is usually performed assuming that the chip will be fabricated exactly according to the idealized geometrical description.

This convenient abstraction is only one of many that arise in layout analysis. It involves an approximation since processing variations mean that most of the information obtained by layout analysis is only roughly true of any real chip.

The raw material for layout analysis is the geometrical information contained in a collection of mask-level polygons. In practice, the layout often contains other information used by designers, such as textual labels that tell what electrical signal is carried on each region. Since the intent of layout analysis is to discover how closely the mask geometry matches some other idea we have about the circuit being designed, the safest course is usually to derive all necessary information from scratch. (Of course, we can then use this derived information to check the correctness of any auxiliary information.) Thus, layout analysis programs must solve geometrical problems that involve polygons, such as finding the intersection of polygons, discovering which polygons contain a given point, or finding polygons that lie closer together than a given distance.

In many computational applications, a naive approach to algorithm design can render a system unworkable. This observation holds true *a fortiori* for layout analysis, because the problems are so large. An algorithm that processes input of size $n$ in $O(n^2)$ time can be acceptable for $n \approx 100$; it is unlikely to be practical on chip-sized problems with $n \approx 10^6$. Indeed, the size of layout analysis problems can force us to consider other complexity measures besides runtime. A common one is the amount of main memory required, since even an $O(n)$-time algorithm will perform unacceptably if it causes a large amount of thrashing on a computer with large virtual memory but small main memory. The time and space complexity of layout analysis operations depends on the way mask geometries are stored; we discuss several possibilities in the next section.

## 8.3  Geometric Representation Schemes

Mask geometries can be represented using a variety of schemes. No single scheme is best suited to all of the different kinds of operations that are involved in layout analysis. If it were practical to maintain several representations of a chip layout, this point would barely be worth mention. However, the amount of data required to describe a chip of moderate or large size is so great that designers of layout analysis programs usually must choose a single representation and tolerate more complicated algorithms or suboptimal performance in some parts of their systems.

One key to deciding among representation schemes is where layout analysis fits into a complete collection of chip design tools. When users want layout analysis to be an independent function that operates on geometry no matter what its source, then layout analysis tools usually accept geometry expressed in a language like CIF [Mea80]; the choice of internal representation is entirely up to the authors of the layout analysis program. On the other hand, when layout analysis is to be an integrated feature in a layout editor, like Magic [Ous85], then it can use the results of computation already done during other parts of the editor's processing; this dictates the use of the same representation as the editor employs.

Some design methods require that the polygons in a layout have edges that are parallel to one of the coordinate axes. This restriction to *Manhattan geometry* (so-called because of the layout of streets in midtown New York City) makes it easier to state layout analysis problems and to write programs that solve them, principally because the intersection of two lines in Manhattan geometry can be computed quickly and with no loss of precision. On the other hand, neither Manhattan geometry nor the variant in which edges lie at angles to the horizontal that are a multiple of 45° changes the fundamental complexity of most layout analysis problems. When possible, we avoid the assumption that layout geometry is Manhattan, because we believe that layout analysis programs should not unnecessarily restrict the freedom of circuit designers to use the best method available to do their jobs.

In our discussion of different representation methods for layers of polygons, we shall consider two tasks that are simple versions of layout analysis problems:

(1)    Form the intersection of the polygons on two layers.

(2)    Find the maximal connected region on a given layer that contains some specified point.

Problem (1) is an instance of the more general problem of forming *derived layers*, which are a Boolean combination of several original layers. Problem (2) is a simple version of finding all points to which a given point is electrically connected. Thus, the performance of different methods on these two problems offers some indication of how suitable the methods are for layout analysis.

## The Polygon Representation

Since chip artwork is described as layers of polygons, the obvious representation scheme is to use a list of polygons for each layer of a chip, with each polygon represented by a list of its vertices. It is common to insist that each polygon's vertices be given in a particular order — for example, clockwise around the boundary, so that the polygon interior lies to the right of each edge as it is traversed. This constraint simplifies processing in many parts of a system and makes representing polygons with "holes" natural: the vertices of hole boundaries appear in counterclockwise order around the hole. (See Figure 8.1.)

Since each polygon is represented by a list of its vertices, a collection of polygons with a total of $n$ vertices is represented in $O(n)$ space. The polygon representation is not biased toward Manhattan geometric features: line segments of arbitrary slope fit into the polygon framework as readily as horizontal and vertical edges do. Despite these advantages as a representation, polygons are inconvenient for computation.

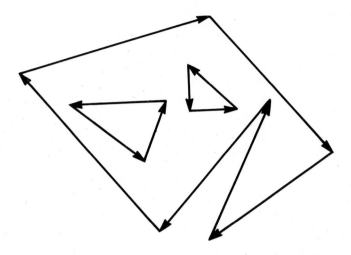

**Figure 8.1** A Polygon with Two Holes

First of all, routines that use the polygon representation must cope with degeneracies [Van84] and nonsimple paths [New80]. Even when a polygon boundary is well behaved, a simple problem like determining whether a given point lies inside the polygon requires examining all edges of the polygon. Finding the maximal connected region that contains a given point on a layer requires finding all polygons on the layer that contain the point, then forming their union; the latter problem is complicated even for two polygons [Yam72]. Finding the intersection of two sets of $k$ polygons requires finding the intersection of $O(k^2)$ pairs of polygons in the worst case; schemes that first check whether the polygons' bounding boxes intersect, in the hope of avoiding this quadratic behavior [Bar80], fail to separate the interlocking collections of polygons that are common on integrated circuits.

We know of no working system that uses the polygon representation internally in its computation, and indeed of only one proposal to do so [Bar80]. Some systems store mask information using polygons but convert them to another form before further processing [Lin76, Swa83].

### The Pixelmap Representation

The *pixelmap* representation is defined for regions whose corners have integer coordinates and whose edges are either horizontal or vertical. Position $(x, y)$ in the pixelmap corresponds to a unit cell or *pixel* $[x, x+1] \times [y, y+1]$ in the plane. Each pixel contains several bits, each of which corresponds to a mask layer; the value of a bit is 1 if the corresponding pixel is covered by a region in the corresponding mask layer or 0 if it is not. (See Figure 8.2.) Other common terms for this representation include *bitmap* and *raster*. Most machines used for constructing lithography masks use the pixelmap representation, because pixelmaps can offer a close approximation to the processing the chip will actually undergo.

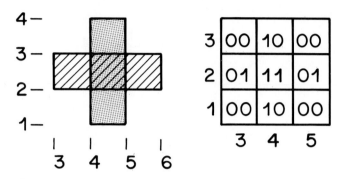

**Figure 8.2**    The two-level layout on the left is represented by the pixelmap on the right.

By using the appropriate masking pattern of bits, we can find the intersection of two layers in a pixelmap, or indeed any Boolean combination of any number of layers. In this sense all derived layers are contained explicitly in the pixelmap representation and hence are so easy to compute that layout analysis systems using pixelmaps rarely form a separate representation of derived layers.

There are two common techniques for finding the maximal connected region that contains a given point. One uses a graphics technique called *flooding* in which neighbors of the given point are visited, then their neighbors are visited, and the computation proceeds recursively until the whole region has been visited [Pav81]. (See Figure 8.3.) This method assumes that the whole pixelmap is accessible in main memory, which may not be realistic.

The other method is to process the entire pixelmap one column at a time, assigning distinct labels to pixels not known to be connected. This method requires only enough main memory to hold two columns of pixels, because each pixel column can be written to secondary storage as soon as the column after it has been labeled. However, regions that are concave with respect to the sweeping column contain pixels that are written out with different labels and are only later discovered to belong to the same region. Thus, further processing is required to adjust all labels to the correct values [Bak80a]. (See Figure 8.4.) The problem to be solved is an instance of finding connected components, for which standard algorithms are well known [Tar83a]; however, the memory required to solve the problem can be much larger than that required for the first pass, requiring space proportional to the number of pixels in the pixelmap in the worst case.

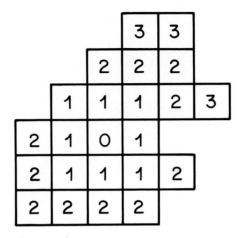

**Figure 8.3**    Flooding: the given point is labeled "0"; points first visited during the $i^{\text{th}}$ recursive call are labeled "$i$."

By definition, pixelmaps can represent only Manhattan geometries. This is a limitation on their expressive power, despite the fact that all chips are ultimately treated as if they contained only Manhattan features. The limitation arises because of a difference in resolution: while the circuit designer might work with a grid spacing of one or two microns, the fabrication machine is apt to work on a grid spacing of one or two tenths of a micron. Since diagonal lines of moderate slope on a coarse grid spacing can be closely approximated as Manhattan paths on a finer grid, features on the manufactured chip are close to the desired geometry. However, in the pixelmap representation, the cost of making the grid finer by a factor of $k$ is an increase in the number of cells by a factor of $k^2$; thus, doubling resolution quadruples the space required to store a chip description in a pixelmap, while a tenfold increase in resolution costs a factor of 100 in space.

Even if a design contains only Manhattan features, the space required to store a chip design is impressively large. For example, a one-half-centimeter square chip designed on a one-micron grid requires 25 million bits to represent each mask layer. A typical chip has several layers, so representing all of the geometry on the chip with a pixelmap is apt to require 25 to 50 megabytes. The space requirement is further increased if other information like signal names must be stored at each cell. Thus, although pixelmap algorithms are usually simple to understand and to implement, their large space requirement often makes using them uncomfortably expensive.

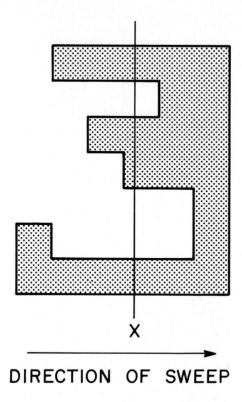

**X**

## DIRECTION OF SWEEP

**Figure 8.4**    This region is concave to the given direction of sweep: at point $x$, when only the half-plane to the left of the scanline has been examined, there appear to be three separate regions.

The pixelmap representation has been used in some systems for many years [Los79] and in other, newer systems as well [Bak80b]. The HIBAWL system uses a compressed pixelmap representation intended to take advantage of the hierarchy present in many chips [Wil80, Wil81]. Special-purpose hardware engines for accelerating the processing of pixelmaps have been described [Bla81, Sei82, Sei85]. Unfortunately, such engines impose tremendous I/O requirements on their host computers. Although several such engines have been designed, to the best of our knowledge, none has ever been built.

### The Tile- or Corner-Based Representation

Like pixelmaps, tile-based representations are usually restricted to Manhattan layouts. On a single layer, the corners of regions induce a partition of the plane into rectangular *tiles*. The partition can be formed by drawing horizontal lines across the plane through each corner and then merging adjacent tiles of the same color (*i.e.*, both inside or both outside a region) whenever the result will be rectangular. A data structure is constructed by storing information at each corner that permits access to the tiles it touches, a process called *corner-stitching* [Ous84]. (See Figure 8.5.) In the simplest case each tile simply indicates whether or not it belongs to the layer being represented; in various applications, tiles also contain information about the electrical signals they carry. This scheme is readily extended to represent $k$ layers. In this extended scheme, a *corner* is any point where two or more edges intersect. Each tile is labeled to indicate which of the $2^k$ possible subsets of layers is present.

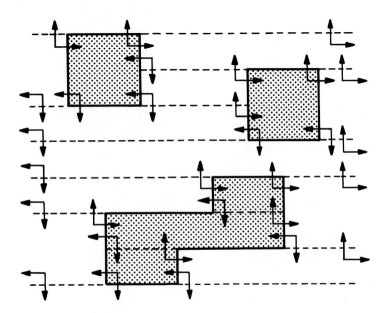

**Figure 8.5**   Corner-stitched representation of plane containing three objects. Both the objects and the space around them are partitioned into tiles. Each tile contains pointers to those tiles that share a border with it at its top right or bottom left corner.

As in the case of pixelmaps, the Boolean combination of any number of layers is available using a search for tiles whose color matches the appropriate mask. A close analog of the flooding technique for pixelmaps can be used to find the maximal connected region that contains a given point; since the recursion occurs on tiles, rather than pixels, the memory requirements for flooding on tiles are reported to be less forbidding than those for flooding on pixelmaps [Sco85]; nevertheless, the expected memory requirement is still linearly proportional to the number of tiles in the layout.

Since every tile is represented in the same amount of space (regardless of area), a tile-based representation can occupy much less space than would be used by a pixelmap. The number of tiles required to represent a layer is directly proportional to the complexity of the features on that layer: a single layer with $n$ corners occupies $O(n)$ space. However, this relationship does not mean that an entire layout can be described in linear space as in the polygon representation, since two intersecting layers contain new corners that are not present in either of the original layers. (See Figure 8.6.) In the worst case, $k$ layers of $n$ corners each require $O(k^2 n^2)$ corners in the overall representation. Under the more realistic assumption that the average edge on one layer crosses at most a constant number of edges on each other layer, the overall representation has $O(k^2 n)$ corners.

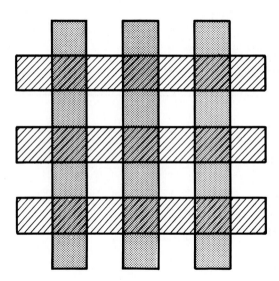

**Figure 8.6**    Each layer contains 12 corners. The tile-based representation of both layers contains 50 corners.

The tile-based representation has many attractive features for layout analysis. Much of the computation required for layout analysis using other representations is performed implicitly while a corner-stitched data structure is built. Tile-based algorithms are especially well suited to interactive applications. Extension of corner-stitching to non-Manhattan geometry has been proposed, but the feasibility and additional cost of its implementation have not been fully evaluated [Arn85].

Several chip-design programs from the University of California at Berkeley use corner-stitching and tile-based algorithms, including the interactive editor Caesar [Ous81], the design-rule checker Lyra [Arn82], and the Magic system [Ous85].  In these systems, not all mask layers are placed in the same corner-stitched structure.  Instead, only mask layers that interact strongly, for example, polysilicon and diffusion, are placed in one structure, with a separate structure used for each level of metal.  This method prevents the data structures from being polluted with irrelevant information that is an artifact of the representation.

### The Edge or Trapezoid Representation

Like the polygon representation, the edge-based representation of a mask level uses a list of edges to represent geometric features.  However, the edges are not stored in order around the boundary of each polygon but are rearranged and tagged with useful information.  An *edge* is represented by an ordered pair of its endpoints $((x_1, y_1), (x_2, y_2))$, with $x_1 < x_2$ or $x_1 = x_2$ and $y_1 < y_2$; each edge is also marked as a bottom (if the region it bounds lies to its left as one goes from $(x_1, y_1)$ to $(x_2, y_2)$) or a top.  Edges appear in the list sorted by slope within $y_1$ within $x_1$; in other words, edge $((x_1, y_1), (x_2, y_2))$ is less than edge $((\bar{x}_1, \bar{y}_1), (\bar{x}_2, \bar{y}_2))$ if

(1)    $x_1 < \bar{x}_1$, or

(2)    $x_1 = \bar{x}_1$ and $y_1 < \bar{y}_1$, or

(3)    $(x_1, y_1) = (\bar{x}_1, \bar{y}_1)$ and $(y_2 - y_1)(\bar{x}_2 - \bar{x}_1) < (x_2 - x_1)(\bar{y}_2 - \bar{y}_1)$.

The ordering is a partial order because edges with the same slope and left endpoint may appear together in any order, as illustrated in Figure 8.7.

To compute the edge representation of the intersection of two layers represented by their edges, we can use an algorithm based on *scanlines* [Ben79, Pre82, Sha76].  Scanlines have been aptly described as a reduction of a two-dimensional static problem to a one-dimensional dynamic problem: rather than consider at once all edges on both layers, we consider the edges as they cross a vertical line that scans across the plane from left to right; the sorted order of the edges on each list makes them available for this processing at exactly the right time.  A scanline algorithm uses a data structure to represent the edges that cross a vertical line as it scans across the plane; the data structure starts out empty and changes to reflect the events when a scanline encounters the endpoint of an edge (beginning or ending) or an intersection between two edges.  (See Figure 8.8.) Many choices of data structure permit a scanline that contains $k$ elements to be updated in time $O(\log(k))$ [Tar83a].  To determine for any point $p$ on the scanline whether $p$ lies in an input region, examine the edges of the desired layer that lie below $p$; $p$ lies in an input region if and only if the number of bottom edges exceeds the number of top edges.  A point lies in an output region if and only if it lies in an input region with respect to both input

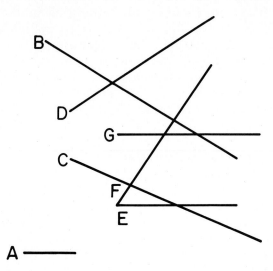

**Figure 8.7**    The edges are labeled so that their sorted order corresponds to alphabetical order.

layers. Observe that all *output edges* are subsegments of input edges and have endpoints that are either endpoints of input edges or else points where input edges intersect.

The ordering rule above makes it possible to implement scanlines to run in a space-efficient manner. An input edge is brought in from secondary storage when its left endpoint touches the scanline and is dropped from main memory when its right endpoint touches the scanline. Thus, scanlines use $O(m)$ space, where $m$ is the maximum number of edges in the layout that cross any vertical line [Lau81]. Since the expected value of $m$ is $O(\sqrt{n})$ [Ben80], the expected main memory requirement for scanline algorithms is $O(\sqrt{n})$. Although it is usually most convenient to state scanline algorithms for combinations of two layers, it is straightforward to extend them to compute arbitrary Boolean combinations of more than two layers [Bro86, Hof83, Ott85, Szy85].

Computing the maximal connected region that contains a given point is not easy in the edge representation. The objective is to assign a label to each edge so that two edges belong to the same connected region if and only if they have the same label. As was the case when sweeping pixelmaps, the fact that two edges should have the same label might not be discovered until long past the time when both edges have been written out to secondary storage. This situation is exemplified by the two horizontal edges that lie completely to the left of the scanline in Figure 8.4. Most algorithms that have been proposed to remedy this difficulty use an amount of main memory proportional to the

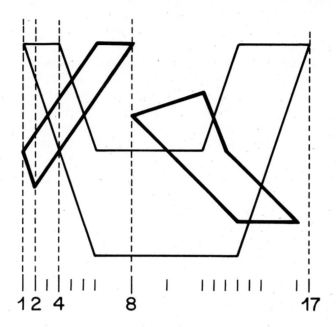

**Figure 8.8**    Illustrating a scanline running on two layers. One or more events occur at each of the marks; numbered marks are explained below. Before 1, the scanline is empty; afterward, it contains four edges. At 2, one edge ends and another starts. At 4, two edges cross, one edge ends and another starts. At 8, two edges end and two others begin. Before 17, the scanline contains two edges, afterward, the scanline is empty.

number of output edges [Bai78, Los79, Mit80, Noi81]. However, the algorithm of Szymanski and Van Wyk [Szy85] uses $O(\sqrt{n})$ space.

A *region-oriented* variation of the above edge-oriented scanlines is sometimes convenient. A set of $k$ layers (primary colors) induces a partition of the plane into areas of $2^k$ possible composite colors. A maximal connected area of uniform composite color is called a *region*. A region-oriented scanline is implemented by storing in the bottom-most edge of each region its composite color, the electrical node associated with each of its primary colors, its accumulated area, and its accumulated perimeter with regions of each of the other $2^k - 1$ composite colors. Regions are linked together so that regions that abut across an edge cut by a scanline can be found in constant time. At each stop of the scanline, a finite automaton traverses the scanline to update area, perimeter, and link information. Processing region-oriented scanlines requires roughly the same time and space as edge-oriented scanlines [Szy85].

We can make the space requirement for the edge-based representation approach that for the polygon representation by omitting vertical edges from the representation [Lau81]. Moreover, if Manhattan geometry predominates (as it does even in all-angle layouts), we can achieve a further reduction in storage requirements by creating a third kind of "edge" that represents a rectangle with horizontal and vertical edges (using the "edge" $((x_1, y_1), (x_2, y_2))$ in place of the two edges $((x_1, y_1), (x_2, y_1))$ and $((x_1, y_2), (x_2, y_2)))$ [Szy85]. Implementation of scanlines is simpler if input is restricted to Manhattan geometry [Gup83, Phi85, Wid86]; however, scanlines can be implemented to accept all-angle geometry while incurring little performance penalty when the input contains mostly Manhattan edges [Szy85].

The edges output by a scanline algorithm are sorted in increasing order of the $x$-coordinate of their right endpoints, instead of the original ordering on their left endpoints. This means that the output of one geometric operation cannot be directly input to a subsequent operation. It was suggested [Bro86] that a right-to-left scanline can be used for this subsequent operation, but this observation does not help when some operands are sorted left-to-right and others are sorted right-to-left. It is therefore necessary to sort the output edges between operations. The right endpoint order of the output edges can be exploited to perform this sort in expected $O(\sqrt{n})$ space [Szy85].

A related representation uses trapezoids whose parallel edges are parallel to one of the coordinate axes — for example, trapezoids with two vertical edges [Mit80]. Algorithms that work on edge-based representations are easily transformed into algorithms on trapezoids.

Many layout analysis systems use the edge representation and scanline algorithms. Baird is usually credited with the earliest work [Bai78]. ACE [Gup83], Goalie [Szy85], HEX [Hof83], the Polygon Package [Noi81], and NCA's system [Per85] all use some variation of these techniques. Carlson and Rutenbar have proposed special-purpose hardware to accelerate the processing of artwork represented using edges [Car86].

## 8.4  Circuit Extraction

The title *circuit extraction* includes a broad class of layout analysis problems. The fundamental problem is *connectivity extraction*, which derives a list of interconnections among the terminals of transistors from a layout description. There are several kinds of *parameter extraction*, which augment the basic connectivity information with measurements of features that are related to the (analog) electrical characteristics of the chip. In the discussion of circuit extraction, the examples are in MOS technology. For the most part, modifications to handle bipolar circuits are straightforward.

Consider the problem of finding transistors. Transistors are formed by intersecting the polysilicon and diffusion layers; their type depends on the presence or absence of different kinds of implant or tub. Thus, finding transistors is analogous to example geometric operation (1) in Section 8.3 above.

Most circuit extractors treat two points (on the same or different layers) as electrically connected if they lie in the same region of a single layer or if they can be joined by a sequence of regions on several layers that are connected explicitly by contact windows. A common circuit extraction operation is to find maximal regions of electrically connected points, more commonly called *nodes*. This operation involves labeling the contents of each layer so that items belong to the same node if and only if they have the same label. This is a generalization of geometric operation (2) in Section 8.3 above.

**Connectivity Extraction**

The output of connectivity extraction is a list of transistors on the chip, together with the node numbers on each transistor's gate, source, and drain. This transistor list is adequate for checking the logical correctness of the circuit. In order to check analog characteristics of the circuit, it is necessary to extract parasitic capacitances and resistances and transistor size information, as described in succeeding sections.

The first step in connectivity extraction is to create derived layers that correspond to transistors of different kinds and to electrically connected regions on single layers. In the bitmap and tile-based representations, these derived layers are immediately available. To illustrate the creation of derived layers using the edge representation, suppose the artwork for an nMOS chip includes the following six levels: *Dmask*, the diffusion mask, *Pmask*, the polysilicon mask, *Mmask*, the metal mask, *Cmask*, contact windows from metal to an underlying layer, *Bmask*, buried contact windows between polysilicon and diffusion, and *Imask*, the depletion transistor implant. Then we would create five derived layers as follows:

$$trans \leftarrow Dmask \textbf{ and } Pmask \textbf{ and not } Bmask$$

$$dwires \leftarrow (Dmask \textbf{ and } Bmask) \textbf{ or } (Dmask \textbf{ and not } trans)$$

$$PDcuts \leftarrow Pmask \textbf{ and } Dmask \textbf{ and } Bmask$$

$$MPcuts \leftarrow Mmask \textbf{ and } Pmask \textbf{ and } Cmask$$

$$MDcuts \leftarrow Mmask \textbf{ and } Dmask \textbf{ and } Cmask \textbf{ and not } Pmask$$

Regions in layer *trans* are transistor channels, that is, places where polysilicon crosses diffusion outside of a buried contact region. Conducting diffusion regions are represented in layer *dwires*. Files *PDcuts*, *MPcuts*, and *MDcuts* contain precisely the places where materials of the appropriate types make electrical contact.

The next step is to assign globally consistent signal labels to the items on each conducting layer that belong to a node, using the contact windows to merge signals between layers. Algorithms for this step are generalizations of the algorithms described in Section 8.3 for labeling connected regions on a single layer.

The final step in connectivity extraction is to find for each transistor the signal labels on the nodes that are its terminals. This requires examining all regions that abut a transistor region. (See Figure 8.9.) This operation is straightforward in pixelmap and tile-based representations. It can be performed in the edge representation using region-oriented scanlines. An alternative is to find edges in the diffusion and polysilicon edge lists that overlap edges of transistors [Szy83], which is simpler to implement and can run faster than region-oriented scanlines, while still obeying the expected space bounds for scanlines [Van86]. Either method must also take account of the contents of *Imask*, which tells whether a transistor is depletion- or enhancement-mode.

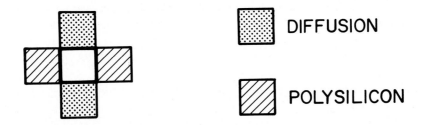

**Figure 8.9**    If edges for transistors, diffusion less transistors, and polysilicon less transistors are available, then connectivity can be extracted from overlapping edges. Each edge of the heavy-bordered transistor region overlaps either a diffusion edge or a polysilicon edge.

## Parasitic Capacitance Extraction

To extract capacitance we still treat each node as equipotential but also consider it as the terminal of one or more capacitors. Each region has a capacitance between itself and the chip substrate and also *internodal* capacitances between itself and other overlapping or nearby nodes.

Substrate capacitance can be accurately approximated as a function of the area and perimeter of each region on each layer [McC84]. Since different circuit simulators may use different models for this approximation, extraction systems merely report geometrical statistics. This is an easy job in all three representations: the information is essentially present in bitmap and tile-based systems, and it can be computed using region-oriented scanlines in the edge representation.

The output of the operation just described consists of one area-perimeter record per non-empty region (or tile) of the layout. Thus, substrate capacitance for a single node may be reported in hundreds or thousands of small pieces. If the simulator is known to use a model that is linear in area and perimeter then the program that computes geometrical statistics can accumulate the area and perimeter for regions on the same layer that belong to the same node [Szy85]. This step reduces substantially the size of the data with which the simulator must cope.

Capacitance between two nodes of the circuit is much harder to compute accurately. Internodal capacitance is not a simple function of area and perimeter; nevertheless, many systems report, and many simulators accept, geometrical statistics about regions of overlap between two nodes [Bas83, Sco85, Szy85]. Some tile-based systems seek internodal capacitance between regions that have long parallel edges close together [Sco85]. (See Figure 8.10.) Because scanlines contain only a one-dimensional slice of the chip, this operation can be awkward in edge-based systems; one possible approach is to use a scan-window [Koz81]. Careful calculation of internodal capacitance requires methods similar to those needed for resistance calculation [McC84].

**Figure 8.10**  The internodal capacitance between the leftmost two rectangles is greater (and hence more important) than that between the rightmost two.

**Transistor Size Extraction**

Analog characteristics such as the drive of an MOS transistor are a function of its channel length and width. For a rectangular transistor formed by polysilicon that completely overlaps diffusion, length is one-half of the transistor's perimeter with polysilicon, and width is one-half of the transistor's perimeter with diffusion. (See Figure 8.11.) Non-rectangular transistors are usually considered to have an "effective" length and width, which can be carefully computed using techniques like those for resistance extraction [McC84]. A simpler alternative used by many systems is to take one-half of the transistor's perimeter with polysilicon as its effective length and to divide its area by this effective length to arrive at an effective width [Gup83, Sco85, Szy85].

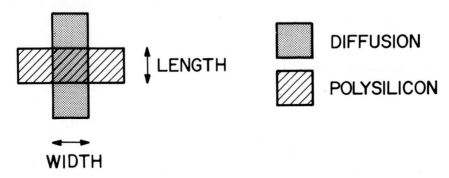

**Figure 8.11**  Channel Length and Width of a Transistor

## Parasitic Resistance Extraction

When we consider the problem of extracting resistances from a layout, the abstraction of transistors connected by equipotential nodes breaks down completely. It does not make sense to associate resistance with a node: resistance is defined between pairs of points. Thus, a node attached to the terminals of $k$ transistors gives rise to $k(k-1)/2$ resistances, one between each pair of terminals. Each of these resistances can be computed by finding the resistance between the two relevant perimeters of the region associated with the node. This computation involves solving Laplace's equation on the region, which can be done using conformal mapping [Cha70, Tre84]. The expense of this computation makes it completely impractical as a solution to the problem of resistance extraction. Moreover, it ignores interactions among currents flowing between different pairs of terminals. These considerations force us to consider approximate schemes.

One idea is to reduce the number of resistances we must compute by chopping the region into electrically isolated regions. If we add the appropriate $k-2$ junctions to a node attached to $k$ terminals, then we need to compute only $O(k)$ resistances, instead of $k(k-1)/2$. (See Figure 8.12.) We must add some correction to account for the resistance of the regions chopped out if we use this scheme.

A second way to approximate resistances uses the special case that the resistance between the two sides of length $w$ of an $l \times w$ rectangle is proportional to $l/w$. (See Figure 8.13.) Thus, if we break nodes into rectangles by

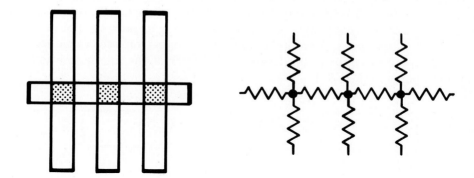

**Figure 8.12**   A region with eight terminals has 56 interconnection resistances. Making the cross-hatched junctions into new nodes splits the region into 10 electrically isolated regions and reduces the number of interconnection resistances to 10.

introducing artificial junctions, then we can compute resistance more easily. (Since artificial junctions can also chop up multi-terminal nets into several two-terminal nets, this idea subsumes the one described in the preceding paragraph.) This general scheme is used by many resistance extraction programs [Bas83, Hor83, McC84, Oza80, Put82].

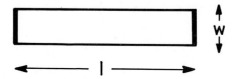

**Figure 8.13**   Resistance between the $w$-length edges of an $l \times w$ rectangle is proportional to $l/w$.

While the idea of introducing junctions at corners and places where paths branch has considerable intuitive appeal, describing it formally so that it works correctly for all cases is difficult. Furthermore, the addition of junctions creates an error in the approximation, because, in general, resistance does not obey simple composition rules. (See Figure 8.14.) Of course, if a region is decomposed along equipotential lines, then the components do obey a simple composition rule [Hor83]; since finding such equipotential lines means solving Laplace's equation, however, this observation is not helpful. A rule of thumb used by some systems is to subtract from long rectangular components a rectangle whose side is half as wide as the path [Hor83, McC84]. This rule leaves oddly shaped junction regions whose resistance must be computed by solving Laplace's equation. (See Figure 8.15.) Since regions of these shapes appear many times during resistance extraction, some systems save them in a library so that the results can be used many times [Hor83, McC84]. One paper on resistance extraction contains detailed analytical results about the error of the approximations made using this approach [McC84].

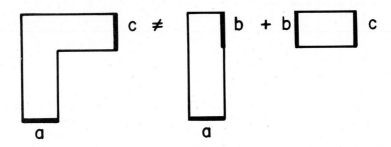

**Figure 8.14**    Simple composition rules are not accurate. The resistance from $a$ to $c$ on the left is not the sum of the resistances $a$ to $b$ and $b$ to $c$ on the right.

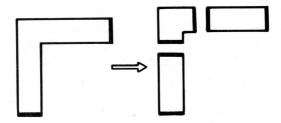

**Figure 8.15**    Subtracting half-width regions from wires leaves oddly shaped corners.

Careful resistance extraction is the hardest and most expensive problem we have discussed in circuit extraction. Indeed, most chips are manufactured without ever undergoing a complete resistance extraction because such an extraction would result in a prohibitively large network of resistors. What is necessary here is more sophisticated than simply discarding small resistances, because even small resistances can give rise to large delays if they drive large capacitances. Unfortunately, attempting to associate $RC$ delays with wires requires some knowledge of how the circuit operates. Su, Rao and Trick [Su86] have recently suggested a heuristic technique for reducing the size of the networks produced by resistance extractors.

## 8.5 Design Rules

*Design rules* are constraints that are imposed on the geometry of an integrated circuit layout in order to guarantee that the circuit can be fabricated with an acceptable yield. The rules originate in several sources.

First, there are considerations that arise from the basic physics of semiconductor circuits. For example, a wire that is too narrow will act like a fuse and might melt during normal circuit operation. Diffusion wires are surrounded by depletion regions that will cause a short circuit if they overlap. A MOSFET transistor channel that is too short will leak current and be impossible to turn off. Thus, we need rules that specify the minimum width and separation of wires and the minimum channel length of transistors.

The fabrication process also gives rise to many constraints. The photographic resist has some maximum resolution that determines the minimum size of features that can be reproduced on the wafer. The various etching techniques used to transfer patterns to the wafer can only be controlled within a certain tolerance. Features created by different mask layers can be misaligned by an amount that depends on not only the alignment tolerance of a mask with a target but also the sequence of alignments used during fabrication. For example, suppose that the contact cut mask is aligned to a polysilicon feature, and then the metal mask is aligned to a contact cut feature. If each alignment can introduce an error of $a$, then the relative position of a metal feature to a polysilicon feature could be off by as much as $2a$.

By considering all of these factors simultaneously, we can derive a set of guidelines to be observed in a layout. For example, consider a contact window between a metal wire and a polysilicon wire. If the window misses the polysilicon wire, it might contact some lower level or the circuit substrate, creating a fatal fabrication defect. We therefore require that such a window be enclosed, in all directions, by at least $a_{pc} + w_{poly} + w_{cut} + s$ polysilicon, where $a_{pc}$ is the alignment tolerance between the polysilicon and contact window masks, $w_{poly}$ and $w_{cut}$ are the maximum expected variations in the width of features created on the corresponding levels, and $s$ is a safety factor thrown in for good measure. Guidelines produced by such an analysis are overly pessimistic, because they assume that all manufacturing parameters are simultaneously maximally deviant; it is common practice to reduce the requirement in the interest of improving layout density at some expense in chip yield.

The reader is undoubtedly familiar with basic width, spacing, enclosure, and extension rules, such as those depicted in Figure 8.16. These basic rules, which only use local context and involve but one or two mask layers at a time, are necessary parts of every set of design rules and are sufficient to express such simplified rule sets as the Mead-Conway rules.  More aggressive rule sets frequently include *conditional design rules*, which involve either non-local context or multiple mask layers.

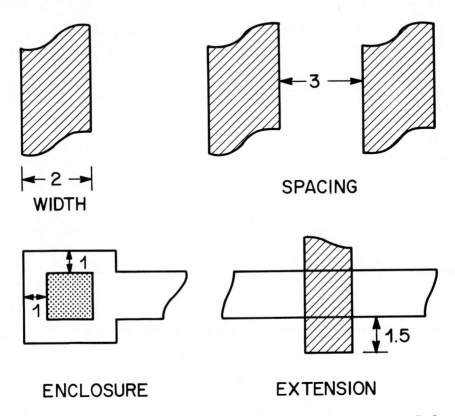

**Figure 8.16**  Typical Width, Spacing, Enclosure, and Extension Rules

Some conditional rules depend on electrical connectivity information.  Consider two metal wires.  If these wires are part of the same electrical node, then a short between them would not affect the operation of the circuit.  Therefore, the spacing requirement between electrically connected wires can be smaller than that between disconnected wires.  However, the spacing requirement cannot be completely removed, because doing so would create small features, or slivers, on the mask, in the resist during processing, or on the wafer.  These slivers are forbidden because they have a tendency to break off and come to rest in inconvenient places during processing.

Other conditional rules contend with the non-planar nature of the wafer surface during fabrication. Hills and valleys caused by the presence or absence of material in lower layers affect the precision with which structures can be built on higher layers. For example, a contact cut between wires on different layers might be manufacturable only when both wires are flat near the cut. This situation can lead to a design rule that prohibits edges of polysilicon regions near a cut between two layers of metal. The *metal reflection* rule is similarly motivated. Metal is typically patterned by coating the entire surface of a wafer with metal, covering it with a photographic resist, exposing the resist to ultraviolet light through a mask, developing the resist, and then removing whatever metal is not protected by resist. If the surface of the metal layer is uneven, light can reflect sideways off the side of a metal hillock, causing the unintended exposure of some portion of the resist as shown in Figure 8.17. In effect, this phenomenon makes metal wires wider wherever a metal edge lies close to a polysilicon edge. We can compensate for this phenomenon by making the spacing requirement for metal wires be a function of the proximity of polysilicon edges as shown in Figure 8.18.

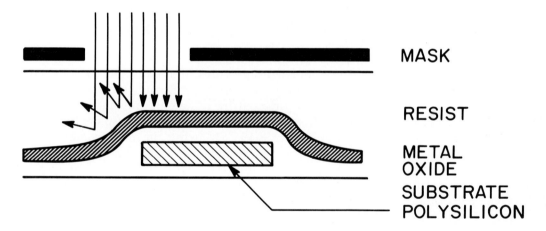

**Figure 8.17** Metal reflection during processing. Light shining through the photomask glances off the side of the metal hillock, thereby exposing part of the resist that should have been unexposed.

Yet another non-local design rule depends on the length of parallel wires. Such a rule might require that parallel wires have to be spaced apart more widely if their common length exceeds some threshold.

These and many other considerations give rise to a tremendous variety of design rules. Fortunately, the full complexity is only necessary for such critical layouts as memories. For most other designs we are usually willing to sacrifice

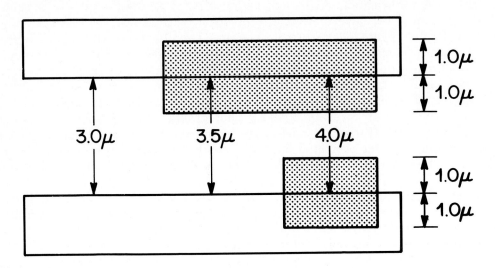

**Figure 8.18**   Metal reflection rule.  The presence of a polysilicon edge parallel to the metal anywhere in the shaded region increases the spacing requirement.

layout density in the interest of simpler design rules.  For example, in nMOS circuits the size of the depletion region surrounding a diffusion wire grows with the potential of the wire and is also affected by a surrounding buried contact region.  This gives rise to nine different spacing rules between diffusion wires as shown in the table at the left of Figure 8.19.  By using the most conservative numbers in each row, we can produce the simpler set of spacing rules in the middle table.  Alternatively, we could use the drastically simplified rule shown in the rightmost table.  The trade-off of rule set complexity against layout density affects design time and cost because more complicated design rules are harder to check automatically.

The representation, nature, and origin of the mask data are major factors in selecting a design rule checking method.  The data might be expressed as raw polygons, edge files left over from a circuit extraction, or tiles from an interactive layout editor like Magic.  The data might be Manhattan or all-angle and might or might not be annotated with electrical node numbers.  Another major factor in the choice of method is the kind of check to be performed.  More intricate industrial design rules require greater programmability than the simpler Mead-Conway rules.  Even the metric to be used is relevant.  Some rules are most naturally expressed in the *Euclidean metric*, that is, the distance between

| # wires in buried regions | # wires at gnd | | |
|---|---|---|---|
| | 0 | 1 | 2 |
| 0 | 4.00 | 3.50 | 3.00 |
| 1 | 4.75 | 4.25 | 3.75 |
| 2 | 5.50 | 5.00 | 4.50 |

| #wires in buried regions | |
|---|---|
| 0 | 4.00 |
| 1 | 4.75 |
| 2 | 5.50 |

| |
|---|
| 5.50 |

**Figure 8.19** Three Different Wire Spacing Rules for Diffusion Wires

two points $(x_1, y_1)$ and $(x_2, y_2)$ is $\sqrt{(x_1-x_2)^2+(y_1-y_2)^2}$. Other rules, particularly those that involve inter-layer alignments, require the *orthogonal distance metric*, that is, $\max(|x_1-x_2|, |y_1-y_2|)$.

One principle, however, is common to all the methods. As shown in Figure 8.20, design rule checks must be based on geometric regions, not mask polygons. This principle implies, at the very least, that a design rule checking system must be able to compute the union of overlapping polygons. Since it is scarcely any harder to perform arbitrary sets of Boolean operations, most design rule checkers have this capability and use it to good advantage.

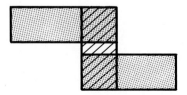

**Figure 8.20** False error reports can be caused by checking polygons rather than regions. Although the three rectangles all lie on the same mask level and form a legal jogged wire, a pairwise comparison between the two shaded rectangles appears to be a spacing violation.

Algorithms for checking design rules are frequently based on the observation that if a rule is satisfied at all the corners* of a design, then it is satisfied everywhere. This observation follows from the geometrical fact that the shortest line between two non-intersecting polygons must pass through a vertex of one of

---

*Recall that a corner is the intersection of two or more edges of the layout.

the polygons. Thus, most design rules can be checked by examining only the corners of the layout.

In the rest of this section, we shall discuss the various approaches that have been used in design rule checking systems. Roughly two distinct classes of methods are in common use: polygon algebra methods and pattern-driven methods. As a running example, let us consider a rule that requires that every contact window to the polysilicon level be surrounded by at least $2\mu$ of polysilicon under the orthogonal distance metric. We will use the symbols $P$ and $C$ to denote the polysilicon and contact window levels, respectively.

## Polygon Algebra Design Rule Checking

A popular method for checking design rules employs sequences of geometric operations. Just a few of the many systems using this approach are described in [Hak80, Lin76, Koz81, Swa83, Szy85] and [Yos77]. The Boolean operations described earlier are important components of these sequences as are the operations described below. For the moment, we shall provide only loose definitions. The *grow* operation can be used to increase the size of all the regions in some set by a specified amount. The *clearance* operation can be used to verify that all edges of a given type are at least some specified distance from all edges of some other type; its output contains a region marking the site of any such violation.

With these additional operations, we can code our enclosure test for polysilicon contacts in two ways. One way involves the sequence

$$PC \leftarrow P \textbf{ and } C$$
$$GPC \leftarrow \textbf{grow}(PC, 2)$$
$$E \leftarrow GPC \textbf{ and not } P$$

at the end of which the set of regions $E$ gives the precise locations where enclosure errors occur. An alternative way is with the simpler sequence

$$PC \leftarrow P \textbf{ and } C$$
$$E \leftarrow \textbf{clr}(PC, P, 2)$$

in which the clearance operator is invoked to check that every $PC$ edge is at least $2\mu$ from every $P$ edge.

It is very useful to express the violations detected by a checking operation as a set of geometric regions. This method allows the errors to be displayed graphically and also permits spurious error regions to be filtered out by subsequent processing. For example, although the spacing required between diffusion regions that belong to different electrical nets might be $3\mu$, the minimum length of a transistor channel might be only $2\mu$. Thus, the source and drain regions of transistors appear to violate the diffusion wire spacing rule. Rather than deluge the user with false error messages, we can filter the errors with the sequence shown below.

$$E \leftarrow \textbf{clrn}(DW, 3)$$

$$F \leftarrow E \textbf{ and not } T$$

Here the sets $T$ and $DW$ are, respectively, transistors and diffusion wires with electrical node numbers. The **clrn** operation checks for electrically distinct regions that are too close. The regions of set $E$ contain every point that lies on a line segment of length less than 3 whose endpoints lie on the edges of $DW$ regions that belong to different electrical nodes. The **and not** operation removes false errors across transistor channels.

**Implementing the Grow Operation.** There are at least three distinct ways to define the grow operation. Shape $A$ in Figure 8.21 was obtained by bisecting all angles in the original shape and moving all edges outward by the desired distance, while keeping all vertices of the shape on the original angle bisectors. This apparently natural concept is not particularly useful for design rule checking because it overextends acute angles in a way that has no correspondence to any of the physical origins of design rules. Shapes $B$ and $C$ were obtained by taking the set of points that are within distance $d$ of the original region according to the orthogonal distance metric and the Euclidean metric. Let us refer to these as the orthogonal and Euclidean grow operations.

The result of an orthogonal grow operation depends on not only the shape of the region being grown but also the region's orientation. The operation captures quite well the distortions induced in a pattern by mask misalignments because masks are typically aligned independently in $x$ and $y$. An easy way to implement the orthogonal grow is to decompose the input region into trapezoids, orthogonally grow the trapezoids, and then remove overlaps from the result.

Although the Euclidean grow is significant in design rule checking, it has the unfortunate property that the result is a circular arc polygon. Although representations can be devised for such shapes and algorithms can be constructed to perform geometric operations on them, the performance penalties over straight line polygons are quite severe. Fortunately, most tests can be expressed in terms of the clearance operator, and so we rarely need to perform a Euclidean grow. In those situations in which it cannot be avoided, a polygonal approximation can be substituted after decomposing the figure into trapezoids as described above for the orthogonal grow.

It should also be noted that a grow operation is frequently needed to produce the actual physical masks used in fabrication. In this context, the operation is called a *mask compensation* and is necessitated by processing considerations. For example, in order to make an aluminum wire $w$ microns wide using an isotropic etching process, we might cover the wafer with aluminum, lay down

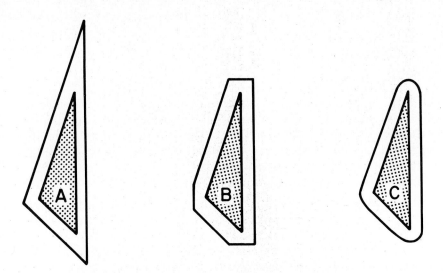

**Figure 8.21**  Three definitions of grow.  *A* is produced by extending parallel lines, whereas *B* and *C* are based on the orthogonal and Euclidean distance metrics, respectively.

a strip of resist $w + 2c$ microns wide, and then etch away the unprotected metal. Because of boundary effects, the wire turns out to have the desired width.

**Implementing the Clearance Test.**   The clearance check operation can be viewed as a test applied to all pairs of edges from the layout.  Of course, implementing the operation this way would be impossibly slow, so alternatives must be considered.  The most powerful method uses *bins*.  Suppose that we partition the plane into a grid of $c \times c$ bins, where $c$ is the desired clearance distance, and place a copy of every region edge from the layout in every bin that it crosses. Any pair of edges that come within distance $c$ of each other will, at some point, either pass through the same bin, or else pass through adjacent bins, as shown in Figure 8.22.  Since the number of edges in any bin is expected to be quite small, a reasonable way to look for clearance violations is to compare each copy of an edge in bin *B* with all the other edges in *B* and the eight bins adjacent to *B*.  This primitive bin-based algorithm has two problems.  It requires that all edges be resident in main memory simultaneously and it makes many copies of long edges.

The memory requirement can be reduced by using the basic scanline idea. Only three adjacent columns of bins are needed in memory at one time.  Checking is restricted to endpoints that lie in the bins in the middle column.  Each endpoint is compared with edges in the nine relevant bins.  When a column has

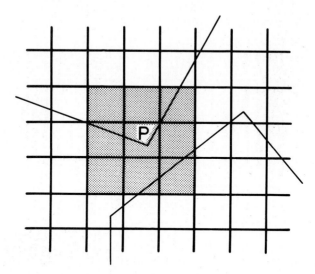

**Figure 8.22**  Clearance checking with $c \times c$ bins.  Any edge that passes within distance $c$ of corner $P$ must pass through one of the nine shaded bins.

been completely examined, we discard the leftmost column, read in a new rightmost column, and repeat the process.

Edge replication can be avoided by representing the three columns of bins with a binary tree.  Each horizontal row of bins constitutes a leaf of the tree. An internal tree node corresponds to the vertical interval of bins spanned by its descendants.  An edge is placed in the lowest tree node that contains the $y$ value of its two endpoints, as shown in Figure 8.23.  As a consequence, horizontal edges are stored at the leaves of the tree, but all other edges can appear anywhere.  Shifting one column out of memory while bringing the next column in is implemented by removing the old edges (whose right endpoint lies in the left column before the shift) from the tree and inserting the new ones.  Each edge is handled twice, once at each endpoint.  To check an endpoint of an edge, we start at the leaf of the tree that contains the endpoint and examine every edge stored at every tree node along the path from that leaf to the root of the tree; in addition, edges in some nodes that are immediately adjacent to nodes on the path must be examined; further optimizations and details for maintaining this structure have been given by Szymanski and Van Wyk [Szy85].  A clearance check performed on $n$ edges has an expected storage requirement of $O(\sqrt{n})$ and an expected running time of $O(n \log n)$.  In practice, the algorithm is I/O bound, checking edges faster than they can be read into memory.

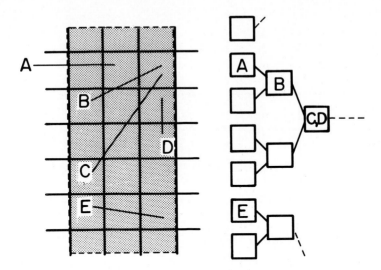

**Figure 8.23**   A binary tree representing nested vertical intervals of bin-rows. An edge is stored at the lowest node that contains both of its endpoints.

### Pattern-Driven Design Rule Checking

Many design rule checkers operate by matching geometric patterns to regions of the layout. Each pattern is associated with a set of constraints that must be satisfied at every location matched by the pattern. Most pattern-driven design rule checking systems are restricted to Manhattan geometry. Although the extension to all-angle geometry is possible, it requires cumbersome notation and is expensive to implement. Pattern-driven systems are most naturally classified in terms of the kinds of geometric patterns supported. In this section we shall discuss pixel-based patterns, corner-based patterns, and edge-based patterns. This order reflects not only the historical order of their development but also their power and utility as well.

**Pixel-Based Patterns.**   Most design rule errors can be detected using only local information from the layout. Accordingly, consider a pixelmap representation of the layout, and pass a small square window over the pixelmap so that every pixel is at some time the lower left corner of the window. At each such location some combination of pixels appears in the window. Since the number of such combinations is finite, a table lookup can immediately reveal whether the window contains a rule violation.

Consider our rule for a contact to the polysilicon level, and suppose that the resolution of our bitmap is $0.5\mu$. Then this design rule can be checked with a 5×5 window. If both polysilicon and contact are present at some pixel within the window, then polysilicon must be present at all the pixels of the window.

The most extensively used program using bitmap patterns is probably the program written by Clark Baker [Bak80a, Bak80b]. One of the most serious shortcomings of the method is that the amount of storage needed for representing the table of legal patterns grows exponentially with the number of pixels in the window. In practice, windows can be no larger than about 4×4. This operation implies that all edges in the artwork must lie on a rather coarse grid, and that all distances that appear in the design rules must be small multiples of this grid size.

Eustace and Mukhopadhyay proposed a way to avoid the use of a window [Eus82]. Their idea was to label the pixels of the pixelmap with the states assumed by a two-dimensional finite automaton that traverses the pixelmap in a regular fashion. In essence, the state associated with a pixel is determined by its mask layers and the states of its neighbors below it and to its left. This allows information to be propagated over a much wider area than is practical with a window. Although this technique does keep table sizes manageable, a pixelmap representation requires such high resolution in order to be useful that the number of pixels is extremely large and computing times are unacceptable.

**Corner-Based Patterns.** The Lyra [Arn82], Mart [Nel85], and Leo [Arn85] systems specify design rule patterns in terms of the mask layers present in the immediate vicinity of corners of the layout. A pattern for the polysilicon contact enclosure rule is shown in Figure 8.24. Each rule has two parts, a *pattern part* and a *constraint part*. A pattern matches a corner if its origin can be placed at the corner in such a way that the layers present in each quadrant satisfy the predicates given in each quadrant of the pattern. If a rule's pattern matches a corner of the layer, then the rule's constraint part is checked against the corner. In general, pattern predicates can specify any Boolean combination of mask layers. Constraints can check for the presence or absence of any desired Boolean combination of layers within a rectangle of fixed specified size that is located at some specified position relative to the corner. The system automatically considers rotations and reflections during the matching process.

Unfortunately, the pattern given in Figure 8.24 is not enough to enforce our example design rule. The layout shown in Figure 8.25 satisfies the pattern but violates the design rule. The problem here is that the rule we are trying to enforce is more naturally thought of in terms of edges of regions than corners. Fortunately, the addition of another pattern remedies the immediate problem. Together, the two patterns specified in Figures 8.24 and 8.26 suffice for our example rule. Evidently the patterns needed for checking design rules can be quite subtle. All of the corner-based systems mentioned in this section include powerful preprocessors to assist in encoding design rules.

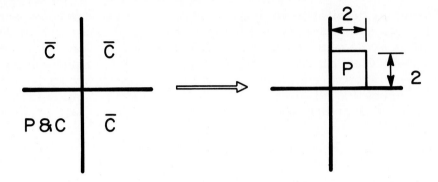

**Figure 8.24**  Corner pattern for polysilicon contact enclosure. The pattern matches any outside corner of a contact window polygon provided that polysilicon is also present under the cut.

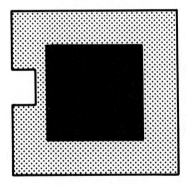

**Figure 8.25**  Although all four corners of the contact cut in this layout satisfy the rule given in Figure 8.24, the intent of the enclosure rule is violated in the vicinity of the notch.

A corner-based design rule checking program needs to find the corners of a layout, determine which patterns apply at each corner, and check the corresponding constraints. Lyra divides the plane into small bins and associates with each bin a list of mask rectangles that intersect it. Corners are found by intersecting the rectangles that overlap the bin in question. Constraints are checked by examining those rectangles within the bin that have the type specified by the constraint and overlap the constraint rectangle in the rule. Mart is driven by a scanline and discovers corners in the usual manner. All corners that are immediately behind the scanline, that is, within a distance

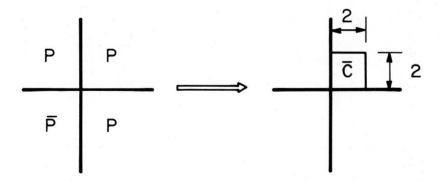

**Figure 8.26** This pattern matches inside corners on the polysilicon layer, thereby preventing notches from occurring too close to a contact window.

equal to the largest distance in any design rule, are also retained in memory. These saved corners are used in constraint checking whenever a pattern is matched. A list of pending checks in front of the scanline is maintained and applied as the scan progresses.

Fast pattern matching at corners is the key to speed in all corner-based systems. Both Lyra and Mart employ a rule compiler to generate various tables that may then be indexed by bit vectors to find applicable rules. These bit vectors are constructed at each corner and specify the mask layers present in each quadrant. The quadrant bit vectors are then OR'ed together to give a corner bit vector. The corner bit vector is used to index a primary table, and the quadrant bit vectors are used to make further decisions.

Corner-based systems easily handle directional rules like transistor extensions. On the other hand, they have a much harder time with conditional rules that involve the interaction of multiple non-coincident edges, such as metal reflection rules. In order to correct this deficiency, the Leo system includes a grow operator that can be used to create additional derived layers for use in pattern checking. Although corner-based systems traditionally employ the orthogonal distance metric, the Euclidean metric could be implemented at small additional cost.

**Edge-Based Patterns.** The Magic [Tay84] layout system extends the corner-based methods described above by allowing constraint checks to be performed perpendicular to the entire length of one of the edges that runs into a corner. The viewpoints of both the implementer and rule specifier focus on this edge. A generic pattern is shown in Figure 8.27. A pattern matches any edge in the layout that separates a region of type 1 from a region of type 2. Rectangle *B*, whose height is the same as that of the matched edge, but whose width is specified in the rule, is then checked for the presence or absence of some other material. In addition, if the predicate $P$ associated with the quadrant above and to the left of the edge is satisfied, then rectangle $A$ (both of whose dimensions are part of the rule) is also checked for the presence or absence of various materials. A pattern suitable for checking our example rule is shown in Figure 8.28.

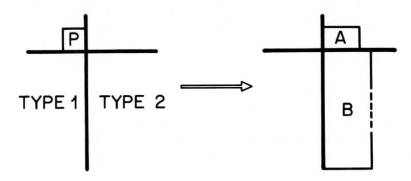

**Figure 8.27**  Generic Edge-Based Pattern as Used in the Magic System

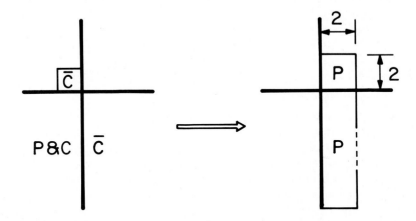

**Figure 8.28**  An Edge-Based Pattern for the Polysilicon Contact Example

Rule checking in Magic is facilitated by the corner-stitched data structure used to represent the layout. The checking algorithm is to examine every tile of the layout and partition its border into edges having a single material combination on either side. A bit vector representing this combination is then used to index into a table of patterns. Region *B* and possibly region *A* are then checked using the standard traversal algorithms for corner-stitched structures [Ous84].

The advantages and disadvantages of edge-based pattern matching are quite similar to those of corner-based patterns.

### Comparison of Design Rule Checking Methods

The polygon algebra approach handles all-angle geometry naturally. The chief advantage of the method is flexibility. Intricate rules that involve the interaction of many layers can be expressed in terms of just a few primitive operations. Certain rules that involve directional considerations are sometimes tedious to represent and require lengthy sequences of operations to check exactly. Indeed, the biggest disadvantage of the method is probably the CPU time required to perform all the operations needed to check a design thoroughly.

The pattern-driven checkers, although effectively restricted to Manhattan designs, promise attractive performance when driven by a layout system whose basic data structures correspond closely to those needed by the checker. The specification of patterns for the checker is an exacting task and is usually relegated to an expert. Fortunately, this job is infrequently needed and can be facilitated by macros. Sophisticated preprocessing of the generated patterns is necessary to reduce overhead in applying patterns. Although pattern-driven checkers can test directional rules more naturally than polygon algebra systems, they are unable to express some of the more intricate industrial design rules.

## 8.6 Hierarchical Layout Analysis

The artwork that describes a VLSI layout typically employs repetition and hierarchy. Hierarchical methods attempt to reduce the cost of layout analysis by exploiting this fact. The basic idea is to analyze each cell once and then reuse the analysis with appropriate modification wherever the cell is instantiated. The challenge in exploiting hierarchy in this manner is to keep the overhead low enough to justify the added implementation complexity. The difference in analysis time obtained by using hierarchical methods can vary from gaining a factor of several hundred on highly regular designs to losing a factor of five on highly irregular designs. It is unclear at the moment whether future designs will be more or less regular than those of the present. On the one hand, increasing the regularity of a design is a powerful way to produce larger designs. On the other hand, increasingly powerful synthesis tools provide a means of quickly producing designs in which each instance of a subcircuit is customized rather than replicated.

Probably, the fundamental difficulty in hierarchical analysis is caused by the fact that the properties of a cell may be drastically altered by the context in which it is instantiated. For example, a polysilicon wire run through a cell could create new transistors where it crosses diffusion wires, short nodes together where it crosses polysilicon wires, and break connections where it passes through a metal-to-diffusion contact.

We shall therefore classify hierarchical methods according to their treatment of overlapping. Two cells overlap if their boundaries intersect. In the simplest case, the *boundary* is taken to be the minimum bounding rectangle of the cell. More generally, the boundary is the minimum area, simply connected polygon that encloses all of the geometric figures that belong to the cell, including all of the subcells of the cell.

## Methods Restricted to Non-Overlapping Designs

A non-overlapping design has the property that, for every cell $C$ in the design, no subcell of $C$ overlaps either another subcell of $C$ or a primitive geometric object in $C$. Electrical connections in a non-overlapping design can only be made by abutment. Systems that require designs to be non-overlapping invariably also require that transistors not be split across cell boundaries. Designs that satisfy these restrictions can be processed in a bottom-up fashion, that is, in an order in which no cell is processed until all of its subcells have been processed.

The absence of overlaps makes hierarchical circuit extraction straightforward. Figure 8.29 depicts how the artwork for a cell that contains a few wires and transistors can be reduced to a circuit and a connection frame. The *connection frame* is the boundary of the original circuit and indicates the locations and layers of the cell's *terminals*, that is, the locations and layers where electrical nodes in the cell touch the cell's boundary. A non-leaf cell can be extracted by examining its geometric primitives and the connection frames of its subcells. The resulting circuit is the union of the circuits of the subcells and the circuit of the cell itself, possibly with some of the nodes connected together.

There is more work to do when the parasitics of the composition must also be computed. We express the capacitance of a net in terms of its area and perimeter on each conducting level. The appropriate adjustment can then be made when terminals abut by adding the areas and perimeters of the two constituents and then subtracting the common perimeter length. Analogous calculations can be used to give a crude approximation to lumped parasitic resistance.

The key to design rule checking of non-overlapping designs is the fact that only objects near a cell boundary can interact with other cells. Let $I$ be the maximum interaction distance of any design rule; that is, two objects that are at least $I$ apart cannot by themselves violate a design rule. Figure 8.30 shows a cell $A$. Two *interaction frames* of width $I$ are shown immediately inside the boundary of $A$. The purpose of interaction frames is to guarantee that enough context is always provided to avoid false error reports.

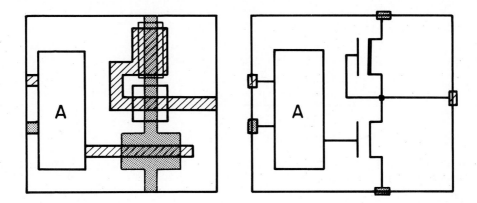

**Figure 8.29**  The cell at left contains some geometric primitives as well as one instance of a subcell.  At right is shown the result of extracting the cell: a connection frame and a circuit that includes the subcircuit corresponding to the subcell.

Cell $A$ is checked by examining all of its geometry.  Any errors found in its outer interaction frame are suppressed at this point because they could be rectified by the context in which the cell is instantiated.  For example, suppose that the polysilicon that forms the gate of a transistor extends a distance $e$ to a terminal on the cell boundary.  Although $e$ might be less than the amount required by the transistor gate extension rule, the construct would still be legal if every connection to this terminal contained enough polysilicon.  When any parent $P$ of $A$ is checked, all of $P$'s geometric primitives along with both interaction frames of $A$ are examined.  Any errors that appear in the inner interaction frame of $A$ are suppressed.

Design rules that use connectivity information can be partially accommodated by propagating connectivity upwards in the hierarchy; that is, connections made in subcells or their descendents will be known to the system, but not connections made in ancestor cells.

Scheffer [Sch81, Sch83] and Tarolli and Herman [Tar83b] have built systems that process non-overlapping designs hierarchically.  However, forbidding all overlapping is inconvenient and unnecessarily restrictive.  Consider a standard cell design in which power and ground run along the top and bottom boundaries, respectively, of each cell.  We would like to allow designers to share power rails between adjacent cells.  Overlapping is the natural way to achieve this but is forbidden to us.  Including only half of each power bus in a cell is difficult because contacts to the bus would also have to be split.  Moreover, any cell that is not paired then needs to have its rail padded out by the abutment of

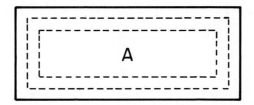

**Figure 8.30** Interaction Frames for Hierarchical Design Rule Checking

additional material. Such other recourses as putting the buses outside the cells or creating several variants of each cell are also unappealing.

Wagner [Wag85] has suggested that the user designate certain cells of a design as *checking cells*. Checking cells must be non-overlapping and may not split transistors, but all other cells are unrestricted. Hierarchical processing is performed by flattening the hierarchy between checking cells. A checking cell is processed by recursively instantiating its subcells and checking the result. For this recursion, a non-checking cell is instantiated in the usual fashion, but a checking cell is instantiated by substituting its interaction frame. This approach requires extra planning and work on the part of its users.

## Methods That Handle Overlaps by Transformation

One of the earliest programs to transform layouts by removing overlaps is the hierarchical design rule checker of Whitney [Whi81]. The program flattened a hierarchical design and fed the result to a conventional, non-hierarchical checker. The hierarchy was flattened selectively in the sense that the context of each instance of a cell was checked before expansion to determine whether some already expanded instance of that cell would produce exactly the same set of design rule violations. Figure 8.31 illustrates the technique. According to the hypothesized circumstances, $C_3$ would not have to be instantiated. Unfortunately, the method does not seem to be applicable to circuit extraction.

Several authors have given mechanical techniques for transforming a design with overlaps to an equivalent one that is free of overlaps. The technique of "disjoint transformations" suggested by Newell and Fitzpatrick [New82, Fit83] is illustrated in Figure 8.32. The original hierarchy consists of three cells $A$, $B$, and $C$ instantiated in the pattern shown on the left. The primes indicate two instances of cell $B$. The transformed hierarchy contains five cells $P$, $Q$, $R$, $S$, and $T$ as shown on the right. All geometric primitives in $B$ are copied into the cell $T$. Another copy of the geometric primitives of $B$ is partitioned by clipping to the bounding polygons of $Q$ and $R$ and distributed into those cells. Similarly, the geometric primitives of $C$ are distributed into $R$ and $S$, and the primitives of $A$ that intersect the subcells are distributed among all five new

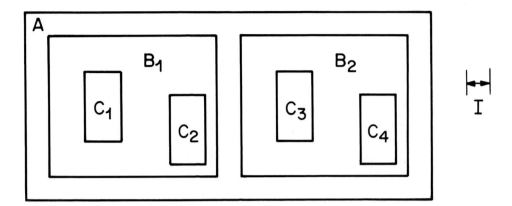

**Figure 8.31** Hierarchical design rule checking by selective flattening. Instances $C_1$ and $C_3$ have identical interactions with their environments. Instances $C_2$ and $C_4$ are within an interaction distance $I$ of the boundary of their containing cell and could react differently with the contents of $A$.

cells after appropriate clipping. In addition, any subcells of $B$ or $C$ that intersect the overlap region between $B$ and $C$ must be recursively split. At this point the first level of hierarchy has been made non-overlapping, but the interiors of the subcells may still contain overlaps. Therefore, we apply the disjoint transformation technique recursively to $Q$, $R$, $S$, and $T$.

The chief advantage of the disjoint transform method is that it allows the straightforward algorithms for the non-overlapping case to be applied to unrestricted layouts. The method has several disadvantages as well, the most significant of which is that the new hierarchy can be much larger than the original. In the worst case, the method could be equivalent to completely flattening the hierarchy. The geometric operations needed to perform the transformation can be quite involved because the new cell boundaries are generally polygons of high order. An additional complication occurs when a transistor straddles the boundary of an overlap. In this case the new boundary must jog around the transistor in order to avoid splitting it. Reporting design rule violations is awkward because errors are detected in the derived hierarchy and must be translated back to the original. A similar comment applies to assigning internal node names to an extracted circuit.

A heuristic method for removing overlaps has been proposed by Gupta and Hon [GuH83, Hon83]. It has basically the same advantages and disadvantages as the disjoint transform method.

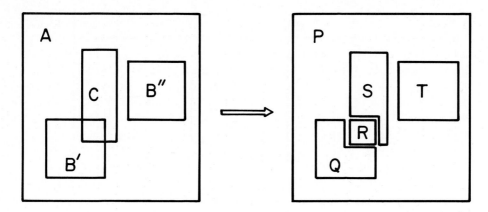

**Figure 8.32**  Disjoint transformation technique. The cell with overlapping sub-cells is transformed into another cell with non-overlapping sub-cells.

## Methods That Resolve Overlaps Directly

Johnson [Joh82] and Weise [Wei82] have proposed that design rule checking be performed by a recursive walk of the hierarchy. Suppose that cells of the design are represented by lists of rectangles and subcells. As always, we assume that by the time we need to check a given cell, all of its subcells have already been checked. The basic checking loop is to compare each element of the list against every other, using appropriate algorithmic techniques to avoid quadratic running times. Comparing two rectangles is easy, and the comparison of a rectangle with a subcell $S$ is done by comparing the rectangle recursively against all rectangles and subcells of $S$. Comparing two subcells requires recursively comparing each component of one cell against every component of the other. This method is quite fast in practice, particularly when the correct heuristics for pruning searches are employed, but it can give erroneous results when more than two objects overlap. For example, two metal rectangles might appear to be too close, whereas in fact they are both covered by a third. Evidently, multiple overlaps must be handled simultaneously.

Several systems originally written at the University of California at Berkeley describe a method that recursively instantiates areas of overlap and then processes the flattened result. These systems include the design rule checkers of Lyra [Arn85] and Magic [Tay84] and the circuit extractor of Magic [Sco85a, Sco85b]. In this method, cells are allowed to overlap so long as the number and type of transistors are not affected. Overlapping is allowed to create connections but not to break them. These restrictions are both natural and mild and could, in principle, be verified as part of design rule checking.

Even if not checked explicitly, their violation would probably be detected as some other kind of design rule error.

The key to this method is an operation that we call *instantiate under mask*. This operation takes a cell and a set of polygons called the *interaction mask* and returns the set of geometric regions formed by intersecting the mask with the geometric figures recursively generated by the cell. If the cell's circuit has already been extracted, then the results of the operation are annotated with hierarchical node names. Instantiation under mask can be implemented using most representation methods but is particularly efficient when implemented in a tile-based system like Magic.

We consider first the Berkeley algorithm for circuit extraction. The restrictions imposed above make it possible to produce a hierarchical circuit description. The circuit extracted from a cell will consist of instances of subcircuits that correspond to the subcells of the cell, primitive circuit elements that correspond to the geometric primitives of the cell, connections between nodes caused by overlaps and abutments within the cell, and adjustments to the parasitics of nodes necessitated by the composition process. Suppose that we are presented with the layout shown in Figure 8.33 and we already have the circuits for the subcells $B$ and $C$. The circuit for $A$ will certainly contain the subcircuits for $B$ and $C$, so we indicate this in the output. The geometric primitives of $A$ can then be extracted, ignoring the subcells of $A$, and added to the circuit being built. A side effect of this extraction is the creation of a representation of the geometry from $A$ with all geometry tagged with node names. Let us call this geometry the *working plane*. Next instantiate $B$ under the mask shown. This mask includes the boundary of $B$, allowing the detection of connections made by abutment. The results of this instantiation can then be merged into the working plane, detecting connections made both by abutment and overlap. Each such connection is noted in the circuit for $A$. At the same time that connections are being detected, the appropriate adjustments to parasitics are also made. This case is somewhat more complicated than in the non-overlapping case because now areas must be subtracted as well. Finally, we can instantiate cell $C$ under the mask shown and merge the result into the working plane.

The Berkeley design rule checking algorithm is very similar to the extraction algorithm, except now the masks used for instantiating portions of subcells must be extended by the maximum interaction distance $I$. The system makes the restriction that a cell, when checked by itself, must be free of design rule violations all the way to its border. Cell $A$ would be checked by initializing the working plane to contain just the geometric primitives of $A$, instantiating subcells $B$ and $C$ under the appropriate masks into the working plane, then checking the working plane for errors.

Two different sets of instantiation masks are possible for subcells. The easier set defines the mask for a subcell to be the area within distance $I$ of the boundary, within distance $I$ of any geometry in the parent cell, or within

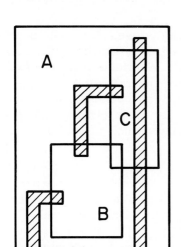
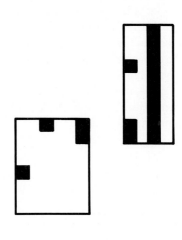

**Figure 8.33**  A cell along with subcell interaction masks for a circuit extraction. Any feature of a subcell that intersects a shaded area of the corresponding interaction mask could cause electrical connections between subcells.

distance $I$ of any other subcell of the parent. A more clever set of masks can be formed by including only those portions of the border that lie within distance $I$ of some geometry or other subcells of the parent. The difference between these two sets of masks is shown in Figure 8.34.

There are situations in which the algorithms described above can exhibit very poor performance. Suppose cell *CHIP* has two subcells called *PADFRAME* and *BODY*, with *BODY* containing almost all of the logic of the design. (See Figure 8.35.) Cells *BODY* and *PADFRAME* are checked individually during a bottom-up walk of the hierarchy. In order to check cell *CHIP*, we need to instantiate the overlapping area of the subcells *BODY* and *PADFRAME*. Since the overlap completely contains the cell *BODY*, this procedure amounts to flattening the entire chip, which both doubles the time needed to check the design rules and leads to a horrendously large main memory requirement. The problem here is that the algorithm resolves overlaps by bringing all geometry to the top level and checking it there. Shand [Sha85] has offered an alternative that he calls *level-order conflict resolution*. Rather than flatten the hierarchy in the area of a conflict, the hierarchy is expanded gradually. In this case, the conflict would be projected down one level, and eventually only the outer subcells of *BODY* and the inner subcells of *PADFRAME* would be examined in detail. Another possibility is to "dice" the area of overlap into regions of some

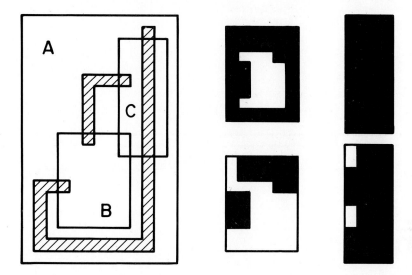

**Figure 8.34**  A cell along with two sets of subcell interaction masks for design rule checking.  The top two interaction masks can be produced by examining the features of the parent cell that intersect the subcell.  To produce the more clever bottom interaction masks, one must also examine all features of the parent cell that lie within distance $I$ of the subcell.

moderate size and restrict flattening to just those regions that contain features belonging to more than one subcell.  Since many representations of geometric data support the ability to test efficiently whether a given area is empty, this scheme can save large amounts of computation while introducing a small amount of overhead.

**Arrays of Cells**

Most layout representation languages include a direct mechanism for constructing two-dimensional arrays of instances of a single cell.  Not only is such a method more natural than the indirect method of using several layers of hierarchy for this purpose, but it can also make it easier for layout analysis tools to process the structure.  Figure 8.36 shows a 4×4 array of overlapping cells along with an interaction mask that captures all design rule errors that could occur between cells in the array, or between cells and the environment of the array.  For circuit extraction, we can consider the overlap regions of the array as being composed of seven distinct interaction tiles, as shown by the shaded regions in Figure 8.37.  Once the cell itself has been analyzed, it is straightforward to use

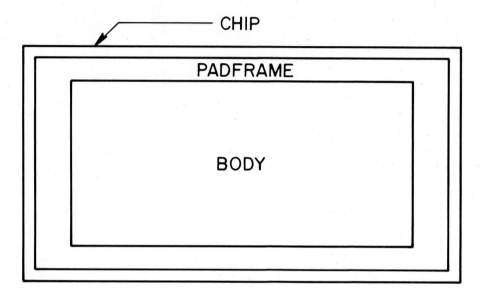

**Figure 8.35**  A Layout That Could Cause a Hierarchical Checker to Perform Badly

the analysis of the interaction tiles to produce a circuit representation for the entire array.

## 8.7  Incremental Layout Analysis

The time needed to check the design rules completely and extract the circuit for an entire VLSI layout is quite long even when hierarchical methods are used. In the final stages of chip assembly, it is not uncommon for every artwork modification to be followed by a multi-hour verification run on a mainframe computer. This tends to be a critical phase of the design schedule because the layout is nearly complete, so simultaneous work by several designers is difficult to coordinate. *Incremental* techniques can achieve great savings by limiting analysis to those portions of the layout that have actually changed since the last design rule check took place.

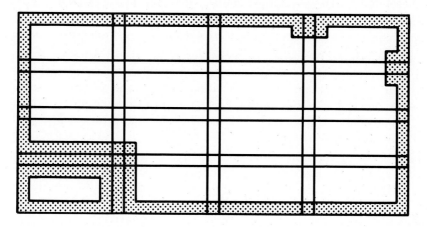

**Figure 8.36** An Array of Cells along with an Interaction Mask Appropriate for Design Rule Checking

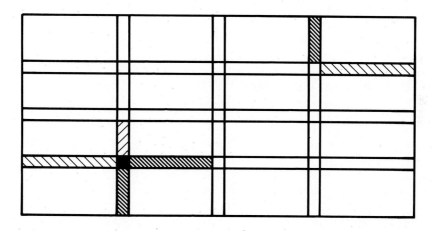

**Figure 8.37** An Array of Cells along with Seven Distinct Interaction Tiles

The local nature of most design rules makes incremental design rule checking possible. The standard technique (for example, see [Swa85] and [Tay84]) involves the cooperation of the layout editing system. Each editing change places an *unchecked region* polygon over the vicinity in which the change took place. When a design rule check is desired, only unchecked regions, extended by an interaction frame, need to be examined. Such checking may even be performed continuously in the background while the designer edits the layout. Although fast feedback is usually considered an asset, instantaneous response can be distracting when the errors that are reported are only temporary side effects of a sequence of changes being executed by the user. Therefore, a good continuous checker signals violations in an unobtrusive manner.

In a hierarchical design, it is necessary to check all parents of a cell when that cell is modified. When doing this, it is beneficial to reverse the usual process of instantiation under mask as described in the previous section. The region being checked in a cell can be projected up into each of its parents to find parental features that might interact with the changed area. The process continues all the way up the hierarchy to the root cell.

The process of incrementally extracting a circuit is considerably harder, essentially because changes to a layout can have far-reaching effects on the connectivity of the circuit. Severing a wire can cause design rule violations only in the immediate vicinity of the break. However, if such a break splits a net into two electrically separate nets, then the circuit connectivity of every transistor connected to the original wire is affected. This means that incremental extraction of a flattened circuit is tantamount to re-extracting the entire circuit. In the case of a hierarchical layout representation, when a cell is changed, not only must its contents be re-extracted, but every ancestor of the cell must also be re-extracted. In particular, this means that the root cell is processed every time any cell in the entire design changes. Unfortunately, cells near the top of the hierarchy often contain large amounts of global routing geometry, which makes extracting them very costly.

## 8.8  Netlist Comparison Algorithms

Designers are frequently confronted with different netlists representing the same design. For example, one netlist might be generated from a schematic representation of a circuit, while the other is produced by an extraction program from a physical layout of that circuit. Inevitably, the two netlists employ different names for the nets and devices of the circuit and list the objects in different orders. A *netlist comparison program* checks whether two netlists represent the same circuit. If they do, then the program produces a mapping that associates each object in one netlist with the corresponding object in the other. More importantly, if the two netlists represent different circuits, the program will pinpoint the differences.

The cross-reference mapping produced by a netlist comparison program is often even more useful than the program's ability to verify that a layout matches its schematics. During circuit design, tuning, and testing, engineers must be able to go easily back and forth between layout and schematics. A cross-reference mapping makes this possible. The mapping can also be used for back-annotating layout parasitics into the schematics.

Technically, the netlist comparison problem is a thinly veiled variant of the well-studied *graph isomorphism problem*: given a pair of graphs $G_1$ and $G_2$, does there exist a one-to-one function $f$ mapping the vertices of $G_1$ onto the vertices of $G_2$ that *preserves adjacency*, that is, $x$ and $y$ are adjacent in $G_1$ if and only if $f(x)$ and $f(y)$ are adjacent in $G_2$. The complexity of the general graph isomorphism problem is currently open [Joh81], although polynomial time algorithms are known for various restricted families of graphs, such as planar graphs or graphs in which all vertices have bounded degree. Unfortunately, the graphs that arise naturally from circuits cannot readily be coerced into any of these easily solved special cases. Moreover, a fast algorithm for graph isomorphism would not necessarily provide a useful basis for a netlist comparison program. The reason is that fast graph isomorphism algorithms tend to propagate distinctions between vertices in very clever and rapid ways. If the netlists in question are only slightly different, then aggressive propagation of distinctions can eliminate the opportunity to issue useful diagnostics. Indeed, the hard part of engineering a useful netlist comparison tool is getting the diagnostics right; the algorithmics of finding an isomorphism are comparatively straightforward.

### Graph Isomorphism Algorithms Based on Partitioning

Although many different approaches to the graph isomorphism problem have been studied in the literature (good surveys appear in [Rea77, Cor80]), the fastest practical netlist comparison programs tend to use some variation of *partitioning* or *partition refinement*.

Consider the pair of graphs shown in Figure 8.38. We first compute a *signature* for each graph vertex based on some property of the vertices that is independent of the names of the vertices, for example, degree. Even though signatures need not be unique, they can be used to partition the vertices of the graph into classes of vertices with the same signatures. For the example in question, if we partition on vertex degree we get three partition blocks consisting, respectively, of the vertices with degree two, $\{A, D, a, d\}$, the vertices with degree three, $\{B, E, F, G, b, e, f, g\}$, and the vertices with degree four, $\{C, c\}$.

This initial vertex partition can frequently be *refined* by considering the edges of the graph. First assign a name to each class of the partition. Then to each vertex $v$, assign a lexically ordered list of the names of the classes of the vertices to which $v$ is adjacent. These lists constitute another kind of signature that can be used to subdivide each partition class into subclasses. For the example at hand, let us call the blocks of the initial partition 2, 3, and 4, respectively, after the degrees of the constituent vertices. Now consider block 3, $\{B, E, F, G, b, e, f, g\}$. Nodes $B$, $G$, $b$, and $g$ would all be associated with the list $<2, 3, 4>$, the vertices $E$ and $e$ would have the list $<2, 3, 3>$, and the vertices $F$ and $f$ would have the list $<3, 3, 4>$. Thus, the block $\{B, E, F, G, b, e, f, g\}$ can be split into three subblocks, $\{B, G, b, g\}$, $\{E, e\}$, and $\{F, f\}$. The refinement process can be continued until no further changes occur.

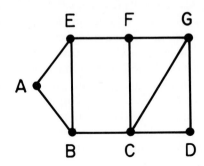 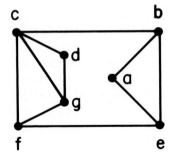

**Figure 8.38**  A pair of isomorphic graphs. Each vertex labeled with an upper-case letter can be mapped into the vertex labeled with the corresponding lowercase letter.

For the graphs of Figure 8.38, the final partition contains the blocks $\{A, a\}$, $\{B, b\}$, $\{C, c\}$, $\{D, d\}$, $\{E, e\}$, $\{F, f\}$, and $\{G, g\}$.

A number of situations may present themselves at the end of the partition refinement stage. One possibility is that each block contains exactly one vertex from each graph. In this case, the graphs are isomorphic, and the pairing provided by the final partition defines the unique isomorphism mapping between the graphs. On the other hand, if any block of the partition contains an unequal number of vertices from each graph, then the graphs are certainly not isomorphic. Unfortunately, there is a third situation that is more difficult to interpret: each partition block contains equal numbers of vertices from each graph, but some blocks contain more than one vertex from each graph. This situation, which we shall term the *ambiguous partition* situation, can arise for two distinct reasons. The first is that one of the graphs in question might have a non-trivial *automorphism*, that is, the graph could be isomorphic to itself by means of a mapping other than the identity. This is illustrated in Figure 8.39. A second reason is that the vertex invariant used for the initial partition might not be sufficiently discriminating for the graph in question. Consider two copies of the graph in Figure 8.40, a graph with no non-trivial automorphisms. If the initial partition is based on vertex degree, then all vertices are placed in the same block of the partition and no refinement can be performed. However, if the initial partition had been based on whether or not each vertex lies on a cycle of length three, then refining this partition would have yielded the unique isomorphism between the copies.

There are essentially two ways to approach the ambiguous partition situation. One is to switch to some other vertex invariant and continue the refinement process. The other approach is to break the impasse by arbitrarily picking some block of the partition, selecting one vertex from each graph,

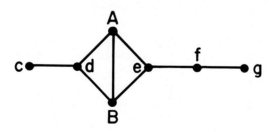

**Figure 8.39** A graph with a non-trivial automorphism. The lowercase vertices must be mapped onto themselves, but $A$ could be mapped to $B$, while $B$ is mapped to $A$. As a consequence, $A$ and $B$ will always occupy the same partition block during refinement.

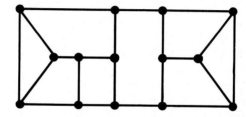

**Figure 8.40** A regular graph with no non-trivial automorphisms. An initial partition on vertex degree yields one large block which cannot be refined. An initial partition based on whether each vertex lies on a triangle, however, will be refined into a partition in which all blocks are singletons.

placing them into a new partition block, and continuing the refinement. Both of these approaches have limitations. The former is limited by the fact that no known vertex invariant is both easy to compute and guaranteed to lead to an isomorphism partition. Moreover, there is no limit on how many invariants might have to be tried before one is found that leads to an unambiguous partition. Finally, if the graphs really do contain a non-trivial automorphism, then no vertex invariant can possibly succeed in separating the vertices in question. The second approach is limited because if it eventually leads to a "not-isomorphic" partition, it is necessary to go back and repeat the process with all other pairs of vertices before concluding that the graphs are indeed not isomorphic. This backtracking can lead to exponential running time.

So long as the ambiguous partition situation does not occur, graph isomorphism can be tested quite efficiently. A straightforward implementation of partition refinement runs in $O(ev)$ time where $e$ is the number of edges in the graph and $v$ is the number of vertices. An implementation [Kub79] that is more sophisticated and that is based on ideas of Hopcroft [Hop71] runs in time $O(e \log(v))$.

### Netlist Comparison Algorithms Based on Partitioning

The partitioning method described in this section is most heavily influenced by the Gemini program [Ebe83] but is also quite close in spirit to the techniques used in many other programs [Kod86, Spi83, Tak82, Tyg85]. Other techniques [All77, Abl81, Bar84] have also been suggested.

A useful undirected graph can be constructed from any circuit. The graph contains two classes of vertices that represent, respectively, the devices and nets of the circuit. A graph edge is placed between a device vertex and a net vertex whenever the corresponding net is connected to a terminal of the device. Notice that this graph is *bipartite* because no edge directly connects two devices or two nets. Various attributes can be associated with graph vertices to correspond to labels in the layout or to specify physical parameters like net capacitance or device size. This information can be used to issue diagnostics, to create cross-reference lists, or to back-annotate circuit schematics. In addition, attributes can be associated with edges to specify the nature of the connection between the corresponding net and device. For example, we might wish to note that a given net is connected to a source or drain terminal of a transistor rather than the gate terminal.

Many different schemes have been proposed for assigning signatures for partitioning the vertices in such graphs. Some of these proposals have been fairly sophisticated such as a vector of distances from some set of labeled vertices [Wan83]. More commonly, the initial signature of a vertex is derived from such local information as the attributes of the vertex and the number and attributes of its incident edges. This suffices because global information is implicitly derived by the partition refinement process. A good choice for the initial signature of a device is simply its device type, that is, enhancement or depletion, nFET or pFET. The initial signature of a net might simply be its degree.

Signatures for refining the partition are typically based on the current signature of a vertex, the attributes of the edges incident on the vertex, and the signatures of the immediately adjacent vertices. Assuming that all signatures are integers, each transistor might be given a new signature based on the formula

$$s_{current} + c_{sd}s_{source} + c_{sd}s_{drain} + c_g s_{gate},$$

where $s_{current}$ is the current signature of the device, $s_{source}$, $s_{drain}$, and $s_{gate}$ are the signatures of the nets constituting the terminals of the device, and $c_{sd}$ and $c_g$

are weights tossed into the formula to ensure that the various device terminals are treated differently based on their function in the device. In the case of a net connected to $n$ devices, a signature might be refined through the formula

$$s_{current} + \sum_{1 \le i \le n} s_i c_i$$

where $s_i$ is the signature of the $i$th device incident on the net and $c_i$ is either $c_{sd}$ or $c_g$ depending on the role played by that connection of the net. Notice that the commutativity of addition guarantees that the above formulas are independent of the ordering chosen for terminals or devices.

At the end of partitioning, if every partition block contains either a single vertex or a pair of vertices, one from each graph, then the pairs represent matched circuit elements and the singletons represent unmatched elements for which suitable diagnostics can be issued. It is far more likely, however, that partitioning will result in the ambiguous partition situation. Backtracking is not a good way to resolve this difficulty, because of the large computing times involved. Fortunately, the most common cause of ambiguous partition blocks is the presence of non-trivial automorphisms in the circuit, rather than the sort of pathology described in Figure 8.40. Such automorphisms arise naturally in most real circuits. For example, a designer might choose to drive a long wire by replicating a gate and tying the corresponding inputs and outputs together. This leads to a non-trivial isomorphism involving the devices and any nets internal to the gates. A strategy that works well here is simply to pick an arbitrary pair of vertices from one of the partition blocks, match them by assigning them a new and unique signature, and then resume partitioning. Alternatively, vertex attributes or labels could be used to select a likely matching pair, or the program could interactively ask the user to select a pair of vertices to be matched.

One of the pitfalls of partitioning schemes as so far described is their behavior when the circuits do not match. Consider a large circuit $C_1$ and create another circuit $C_2$ by deleting just one transistor whose source terminal is $vdd$. The initial partition will separate the nets corresponding to $vdd$ in the two circuits because the nets have different degrees. The first partition refinement will then separate all transistors from $C_1$ that are connected to $vdd$ from their intended mates in $C_2$ because these devices are connected to nets that are in different blocks. Since the $vdd$ net typically has a very large number of connections, in a few more iterations partitioning will terminate with no element in $C_1$ matching any element in $C_2$. The algorithm then halts and prints a totally useless list of unmatched circuit elements.

Gemini contains an elegant method for avoiding this untoward situation. Let us say that a vertex is *suspect* whenever it occurs in a block of the partition that does not contain an equal number of vertices from each circuit. Assign a signature of zero to all suspect vertices. Because the signature functions described above sum the weighted contributions of their arguments, the connections of a circuit element to a suspect vertex have no effect on the refined signature of the element. In the example just given, the two versions of the *vdd* net are suspect in the initial partition and hence do not cause the subsequent separation of their incident devices. Thus, Gemini ignores suspect vertices until no further partitioning is possible. At that point, if more partitioning is desired, the suspect vertices can be "redeemed" and assigned non-zero signatures, thus permitting subsequent partitioning.

The running time of netlist comparison programs can be greatly improved by restricting refinement to those partition blocks that contain vertices that are adjacent to vertices whose signatures have recently changed. This method can be implemented using a queue to contain those partitions that need to be refined. This technique tends to make the partition phase of the program so fast that it takes less time than the program spends simply reading and digesting the input circuits.

### More General Netlist Comparison Algorithms

Although the previous section dealt with netlist comparison of circuits described at the transistor level, it is fairly easy to extend the methods to work at such higher levels of representation as the gate or block levels. When one does this, it is very important to treat properly equivalent or interchangeable terminals on devices. For example, all the inputs to a NAND gate are interchangeable at the logic level, although they might have quite different delay properties when viewed at the transistor level. As another example, consider the and-or-invert gate in Figure 8.41. The four inputs can be permuted in eight different ways without affecting the function computed by the circuit.

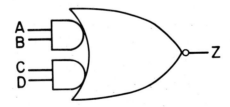

**Figure 8.41**  Inputs to an and-or-invert gate can be permuted without changing the value of the computed function. *A* and *B* can be interchanged, or *C* and *D* can be exchanged, or *A* and *B* can be simultaneously exchanged with *C* and *D*.

The easiest way to deal with permutable terminals in netlist comparison algorithms is to guarantee that the signature functions employed during partition refinement treat all legal permutations of terminals the same way. The transistor signature functions described in the previous section do just this in the case of the source and drain terminals. In the case of our and-or-invert gate, any signature function of the form

$$f(s_{current}, s_z, g(h(s_A, s_B), h(s_C, s_D)))$$

where $g$ and $h$ are commutative functions will suffice.

It is possible to extend these ideas further and consider programs for matching circuits based on functional equivalence. For example, the two circuits shown in Figure 8.42 would be treated as functionally equivalent even though they have quite different implementations at the gate level. Creating such a system is difficult because testing functional equivalence of Boolean formulae is an NP-complete problem for which no efficient algorithms are known. Accordingly, most programs for this generalized form of netlist comparison take the form of rule-based expert systems. Spickelmier and Newton [Spi85] describe one such program. The user is allowed to specify a list of subcircuit transformations such as the equivalence shown in Figure 8.42 along with the input circuits. The system then applies these transformations in an attempt to convert one circuit into the other. In general, programs based on such principles are quite slow, and it is not clear how to use them to produce useful diagnostics when the input circuits cannot be made to match.

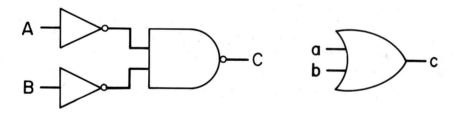

**Figure 8.42**  Functionally Equivalent Circuits

## 8.9  Conclusions

Design-rule checking and connectivity extraction tools are obvious necessities when a layout is created partly or completely by hand. Some authors hope to obviate these tools by layout editors that enforce correctness or by synthesis tools that automatically produce correct layouts with small area and suitable electrical properties. We believe that their efforts can reduce the need for layout analysis programs, but not eliminate it. First, it is difficult or impossible to prove that large programs like chip editors and other synthesis tools are correct; moreover, the individual programs are frequently modified. Thus, layout analysis tools offer an essential line of defense against fabricating chips that don't work.

Moreover, if a synthesis tool eventually does eliminate the need for some phase of layout analysis, then it must incorporate the same knowledge as that contained in a specialized layout analysis program. It is not hard to imagine this happening for design-rule checking and connectivity extraction. However, as we saw in Section 8.4, the extraction of parasitic capacitance and resistance involves many intricate geometric and numerical computations. It would be extraordinarily demanding to require that every synthesis tool be capable of performing such analyses.

We have seen that an appreciation of the methods of computational geometry is necessary for the design of layout analysis tools; however, it is not sufficient. Another important requirement in layout analysis is that we pose the problems carefully. The correct answer to a carelessly posed problem can be useless. For example, if we seek polygons that lie too close together by asking, for each edge, what edges lie within some positive distance of that edge, then the output will be enormous because *every* edge touches at least two other edges.

It is also important that a layout analysis tool produce useful output. A netlist comparator that reports "same" or "different" and then halts is much less useful than one that tries to pinpoint where the difference arises. It is equally obvious that a design rule checker should report the location of violations, and not merely their existence. It is usually useful to have the output available in some graphical form so that the errors can be highlighted on an interactive display.

There are some problems that users would pay dearly to solve, but which in fact have no algorithmic solution. For example, if we learn that a layout contains two electrically connected regions to which the designer has given different signal names, then the layout might contain a short circuit. Unfortunately, a program cannot tell us the location of such a short because it cannot deduce the signal we intended to be carried at each point in the layout; layout analysis can only report what appears in the layout, not why it differs from our intention.

## Acknowledgements

We would like to thank George Entenman, Brian Kernighan, Chi-Yuan Lo, and Michael Williams for their enlightening comments and suggestions on earlier drafts of this chapter. In addition, Mihalis Yannakakis provided us with the graph appearing in Figure 8.40.

## Exercises

**8.1**  Express the design rules for your local process using the edge-based pattern formalism. Repeat using polygon algebra.

**8.2**  Show that the graph in Figure 8.40 has no non-trivial automorphisms.

**8.3†** Show how to use connectivity extraction to check that all tub regions for a CMOS process are properly biased.

**8.4†** *Shrinking* can be defined as growing by a negative amount. Give an example to show that if orthogonal shrinking is implemented by decomposing the region into trapezoids and shrinking each trapezoid, the result can be disconnected even though the correctly shrunken region is connected.

**8.5†** Show how to implement orthogonal shrinking using Boolean operations and growing.

**8.6†** Show how minimum wire width and minimum wire separation tests can be performed using Boolean operations and growing (possibly by a negative amount). Hint: Consider the following sequence: grow by $d/2$, shrink by $d/2$, intersect with original; show that this sequence detects violations of minimum wire separation.

**8.7†** In a Manhattan design, a feature of width $d$ can be defined as one formed by two parallel edges of a region that are closer than $d$. It is sometimes useful to detect features of width less than $d$. Show how to detect such features using Boolean operations and growing.

**8.8‡** Give a consistent definition of *small feature* for non-Manhattan designs.

# References

[Abl81]    Ablasser, I., and U. Jager, "Circuit recognition and verification based on layout information," in *Proc. 18th Design Automation Conf.*, pp. 684-689, 1981.

[All77]    Allgair, R. M., and D. S. Evans, "A comprehensive approach to a connectivity audit, or a fruitful comparison of apples and oranges," in *Proc. 14th Design Automation Conf.*, pp. 312-321, 1977.

[Arn82]    Arnold, M. H., and J. K. Ousterhout, "Lyra: a new approach to geometric layout rule checking," in *Proc. 19th Design Automation Conf.*, pp. 530-536, 1982.

[Arn85]    Arnold, M. H., *Corner-based geometric layout rule checking for VLSI circuits*, University of California at Berkeley Report No. UCB/CSD 86/264, Ph.D. thesis, 1985.

[Bai78]    Baird, H. S., "Fast algorithms for LSI artwork analysis," *Journal of Design Automation and Fault-Tolerant Computing*, vol. 2, pp. 179-209, 1978.

[Bak80a]   Baker, C. M., "Artwork analysis tools for VLSI circuits," Mass. Inst. of Tech. Report No. MIT/LCS/TR-239, M.S. and E.E. thesis, 1980.

[Bak80b]   Baker, C. M., and C. Terman, "Tools for verifying integrated circuit designs," *Lambda*, vol. 1, no. 3, pp. 22-30, 1980.

[Bar80]    Barton, E. E., and I. Buchanan, "The polygon package," *CAD/CAM*, vol. 12, no. 1, pp. 3-11, 1980.

[Bar84]    Barke, E. "A network comparison algorithm for layout verification of integrated circuits," *IEEE Trans. on Computer-Aided Design*, vol. 3 no. 1, pp. 135-141, 1984.

[Bas83]    Bastian, J. D., M. Ellement, P. J. Fowler, C. E. Huang, and L. P. McNamee, "Symbolic parasitic extractor for circuit simulation (SPECS)," in *Proc. 20th Design Automation Conf.*, pp. 346-352, 1983.

[Ben79]    Bentley, J. L., and T. A. Ottmann, "Algorithms for reporting and counting geometric intersections," *IEEE Trans. on Computer*, vol. C-28, no. 9, pp. 643-647, 1979.

[Ben80]    Bentley, J. L., D. Haken, and R. W. Hon, "Statistics on VLSI designs," Carnegie-Mellon Univ. Report No. CMU-CS-80-111, 1980.

[Bla81]    Blank, T., M. Stefik, and W. vanCleemput, "A parallel bit map processor architecture for DA algorithms," in *Proc. 18th Design Automation Conf.*, pp. 837-845, 1981.

[Bon87]    Bonapace, C. R., and C. Y. Lo, "LARC2: a space-efficient design rule checker," in *Digest Intl. Conf. on Computer-Aided Design*, pp. 298-301, 1987.

[Bro86]    Brown, A. D., and P. R. Thomas, "Improving the efficiency of scanline algorithms," *CAD/CAM*, vol. 18, no. 2, pp. 91-94, 1986.

[Car86]    Carlson, E. C., and R. A. Rutenbar, "A data structure processor for VLSI geometry checking," *Digest Intl. Conf. on Computer-Aided Design*, pp. 404-407, 1986.

[Cha70]    Chawla, B. R., and H. K. Gummel, "A boundary technique for calculation of distributed resistance," *IEEE Trans. on Electron Devices*, vol. ED-17, no. 10, pp. 915-925, 1970.

[Cor80]    Corneil, D. G., and D. G. Kirkpatrick, "A theoretical analysis of various heuristics for the graph isomorphism problem," *SIAM Journal on Computing*, vol. 9, no. 2, pp. 281-297, 1980.

[Ebe83]    Ebeling, C., and O. Zajicek, "Validating VLSI circuit layout by wirelist comparison," *Digest Intl. Conf. on Computer-Aided Design*, pp. 172-173, 1983.

[Eus82]    Eustace, R. A., and A. Mukhopadhyay, "A deterministic finite automaton approach to design rule checking for VLSI," in *Proc. 19th Design Automation Conf.*, pp. 712-717, 1982.

[Fit83]    Fitzpatrick, D. T., "Exploiting structure in the analysis of integrated circuit artwork," Univ. of Calif. at Berkeley Report No. UCB/CSD 83/130, Ph.D. thesis, 1983.

[Gup83]   Gupta, A., "ACE: a circuit extractor," in *Proc. 20th Design Automation Conf.*, pp. 721-725, 1983.

[GuH83]   Gupta, A., and R. W. Hon, "HEXT: a hierarchical circuit extractor," *Journal of VLSI and Computer Systems*, vol. 1, no. 1, pp. 23-39, 1983.

[Hak80]   Haken, D., "A geometric design rule checker," Carnegie-Mellon Univ. VLSI Document V053, 1980.

[Hof83]   Hofmann, M., and U. Lauther, "HEX: an instruction-driven approach to feature extraction," in *Proc. 20th Design Automation Conf.*, pp. 331-336, 1983.

[Hon83]   Hon, R. W., "The hierarchical analysis of VLSI designs," Carnegie-Mellon Univ. Report No. CMU-CS-83-170, Ph.D. thesis, 1983.

[Hop71]   Hopcroft, J. E., "An $n\log n$ algorithm for minimizing states in a finite automaton," *Theory of Machines and Computations*, edited by Z. Kohavi and A. Paz, Academic Press, 189-196, 1971.

[Hor83]   Horowitz, M., and R. W. Dutton, "Resistance extraction from mask layout data," *IEEE Trans. on Computer-Aided Design*, vol. 2, no. 3, pp. 145-150, July 1983.

[Joh81]   Johnson, D. S., "The NP-completeness column: an ongoing guide," *Journal of Algorithms*, vol. 4, pp. 393-405, 1981.

[Joh82]   Johnson, S. C., "Hierarchical design validation based on rectangles," *Conf. on Advanced Research in VLSI*, pp. 97-100, 1982.

[Kod86]   Kodandapani, K. L., and E. J. McGrath, "A wirelist compare program for verifying VLSI layouts," *IEEE Design and Test of Computers*, vol. 3, no. 3, pp. 46-51, 1986.

[Koz81]   Kozawa, T., A. Tsukizoe, J. Sakemi, C. Miura, and T. Ishii, "A concurrent pattern operation algorithm for VLSI mask data," in *Proc. 18th Design Automation Conf.*, pp. 563-570, 1981.

[Kub79]   Kubo, N., I. Shirakawa, and H. Ozaki, "A fast algorithm for testing graph isomorphism," in *Proc. Intl. Symposium Circuits and Systems*, pp. 641-644, 1979.

[Lau81]   Lauther, U., "An $O(N\log N)$ algorithm for boolean mask operations," in *Proc. 18th Design Automation Conf.*, pp. 555-562, 1981.

[Lin76]   Lindsay, B. W., and B. T. Preas, "Design rule checking and analysis of IC mask designs," in *Proc. 13th Design Automation Conf.*, pp. 301-308, 1976.

[Los79]   Losleben, P., and K. Thompson, "Topological analysis for VLSI circuits," in *Proc. 16th Design Automation Conf.*, pp. 461-473, 1979.

[McC84]  McCormick, S. P., "Automated circuit extraction from mask descriptions of MOS networks," Massachusetts Institute of Technology Report No. 7-75-86, M. S. thesis, 1984.

[Mea80]  Mead, C., and L. Conway, *Introduction to VLSI systems*, Addison-Wesley, Reading, Massachusetts, 1980.

[Mit80]  Mitsuhashi, T., T. Chiba, M. Takashima, and K. Yoshida, "An integrated mask artwork analysis system," in *Proc. 17th Design Automation Conf.*, pp. 277-284, 1980.

[Nel85]  Nelson, B. J., and M. Shand, "An integrated, technology-independent, high-performance artwork analyzer for VLSI circuit design," *Journal of VLSI and Computer Systems*, vol. 1, no. 3, pp. 271-295, 1985.

[New80]  Newell, M. E., and C. H. Sequin, "The inside story on self-intersecting polygons," *Lambda*, vol. 1, no. 2, pp. 20-24, 1980.

[New82]  Newell, M. E., and D. T. Fitzpatrick, "Exploiting structure in integrated circuit design analysis," *Conf. on Advanced Research in VLSI*, pp. 84-92, 1982.

[Noi81]  Noice, D., J. Newkirk, and R. Mathews, "A polygon package for analyzing integrated circuit designs," *VLSI Design*, vol. 2, no. 3, pp. 33-36, 1981.

[Ott85]  Ottmann, T., P. Widmayer, and D. Wood, "A fast algorithm for the boolean masking problem," *Computer Vision, Graphics, and Image Processing*, vol. 30, no. 3, pp. 249-268, 1985.

[Ous81]  Ousterhout, J. K., "Caesar: an interactive editor for VLSI layouts," *VLSI Design*, vol. 2, no. 4, pp. 34-38, 1981.

[Ous84]  Ousterhout, J. K., "Corner stitching: a data-structuring technique for VLSI layout tools," *IEEE Trans. on Computer-Aided Design*, vol. CAD-3, pp. 87-100, 1984.

[Ous85]  Ousterhout, J. K., G. T. Hamachi, R. N. Mayo, W. S. Scott, and G. S. Taylor, "The Magic VLSI layout system," *IEEE Design and Test of Computers*, vol. 2, no. 1, pp. 19-30, 1985.

[Oza80]  Ozaki, T., T. Yoshida, and M. Kosaka, "PANAMAP-1: a mask pattern analysis program for IC/LSI-centerline extraction and resistance calculation," in *Proc. IEEE Intl. Symposium on Circuits and Systems*, pp. 1020-1026, 1980.

[Pav81]  Pavlidis, T., "Contour filling in raster graphics," *Computer Graphics*, vol. 15, no. 1, pp. 29-36, 1981.

[Per85]  Perry, S., S. Kalman, and D. Pilling, "Edge-based layout verification," *VLSI Systems Design*, vol. 6, no. 9, pp. 106-114, 1985.

[Phi85]   Phillips, C. A., "Space-efficient algorithms for computational geometry," Mass. Inst. of Tech. Report No. 7-83-86, 1985.

[Pre82]   Preparata, F., and J. Nievergelt, "Plane-sweep algorithms for intersecting geometric figures," *Communications of the ACM*, vol. 25, no. 10, pp. 739-747, 1982.

[Put82]   Putatunda, R., "Auto-delay: a program for automatic calculation of delay in LSI/VLSI chips," in *Proc. 19th Design Automation Conf.*, pp. 616-621, 1982.

[Rea77]   Read, R. C., and D. G. Corneil, "The graph isomorphism disease," *Journal of Graph Theory*, vol. 1, pp. 339-363, 1977.

[Sch81]   Scheffer, L. K., "A methodology for improved verification of VLSI designs without loss of area," *Caltech Conf. on VLSI*, pp. 299-309, 1981.

[Sch83]   Scheffer, L. K., "The use of strict hierarchy for verification of integrated circuits," Stanford Univ., Ph.D. thesis, 1983.

[Sco85a]  Scott, W. S., and J. Ousterhout, "Magic's circuit extractor," in *Proc. 22nd Design Automation Conf.*, pp. 286-292, 1985.

[Sco85b]  Scott, W. S., *Compaction and Circuit Extraction in the MAGIC IC Layout System*, University of California at Berkeley Report No. UCB/CSD 86/269, Ph.D. thesis, 1985.

[Sei82]   Seiler, L., "A hardware assisted design rule check architecture," in *Proc. 19th Design Automation Conf.*, pp. 232-238, 1982.

[Sei85]   Seiler, L., "A hardware assisted methodology for VLSI design rule checking," Massachusetts Institute of Technology VLSI Memo No. 85-230, Ph.D. thesis, 1985.

[Sha76]   Shamos, M. I., and D. J. Hoey, "Geometric intersection problems," in *Proc. 17th Annual Conf. on Foundation of Computer Science*, pp. 208-215, 1976.

[Sha85]   Shand, M. A., "Hierarchical VLSI artwork analysis," *VLSI 1985*, pp. 415-424, 1985.

[Spi83]   Spickelmier, R. L., and A. R. Newton, "WOMBAT: a new netlist comparison program," in *Digest Intl. Conf. Computer-Aided Design*, pp. 170-171, 1983.

[Spi85]   Spickelmier, R. L., and A. R. Newton, "Connectivity verification using a rule-based approach," in *Digest Intl. Conf. on Computer-Aided Design*, pp. 190-192, 1985.

[Su86]    Su, S. L., V. B. Rao, and T. N. Trick, "A simple and accurate node reduction technique for interconnect modeling in circuit extraction," in *Digest Intl. Conf. on Computer-Aided Design*, pp. 270-273, 1986.

[Swa83]   Swartz, P. A., B. R. Chawla, T. R. Luczejko, K. Mednick, and H. K. Gummel, "HCAP — A topological analysis program for IC mask artwork," in *Proc. Intl. Conf. on Computer Design*, pp. 298-301, 1983.

[Swa85]   Swartz, P. A., and K. Mednick, "Incremental design rule checking," *IEEE Custom Integrated Circuits Conf.*, pp. 60-63, 1985.

[Szy83]   Szymanski, T. G., and C. J. Van Wyk, "Space efficient algorithms for VLSI artwork analysis," in *Proc. 20th Design Automation Conf.*, pp. 734-739, 1983.

[Szy85]   Szymanski, T. G., and C. J. Van Wyk, "Goalie: a space efficient system for VLSI artwork analysis," *IEEE Design and Test of Computers*, vol. 2, no. 3, pp. 64-72, 1985.

[Tak82]   Takashima, M., T. Mitsuhashi, T. Chiba, and K. Yoshida, "Programs for verifying circuit connectivity of MOS/VLSI artwork," in *Proc. 19th Design Automation Conf.*, pp. 544-550, 1982.

[Tar83a]   Tarjan, R. E., *Data Structures and Network Algorithms*, SIAM Press, Philadelphia, PA, 1983.

[Tar83b]   Tarolli, G. M., and W. J. Herman, "Hierarchical circuit extraction with detailed parasitic capacitance," in *Proc. 20th Design Automation Conf.*, pp. 337-345, 1983.

[Tay84]   Taylor, G. S., and J. K. Ousterhout, "Magic's incremental design-rule checker," in *Proc. 21st Design Automation Conf.*, pp. 160-165, 1984.

[Tre84]   Trefethen, L. N., "Analysis and design of polygonal resistors by conformal mapping," New York University Report No. 112, 1984.

[Tsu83]   Tsukizoe, A., J. Sakemi, T. Kozawa, and H. Fukuda, "MACH: a high-hitting pattern checker for VLSI mask data," in *Proc. 20th Design Automation Conf.*, pp. 726-731, 1983.

[Tyg85]   Tygar, J. D., and R. Ellickson, "Efficient netlist comparison using hierarchy and randomization," in *Proc. 22nd Design Automation Conf.*, pp. 702-708, 1985.

[Van84]   Van Wyk, C. J., "Clipping to the boundary of a circular-arc polygon," *Computer Vision, Graphics, and Image Processing*, vol. 25, no. 3, pp. 383-392, 1984.

[Van86]   Van Wyk, C. J., and J. S. Vitter, "The complexity of hashing with lazy deletion," *Algorithmica*, vol. 1, no. 1, pp. 17-29, 1986.

[Wag85]   Wagner, T. J. "Hierarchical layout verification," *IEEE Design and Test of Computers*, vol. 2, no. 1, pp. 31-37, 1985.

[Wat83]   Watanabe, T., M. Endo, and N. Miyahara, "A new automatic logic interconnection verification system for VLSI design," *IEEE Trans. on Computer-Aided Design*, vol. 2, no. 2, pp. 70-82, April 1983.

[Wei82]  Weise, D. W., "Exploiting hierarchy in the analysis of VLSI systems," Massachusetts Institute of Technology VLSI Memo No. 82-104, M.S. thesis, 1982.

[Whi81]  Whitney, T., "A hierarchical design rule checking algorithm," *Lambda*, vol. 2, no. 3, pp. 40-43, 1981.

[Wid86]  Widmayer, P., and D. Wood, "A time and space optimal algorithm for boolean mask operations on orthogonal polygons," Univ. Karlsruhe Bericht Nr. 161, 1986.

[Wil80]  Wilmore, J. A., "A hierarchical bit-map format for the representation of IC mask data," in *Proc. 17th Design Automation Conf.*, pp. 585-590, 1980.

[Wil81]  Wilmore, J. A., "Efficient boolean operations on IC masks," in *Proc. 18th Design Automation Conf.*, pp. 571-579, 1981.

[Yam72]  Yamin, M., "Derivation of all figures formed by the intersection of generalized polygons," *Bell System Tech. J.*, vol. 51, pp. 1595-1610, 1972.

[Yos77]  Yoshida, K., T. Mitsuhashi, Y. Nakada, T. Chiba, K. Ogita, and S. Nakatsuka, "A layout checking system for large scale integrated circuits," in *Proc. 14th Design Automation Conf.*, pp. 322-330, 1977.

# Chapter 9

# Knowledge-Based
# Physical Design Automation

**Bryan D. Ackland**

*AT&T Bell Laboratories*
*Holmdel, New Jersey*

## 9.1 Introduction

The goal of computer-aided design (CAD) systems is to assist the designer by (1) performing some of the more mundane repetitive design tasks, (2) organizing and structuring large quantities of design data, and (3) verifying the correctness of the design. In order to accomplish these tasks effectively, a system should include components that [Lat77]:

- *Automate design operations.* These are programs that automatically solve design problems. This includes synthesis (for example, PLA generation), analysis (for example, design rule checking), and optimization (for example, layout compaction).

- *Provide user interaction.* A good CAD system is able to blend the best qualities of the designer and the computer. This requires an interface that presents the state of the design and the options available in a form that can be readily interpreted and manipulated by both humans and machines.

- *Maintain the design database.* One of the most important aspects of a computer is its ability to provide fast, reliable and orderly access to large amounts of design data.

Traditionally, these components have been implemented by modeling design tasks as numerical problems, developing algorithms to solve those problems, and then coding the algorithms using a procedural language (for example,

Fortran, or Pascal). As described in the preceding chapters of this book, this approach has been successfully applied to many physical design tasks. As designs become more complex, however, and as design consumes a larger portion of product cost (both in time and money), it becomes necessary to increase the level of design automation. In particular, tool developers are beginning to look at the automation of not just individual design tasks, but also the design process itself.

There are, however, many aspects of physical design that do not lend themselves to an algorithmic style of programming. In studying the human designer, we can identify a number of design techniques which require a quite different approach if they are to be automated. For example:

- Much of the human design process is exploratory or evolutionary in style [Shr83]. Initially, the designer has only a vague idea of the design plan. As design proceeds, this plan is refined and frequently modified. The complete design plan is not known until the design is finished and is usually not applicable, in detail, to the next design problem. This is in contrast to conventional programs, which are usually built around a sequence of operations that is only very weakly influenced by the design data.

- Design can be thought of as a search over a huge solution space. Any significant search of this space is impractical because of the time required. The search process must therefore be guided by an understanding of what is to be designed, the constraints on the design, and the possible interactions between components of the design. This understanding (or knowledge) is often based on design experience and cannot be easily expressed in procedural programming constructs.

- As design proceeds, the choices made by the designer influence other portions of the design. Constraints must therefore be generated and propagated to maintain consistency. Exhaustive low level constraint propagation is, however, impossible because of the large quantities of data involved. Constraints need to be propagated and applied in an "intelligent" fashion — with an understanding of which constraints are really important at any particular stage of the design process. This informal editing of information requires high level knowledge of component interaction within the context of the system actually being designed.

- The human designer will often simultaneously explore multiple lines of reasoning, comparing and modifying design decisions along the way. This is in contrast to the single thread of control normally found in procedural code. In addition, the designer will propose solutions and then "backtrack" when problems arise as a result of those solutions. Sometimes this backtracking will be to an earlier design state; other times it will be to a new state that is somehow extrapolated from those design states already visited. This "creative leap" is hard to represent using any existing programming technology.

- If the CAD system is to act as a high level assistant, it must be able to communicate using semantics that match the concepts and high level abstractions understood by the human designer. It is much more useful, for example, for a designer to specify that he wants "cells $A$ and $B$ to abut" rather than having to prepare the list of pin matchings required to achieve abutment. A system that is making complex design decisions also needs to be able to explain to the user why those decisions were made. In algorithmic code, such reasons are frequently buried within the model on which the code is built.

- In some design tasks, a solution can frequently be obtained using a simple algorithm. There remains, however, a small yet important set of problems which do not yield to the algorithmic solution and get categorized as "special cases." These are handled by the human designer using intuitive techniques often based on experience (for example, standard cell placement and routing). It is usually quite difficult to modify an algorithm to deal with more than a few of these special cases.

As a result, CAD tool builders have begun to look for new software techniques with which to tackle complex design tasks. One technique which has emerged from artificial intelligence (AI) research and which appears to directly address some of the shortcomings of conventional programming is the use of expert systems. This technique is based on the explicit representation and manipulation of expert design knowledge. In this chapter, we will focus on CAD tools and systems that have been built using this knowledge-based approach. We begin by looking at the role of knowledge in the design process. This is followed by a section on expert systems — what they are, how they evolved, the tools of the trade. We then look at how expert systems can be applied to physical design. A survey of previous work shows how knowledge-based systems have been used to solve many different design problems. This discussion is followed by a closer look at a number of selected examples to demonstrate particular programming techniques. Finally we look at where the field is heading.

## 9.2  Knowledge in the Design Process

Engineers are trained to think of design and its automation in a *procedural* fashion, that is, as a list of operations whose sequence is largely data-independent (as characterized by the flow chart). Design can also be viewed, however, as the application of *knowledge*, knowledge that specifies how and when to refine the design state. In studying the human design process, the following types of knowledge can be identified:

- *Basic principles.* This is the deep knowledge that comes from engineering/circuit theory. At the lowest level, for example, are principles of device physics which define the operation of the basic electrical components. At a slightly higher level are models of transistor operation and transmission line theory. This is knowledge that the designer learned in school.

Although it could, in principle, be used to design a complete circuit, it is rarely applied directly because of the large amount of time that would be required to process a complete design at this level. Rather, it is usually abstracted into a number of higher level design heuristics.

- *Heuristic knowledge.* This is domain-specific knowledge that has been developed (possibly from first principles) to guide the design process at higher levels of abstraction. This higher level knowledge allows the designer either to proceed in larger (coarser) design steps or to prune the design search space. For example, knowing that a MOS gate can be used to store a 0 or 1 logic level (in the form of charge) is a piece of heuristic knowledge. It can be derived by applying basic transistor theory to the circuit configuration. It is a piece of knowledge that is only valid, of course, over a certain range of design parameters (for example, storage time, circuit configuration, temperature, supply voltage), and it is up to the designer to understand when it can be legitimately applied. Human designers also use heuristics to effect trade-offs between design parameters (for example, understanding the performance area trade-off between ripple-carry, carry-save, and lookahead adders).

Heuristic knowledge can range from the low level (for example, a MOS transistor can be modeled as a switch) to the very high level (for example, pipelining increases the throughput of synchronous systems). High level knowledge allows the design to be generated with a small number of large design steps but requires a huge number of heuristics to be available in order to guarantee a good design over a wide range of problems. Low level knowledge, on the other hand, captures the design process in a relatively small set of heuristics but requires their repeated application over a large number of small design steps.

- *Procedural knowledge.* Basic and heuristic knowledge are also used to develop procedures which specify how to perform a certain design task. A procedure may be developed, for example, to specify component values for a differential amplifier given specifications of gain, output impedance, common-mode rejection, *etc.* A logic minimization algorithm is another example of knowledge captured in procedural form.

- *Design experience.* A good designer is invariably an experienced designer who has picked up many "tricks of the trade" along the way. Knowledge of previous design cases and their solutions frequently gives insight into how best to tackle a new problem. A designer may know, for example, from experience that unbuffered dynamic latches are particularly susceptible to induced switching noise on their output. Each designer's experience is, of course, different and, in some situations, very little experience may be required to produce an acceptable result. In a tightly constrained problem, however, design experience will often make the difference between an acceptable and a superior solution.

- *Control knowledge.* All of the above can be classed as *domain knowledge* — knowledge that describes properties of the design domain and simple techniques for refining the state of the design. In addition to this domain knowledge, a human designer requires control knowledge which determines how and when to apply the various pieces of domain knowledge. This kind of knowledge is sometimes referred to as problem solving knowledge [Lat77]. Included in this class is knowledge of how to partition a design problem into manageable fragments (hierarchical decomposition) and how to search the partitioned design space for an acceptable solution. Also included is knowledge of how various pieces of the design interact — how constraints are generated and efficiently propagated.

Much of this design knowledge (particularly the procedural and some heuristic knowledge) can be simply captured in algorithmic form. This is the traditional engineering approach to design automation. Procedural (algorithmic) programs have a structure that is well matched to the architecture of conventional general-purpose computers. They lead to CAD tools that are characterized by efficient, predictable operation over a well-defined class of subproblems.

Other types of knowledge, however (particularly knowledge based on experience, control knowledge, and special-case heuristics), are not easily automated using these techniques. As mentioned previously, much design knowledge is applied in an exploratory fashion, that is, according to incomplete plans that guide the design through a large, constrained solution space. Mapping this type of knowledge into algorithmic control structures leads to an oversimplified model of the design process which must be augmented with many pieces of special-case code before the program can perform any useful design. Much of the knowledge gets buried in the control structures of the program, which become increasingly complex as new knowledge is added in the form of modified code. As a result, tools using these techniques are usually restricted to a narrow class of problems and are unable to adapt to new knowledge or situations that do not match their simplified model of the design process.

An example of the limitations of the algorithmic approach can be seen in the numerous attempts to build a silicon compiler as described in Chapter 7. A silicon compiler is a VLSI synthesis tool for converting functional specification to layout using techniques borrowed from compiler technology. Although a number of such tools have been built and successfully applied to real design problems (for example, [Gaj84, Sis82, Den82]), they are each quite limited in their domain of application and rely heavily on knowledge in the form of pre-defined layout cells or generators. In the following section, we look at techniques for explicitly representing and manipulating knowledge, independent of program control structure.

## 9.3  Expert Systems

*Artificial intelligence* (AI) is a field that is concerned with (a) understanding human intelligent behavior, and (b) building computing systems that simulate intelligent behavior in performing some set of tasks. During the 1960s, AI research focussed mainly on general principles and techniques (for example, logical reasoning, search heuristics) that could be used in solving a wide range of problems [McD81, Joo85b]. Although these techniques were applied with some success in a number of abstract areas (for example, theorem proving), they were found to be insufficient for complex real-world tasks. During the 1970s the focus shifted to developing techniques for capturing, storing and using domain-specific knowledge to guide and minimize the search process. This led to the development of a new class of tools known as knowledge-based expert systems.

An *expert system* can be defined [Wei84] as one that:

- Solves a problem that normally requires human expert interpretation.

- Reaches its solution using a model of computation based on human expert reasoning.

A *knowledge-based expert system* is one whose expertise is contained in domain knowledge that is explicitly represented and manipulated within the system. Note that human experts are usually only proficient in a narrow field. Knowledge-based expert systems are similarly targeted at a specific problem domain; they are built on knowledge that is relevant only to that domain and are not, in general, capable of broad (domain independent) reasoning.

### Production Systems

Knowledge-based expert systems are usually implemented as production (or *rule-based*) systems [Nil80]. A *production system*, as shown in Figure 9.1, consists of three components:

- *Working memory*. This is a global database that contains the objects which describe the current state of the problem. All problem specific data is held in working memory.

- *Rule memory*. This memory contains a set of production rules which specify how the objects in the working memory might be modified in order to advance towards a solution. Rules typically consist of two parts − an *antecedent* (or IF clause), which specifies under which conditions the rule should fire and a *consequent* (or THEN clause), which specifies what action to perform on the working memory.

- *Inference engine*. This engine provides control by determining which rules are able to fire, selecting one according to some strategy and then executing the consequent of that rule.

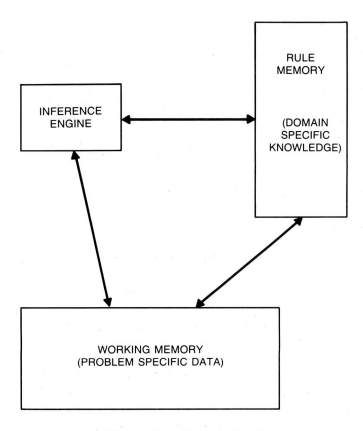

**Figure 9.1** Production System

Initially, the contents of working memory describe the problem to be solved. The inference engine applies rules one at a time to transform these contents. The process terminates either when there are no more rules that can fire, or when one of the rules detects a termination condition and informs the inference engine to stop.

One way to demonstrate the application of rule-based programming is through a simple example. Consider the problem of translating a pull-down tree of $N$ transistors into a reverse polish logic expression as shown in Figure 9.2. Figure 9.3 gives an algorithmic solution to this problem. It consists of a sequence of reduction steps which alternately seek out and simplify series and parallel pairs of transistors, converting them into two-input AND and OR operators. Figure 9.4 lists a set of rules that can be used to achieve the same result. The rules assume a working memory that has objects describing transistors and elements. Each element owns two terminals and a function. The rules also

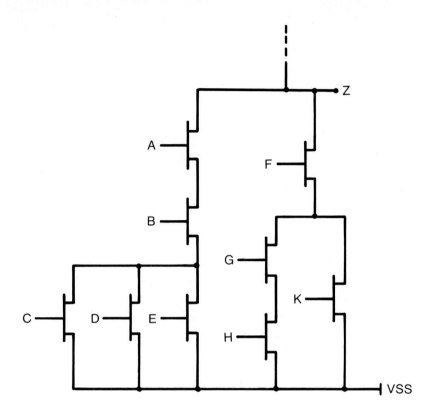

$$Z = K.G.H.and.or.F.and.C.D.or.E.or.A.B.and.and.or$$

**Figure 9.2**  Converting a Transistor Tree to a Logical Expression

assume that there is an inference engine that can perform search operations on the working memory of the form "If there is an $X$ . . . *etc.*"

Some of the clauses in the rules correspond directly to executable statements in the algorithmic version. This is hardly surprising since both systems are solving the same problem using similar object models. There are, however, a couple of differences worth noting. First, much of the searching that is coded into the algorithm in the form of loops is automatically performed in the rule-based version by the inference engine. Second, some of the knowledge of how to perform this task is buried in the loop ordering of the algorithmic version. If another programmer were to subsequently modify this code, there had better be enough comments around to warn about the importance of this ordering. In the rule-based version, this knowledge is explicitly represented in the form of extra clauses in the rules (for example, line 5 of rule 3). The production rule implementation has thus not only exposed the domain specific problem solving

$check = 1$
**until** ($check$ eq 0) **do**
   $check = 0$
   **for** each element $E1$ **do**
      **for** each element $E2$ not eq $E1$ **do**
         **if** ((($E1.term1$ eq $E2.term1$) and ($E1.term2$ eq $E2.term2$))
           or (($E1.term2$ eq $E2.term1$) and ($E1.term1$ eq $E2.term2$))) **do**
           create element $E3$
           $E3.term1 = E1.term1$; $E3.term2 = E1.term2$
           $E3.func = E1.func \mid E2.func \mid$ OR
           delete $E1$ and $E2$
           $check = 1$
         **done**
      **done**
   **done**
   **for** each element $E1$ **do**
      **for** each element $E2$ not eq $E1$ **do**
         **if** (($E1.term1$ eq $E2.term1$) and ($E1.term2$ not eq $E2.term2$)) **do**
           $junction = 0$
           **for** each element $E3$ (not eq $E1$) and (not eq $E2$) **do**
              **if** (($E3.term1$ eq $E1.term1$) or ($E3.term2$ eq $E1.term1$)) **do**
                 $junction = 1$
              **done**
           **done**
           **if** ($junction$ eq 0) **do**
              create element $E4$
              $E4.term1 = E1.term2$; $E4.term2 = E2.term2$
              $E4.func = E1.func \mid E2.func \mid$ AND
              delete $E1$ and $E2$
              $check = 1$
           **done**
         **done**
         **else if** (($E1.term2$ eq $E2.term1$) and ($E1.term1$ not eq $E2.term2$)) **do**
           *etc.*
         **else if** (($E1.term2$ eq $E2.term2$) and ($E1.term1$ not eq $E2.term1$)) **do**
           *etc.*
         **else if** (($E1.term1$ eq $E2.term2$) and ($E1.term2$ not eq $E2.term1$)) **do**
           *etc.*
      **done**
      **done**
   **done**
**done**
**if** (number of elements $E > 1$) output(ERROR) **else** output($E$)

**Figure 9.3** Algorithmic Solution to Logic Generation Problem

(1)  **IF**     there is an element *E1*
                and there is an element *E2*
                and *E1.term1* eq *E2.term1*
                and *E1.term2* eq *E2.term2*
     **THEN**   create an element *E3(E1.term1, E1.term2)*
                and set *E3.func = E1.func | E2.func |* OR
                and delete *E1* and *E2*

(2)  **IF**     there is an element *E1*
                and there is an element *E2*
                and *E1.term1* eq *E2.term2*
                and *E1.term2* eq *E2.term1*
     **THEN**   create an element *E3(E1.term1, E1.term2)*
                and set *E3.func = E1.func | E2.func |* OR
                and delete *E1* and *E2*

(3)  **IF**     there is an element *E1*
                and there is an element *E2*
                and *E1.term1* eq *E2.term1*
                and *E1.term2* not eq *E2.term2*
                and there is no *E4* with *E4.term1 = E1.term1*
                and there is no *E4* with *E4.term2 = E1.term1*
     **THEN**   create an element *E3(E1.term2, E2.term2)*
                and set *E3.func = E1.func | E2.func |* AND
                and delete *E1* and *E2*

(4), (5) and (6)   *similar to rule (3)*

(7)  **IF**     there is an element *E1*
                and there is an element *E2*
                and no other rules can fire
     **THEN**   output error

(8)  **IF**     there is an element *E1*
                and there is no element *E2*
     **THEN**   output *E1*
                and exit

**Figure 9.4** Rule-Based Solution to Logic Generation Problem

knowledge but also suppressed the underlying domain independent utilities (searching).

### Knowledge Acquisition

Rule-based expert systems are frequently developed as a collaboration between a *domain expert*, who knows implicitly how to solve the problem, and a *knowledge engineer*, who understands how to build the system. The knowledge engineer tries to extract domain knowledge from the domain expert and then store that knowledge explicitly within the system.

Rule-based systems store knowledge in two different ways. The first is in working memory. As mentioned previously, this memory contains all the problem specific data. It also contains, however, domain specific knowledge in the form of the objects used to represent the domain. The structure and scope of these objects limit the system's universe of possible states. This has the advantage of simplifying the production rules at the expense of limiting the scope of the system. In Figure 9.4, for example, the domain specific knowledge that there are only two connection points on each pull-down element simplifies the individual rule clauses but prevents the system from being easily modified to deal with more complex structures. The selection of working memory objects around which to build rules is thus a very important task in building production systems. Issues of data representation become very important in design problems. These issues are addressed in more detail later in this section.

The second place where knowledge is stored is in the rule memory. Once the objects in the working memory have been defined, programming a rule-based system consists of adding new rules and refining existing rules. This is normally carried out in an iterative fashion as shown in Figure 9.5. Once the knowledge engineer has gleaned enough knowledge to put together a small set of rules that will lead to some (albeit weak or naive) solution, the system is run on a number of test cases. The domain expert is asked to critique the output of this prototype system. The knowledge engineer uses this new information to modify the rule base. This process is repeated until both are satisfied with the results over a large class of problems. In some systems, this process never completes; knowledge acquisition continues as the system is used in the field.

### Applications

Expert systems find application in solving problems for which either (a) no good algorithmic solution exists, or (b) algorithmic solutions exist but they are too expensive [Bir86]. Problems in this class are characterized by:

- A large problem space with a large number of alternatives at each problem state.

- A large number of requirements, some of which may be conflicting.

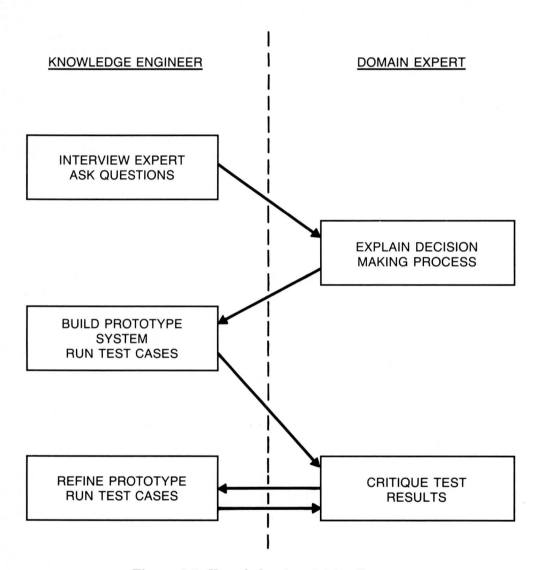

**Figure 9.5**  Knowledge Acquisition Process

- Incomplete knowledge of how to solve the problem under all circumstances.

- A large number of acceptable solutions — lack of definitive evaluation functions.

Rule-based systems have been successfully applied in a number of quite different domains, including chemical analysis [Buc78], medical diagnosis [Sho76], and geological survey [Dud79]. Some of these systems have been shown to outperform human experts. The principal advantages of this type of approach are:

- *Flexibility*. Because knowledge is stored in small modular units (rules), it is simple to modify or expand that knowledge by adding or changing rules. Flexibility is enhanced if the rules are written so as to be (mostly) orthogonal, *i.e.*, independent of one another.

- *Incremental Development*. An implementation can begin long before all the relevant knowledge has been extracted. Prototype implementations, in fact, aid the knowledge acquisition process. Rule-based programming is not just a technique for building working systems; it is a technique for understanding how to go about building systems.

- *Explicit Knowledge*. Knowledge is stored in a form semantically (if not syntactically) equivalent to forms understood by the domain expert. There is no need to transform the problem into numerical quantities manipulated by algorithmic control structures. This simplifies the knowledge acquisition process during system construction and user interaction when the system is being run in the field.

- *Explanation*. A recorded sequence of rule firings can be used to generate an explanation of how a particular solution has been reached. This aids debugging by the domain expert and gives the user of the system confidence in the result.

Needless to say there are a number of disadvantages associated with the use of production systems:

- Rule-based systems tend to be less efficient in terms of execution time than their algorithmic counterparts. Algorithmic control structures are capable of hiding very complex search paths which are difficult to represent in simple rules. For this reason, it rarely makes sense to apply production rules to a problem for which there exists a good algorithmic solution.

- Like the human expert, rule-based systems do not, in general, yield optimal solutions or produce results of predictable accuracy in predictable time. Sometimes a system will not even guarantee any result at all. This is because these systems are built on incomplete knowledge, and the quality of solution is often a continuously evolving property.

- Although rule-based systems tend to be more flexible in terms of new problems within a particular domain, they are frequently of no value at all outside their domain. This is in contrast to some algorithms which find application in many different problem domains. This can limit how well the system scales. For example, the rules used to lay out transistors in a leaf cell are not very useful when it comes to laying out modules on a chip because the knowledge required by the two problems is very different.

## Other Forms of Knowledge Representation

So far in this chapter, we have used the terms *expert system*, *knowledge-based system*, and *rule-based system* interchangeably. There are, however, expert systems (systems that emulate human expert behavior) that are not knowledge-based but procedural in form (for example, cell layout generators). There are also examples of rule-based systems that one would not describe as domain knowledge-based. CV [Spi85], for example, is a connectivity verification program which translates circuit structures from a connectivity description into rules. A second connectivity description is loaded into working memory. The data dependent rules are then executed and cause structures to be deleted from the working memory. The domain knowledge in this system is not in the rules (these contain data). Rather, it is in the algorithm that translates circuit structure into rules.

There is also an important class of knowledge-based systems that are not rule-based. As mentioned previously, large amounts of domain knowledge can be built into the data representations on which a system is built. One approach to capturing this knowledge that has been widely used is *object-based programming*, a technique that has it origins in the programming language *Simula* [Dah66]. All data is described in terms of objects. Objects belong to *class*es which define the object's attributes (parameters), the values the attributes may assume, and maybe some default values for the attributes. These object classes capture, therefore, a significant amount of knowledge about the domain in which these objects exist.

In addition to storing data, objects can also contain code. Object-based systems execute by passing *messages* between objects (as opposed to procedures operating "on" objects). When an object receives a message, it executes the piece of code that is associated with that message. Object classes define which messages an object may receive and also which piece of code will be activated when the message is received. Thus, object classes also capture knowledge about how different objects will respond (or should be processed) in various situations.

A third characteristic of object-based programming is that it usually includes some concept of *inheritance* between classes. The class of *D-latches*, for example, may be represented as a subclass of the class *storage-elements*. Some of the attributes (not values) of D-latch (for example, stored-data-value) can be inherited from the parent class. Others (for example, data-input-hold-time) are specific to the subclass. Using these techniques, a *classification hierarchy* can be constructed which links the various objects in the domain through inheritance relations. This considerably reduces the amount of programming effort and machine storage required to characterize the objects in a large complex domain.

## Expert System Tools

Over the years, researchers have developed a number of special tools for building AI systems and, in particular, expert systems. Languages such as Lisp and Smalltalk provide powerful facilities for manipulating symbolic information (such as objects and their properties). More recently, with the advent of fast compilers and special machine environments, Lisp has gained acceptance outside the AI community. One of the principal advantages of Lisp is its extensibility [Shr83]. Symbolic expressions can be used to form the vocabulary of a language that describes a problem domain. This makes it possible to trivially build special-purpose languages that retain all the power of the underlying Lisp system.

In particular, a number of object-based languages have been built on top of Lisp. Loops [Bob82] and Flavors [Sym84] are examples of these. *DPL* [Sus80] is an object-based language built on Lisp specifically for developing CAD environments.

Powerful programming environments have been built around the Lisp language. They have their roots in work done at Massachusetts Institute of Technology (MIT) and are now commercially available on a number of Lisp machines (for example, Symbolics, TI Explorer). These systems are much more than a Lisp compiler or interpreter running on a general-purpose machine. They include editors, debuggers, file maintenance, window and graphic systems all integrated into the runtime Lisp environment. They facilitate the building and exploration of new concepts (as opposed to designing, writing and executing programs).

*Prolog* is another language that has been widely used in developing expert systems. Prolog is a logic programming language that describes objects and relations between them using a set of logic clauses [Hor84]. These clauses (like rules) consist of a consequent and a number of antecedents as shown in the example of Figure 9.6. In addition to being declarative, these clauses are procedural (*i.e.,* they can be executed as functions). Clauses without antecedents are always true and can be thought of as known *facts.* Clauses without consequents are *goals.* A *logic program* is a set of rules, facts, and a goal. The antecedents of the goal are matched to the consequents of rules. The antecedents of those rules are then matched to other consequents. This process, called *backward chaining,* continues until the search terminates on facts (in which case the goal is satisfied) or the search space is exhausted (the goal is not satisfied).

*Syntax:*

    <consequent>:- <antecedent-1>,<antecedent-2>,...<antecedent-n>.

*Example:*

    rsflop($R$,$S$,$Q$,$Qb$):- nand2($Qb$,$R$,$Q$),nand2($Q$,$S$,$Qb$).

*Meaning:*

An RS flip-flop exists between any four nodes $R$, $S$, $Q$ and $Qb$ if a 2-input NAND gate exists between $Qb$, $R$ and $Q$ and a 2-input NAND gate exists between $Q$, $S$ and $Qb$.

**Figure 9.6**  Example of Prolog Clause

OPS is a language that was developed specifically for building expert systems [For77]. It provides a working memory, a rule memory, and an interpreter that selects and fires rules according to a simple priority function. As rules fire, the working memory is modified. This, in turn, causes more rules to fire. This process of moving from known facts towards a conclusion (solution) is known as forward chaining.

In addition to these languages, there are a number of problem solving systems that seek to provide a comprehensive rule-based environment, including object-based rule and working memories, inferencing mechanisms, database editors, and interactive graphics support. Examples are KEE, S.1, ESE, and ARTS. These systems allow rapid prototyping but may be limited in the size of the problem they can model and may restrict the developer to think only in terms of those primitives supplied and supported by the system. A more complete review of expert system building tools can be found in [Gev87].

## 9.4  Applying Expert Systems to Design

As mentioned previously, the knowledge-based approach has been shown to be effective in solving a number of complex real-world problems. Much of this success, however, has been in problem areas that one would describe as analytic or *diagnostic* (for example, medical diagnosis, chemical analysis). These problems are characterized by (a) a relatively small number of input facts, (b) a relatively small number of possible outcomes, and (c) a body of deductive knowledge which, when applied to the input facts, will lead to a correct solution regardless of the order in which the knowledge is applied.

Physical design is typically performed as a large number of small design tasks. Some of these tasks (for example, verification) can be easily modeled as diagnostic problems. Others (in particular, synthesis tasks and the overall management of the design process) do not fit this model and, in fact, display quite different properties. These tasks are characterized by:

- A large solution space. There is usually no "right" answer but many good ones. Good solutions are frequently separated by a huge distance, *i.e.*, they are quite different, and it is not obvious how one could be derived from another.

- There is a strong notion of *state* in the design process. The actions taken depend very much on the state of the design. Another way of saying this is that order is important (try to describe, for example, how to build a kite without using the words "then" or "next").

- Design is a sequence of decision making steps, but the actions taken at each step may be quite complex. Routing a wire, for example, between two points cannot be performed merely by changing a couple of values in a working memory.

- A large design problem may require knowledge from more than one domain. Planning a VLSI layout, for example, requires knowledge of floor-planning, transistor layout, routing, performance limitations, *etc.*

There has recently been considerable interest in how best to apply expert systems technology to the physical design process. One technique that has been used with some success is to transform a design problem into a series of analysis problems. Dixon and Simmons [Dix84] describe a model (Figure 9.7) in which an initial design is followed by a series of evaluate/redesign cycles. The initial design is readily generated; it is little more than a rough guess. The design process becomes, therefore, a series of diagnostic evaluations which generate suggestions as to how the design may be improved. *TOPOLOGIZER* [Kol85] and *Schema* [Zip83] are examples of systems that use this approach. TOPOLOGIZER converts transistor schematics into CMOS layout subject to physical constraints on aspect ratio, pin position, *etc.* It works in two phases: transistor placement and routing. In both phases it begins with an initial guess (which is little more than a random placement or connection) and then applies rules to iteratively improve the solution.

Many problems do not, however, yield to this approach. In some cases, this is because the problem is too poorly structured to generate suggestions as to how a design might be improved. In others, the solution space is complex and discontinuous so that it is not possible to chart a continuous series of small modifications that will run from the initial design to a good final solution. Researchers have, therefore, proposed a number of modifications to the basic production system. Unfortunately, there is not broad agreement on how best to

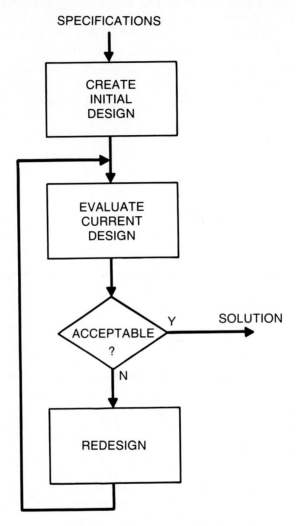

**Figure 9.7**   Redesign Model

tackle this problem, and no single architecture has emerged. In the following sections we will examine some of the critical issues.

### Control

The need for state (or sequence) in a design task leads to a much more complex control structure than is required for diagnostic tasks. Many researchers have recognized the need to separate control knowledge (choosing which action to perform next) from design implementation knowledge (how to perform the action). In complex systems, there may even be a hierarchy of control knowledge. The term *meta-control* is sometimes used to describe the process of controlling the execution of control knowledge. The simplest technique for capturing control knowledge in a rule-based system is to add state variables and rules that manipulate those variables. R1 [McD81], Talib [Kim84] and DAA [Kow85a] are examples of rule-based synthesis systems that use this approach. Tests for the various state variables are included in the antecedents of the rules. A disadvantage of this approach is that it can be very difficult to understand how various rules interact. Also, as the system grows, rules become cluttered with large numbers of state variable tests.

Another approach is to organize rules into a number of separate rule sets and invoke these rule sets through a conventional algorithmic control structure. *Blorec* [Jos85] and *Flute* [Wat86] are systems that employ this scheme. DAA also uses this technique at the highest control level. Alternatively, a rule-based controller can determine when to invoke the various design operations. This type of control structure is seen in *Floyd* [Dic86] and a mechanical design system proposed by Mittal and Araya [Mit86]. In the Cadre [Ack85] and Ulysses [Bus86] systems, control is centralized in a rule-based manager that invokes a number of autonomous expert agents.

A more distributed approach is to cast the controller as one of many knowledge sources interacting through a central working memory known as a *blackboard* [Ens86]. Each knowledge source examines the current state of the design as recorded on the blackboard and posts a proposal as to what should be done next. The controller then examines the various proposals and determines which should be executed. Weaver [Joo85a] is an example of such a system. Another distributed control technique is the use of *demons*. Demons are bodies of code that are automatically executed whenever a specified piece of data is accessed or modified. Demons can be used to provide automatic constraint propagation as in the EL system [Sta77].

Yet another approach is to let the user assume the control function. The design system provides primitive synthesis and analysis operators and maybe offers advice but ultimately allows the user to determine what should occur next. Vexed [Mit84] and Class [Bir85] are programs that employ this type of user interaction. These systems act, therefore, more as a *design assistant*, blending the decision making capabilities of the human with the computational power of the computer.

## Algorithmic Support

There are portions of any complex design task that are best implemented using a conventional algorithmic approach. Numerical procedures, for example, tend to be clumsy and inefficient when coded in terms of rules. A large system will usually consist of some combination of rule-based and procedural programming. Blorec and Flute are two synthesis systems (floorplanning and routing) that are modeled on human expert behavior. Part of the human expertise lies in the ability to recognize and react to visual patterns. This 2-D pattern recognition is most easily performed in the computer using a number of algorithmic primitives. These primitives appear as operators, both in the antecedents and consequents of design rules.

Weaver, DAA and Xsel [McD81] are further examples of synthesis systems that rely on low level algorithmic support. Horstmann and Stabler [Hor84] describe a Prolog-based system for test vector generation and make the observation that efficiency could be dramatically improved by implementing some of the Prolog clauses in a procedural language. In Schema, simulation procedures are used to provide analysis of intermediate results. Test vector generation and analysis of simulation results are performed by the knowledge-based part of the system.

## Backtracking

As previously noted, human designers rely heavily on backtracking and redesign to compensate for their incomplete design knowledge. Systems that emulate human design will likely also include some backtracking mechanism to deal with conflict and dead ends in the search process. The simplest form of backtracking is *chronological, i.e.,* retreat to the previous design state and try a different solution. If no good solution is found here, retreat to the next previous state and modify the choice made there. The problem with this approach is that it is very inefficient — exploring many alternatives that have nothing to do with the original conflict.

Another approach, called *dependency-directed backtracking*, was first proposed by Stallman and Sussman [Sta77] as part of their circuit analysis program. Analysis proceeds as a result of demon-based constraint propagation. As facts are deduced, they are annotated with the source of any constraint that contributed to their deduction. When a conflict or dead end occurs, the system re-examines only those design points that have contributed to the conflict. PLA-ESS [Bre85] is a tool for applying tests to PLAs which uses this backtracking technique. An added benefit of recording constraint data is that it can form the basis of an enhanced explanation facility.

A third alternative is to record design operations in some form of *design history* file and then use high level domain specific knowledge to analyze the conflict in terms of that history and select an appropriate redesign point. This has the advantage that backtracking decisions are being made at a higher level using special-case knowledge similar in form to that used by the human designer. It can also be used to generate a more meaningful explanation for the user. A disadvantage is that it is more difficult to guarantee that such an approach will deal with all possible conflicts. Sussman *et al.* [Sus80] proposed history-based backtracking for their DPL language. Mittal and Araya [Mit86] and Knapp *et al.* [Kna83] both present frameworks for design based on this approach. Cadre and Ulysses are two synthesis systems that are being developed around this scheme.

Other researchers believe that backtracking can be significantly reduced and even eliminated through the use of design plans. They point out that human backtracking is not systematic and that human designers will often backtrack to a point which is not even part of the design history but rather some extrapolation of it based upon design experience. Talib and Floyd are systems that use top-down planning and a principle of *least commitment* to try and reduce any backtracking to a minimum. In DAA and Weaver careful planning has led to systems in which there is no backtracking whatsoever.

**Knowledge Representation**

Complex design tasks require that knowledge from many sources be brought together in a constructive fashion. Often, modularity requires that we keep these knowledge sources separate and, in fact, store these different types of knowledge in quite different ways. In the Palladio design environment, for example, knowledge is stored in three separate ways [Bro83]:

- Knowledge in the form of rule-based expert agents. For example, an agent to convert a switch level transistor circuit into symbolic layout. As mentioned previously, there are a number of other systems that use multiple experts to capture different types of design knowledge.

- Knowledge about the domain (its primitive components, their properties, and legal ways in which they may interact) is stored in a series of *design perspectives*. These perspectives limit the number of possible design states by eliminating meaningless or illegal combinations of data. A number of other authors [Kna83, Wal85, Mit86, Dic86, Sub86] have also proposed techniques for structuring the design domain in this fashion. Stefik and Conway [Ste82] propose similar schemes based on representation languages that by their syntax generate some guarantee of correctness.

- Knowledge about design objects, their properties, and how they are to be designed is captured using object-based programming. Inheritance allows design objects to be structured into classification hierarchies which reduce the volume and complexity of data that needs to be stored. Procedures and demons attached to these objects capture knowledge of the design process (for example, constraint propagation). Many authors (for example, [Kel84, Del85, Sus80, Cla85, Sim85, How85, Gab87, Afs86, Rub83, Wol86]) have proposed or developed systems based on this style of knowledge capture.

Another technique for capturing design knowledge is in the form of design *schemata*. These are implementation methods that recognize a particular design situation and know how to transform it into either a better or a more refined representation. They can be thought of as "super rules." Synapse [Sub86] and Vexed are two systems that have been built around this approach. In the Hitest [Ben84] test generation system, knowledge is stored in the form of known test procedures that are attached to module objects. In *PLA-ESS* [Bre85], also a test system, knowledge is stored in tables that describe the attributes of different test techniques. There are even more *ad hoc* forms of knowledge capture. For example, in a printed wiring board placement tool developed by Odawara *et al.* [Oda85], empirical knowledge is extracted from manual designs and stored in a highly structured database. The placement program clusters the objects to be placed and then searches the database for known patterns that most closely match each cluster.

## 9.5  Knowledge-Based Design Tools

In the previous section we examined, from a knowledge engineering perspective, some issues that arise in applying expert system techniques to physical design problems. In this section we review some knowledge-based design tools that have actually been proposed and/or built.

Most of the work done to date falls into one of three general classes.

- *Synthesis tools* — systems that generate new design data from a higher level specification.

- *Analysis tools* — systems that analyze function, performance, testability, *etc.*; systems that verify correctness.

- *Management tools* — systems that manage the overall design process; decide what action should be taken next in order to advance the state of the design.

In each of these three categories we will describe one tool in some detail. Other tools will then be summarized in tabular form. Finally, we will look at a few miscellaneous systems that do not really fit into these classifications.

### Synthesis Systems

One of the earliest and most successful applications of expert systems to design synthesis was *R1* [McD81], a system developed at Carnegie-Mellon University (CMU) for configuring VAX-11 computer systems. R1 takes, as input, a set of components (disks, CPUs, floating point, *etc.*) and produces diagrams showing what extra components are required (for example, cabinets, power supplies, controllers) and how all the components can be spatially organized. Knowledge for making these decisions is captured in a set of rules coded in OPS5. Control (ordering) of the design is achieved through a number of state variables called goals. Rules are organized into sets that correspond to the different goals and sub-goals. An example of (the English translation of) an R1 production rule is given in Figure 9.8. Note that the first antecedent clause is a state variable (goal) check. R1 contains about 3300 rules. It has been used on a regular basis by DEC for configuring VAX orders since 1980.

**if:**      the most current active subtask is assigning devices to unibus modules
          and there is an unassigned dual port disk drive
          and the type of controller it requires is known
          and there are two such controllers
               neither of which has any devices assigned to it
          and the number of devices that these controllers
               can support is known
**then:**    assign the disk drive to each of the controllers
          and note that the two controllers have been associated
               and that each supports one device

**Figure 9.8**  Example of R1 Production Rule

Table 9.1 summarizes some other knowledge-based synthesis systems and lists a few key features of each.

### Analysis Systems

One of the first applications of knowledge-based techniques to an analysis task was the *EL* system developed at MIT by Stallman and Sussman [Sta77]. EL analyzes the DC operating point of an electronic circuit by applying a set of rules written in a language called ARS. These rules encode familiar approximations to physical laws such as Kirchoff's laws and Ohm's law as well as more complex models for devices such as transistors. Some examples are shown in Figure 9.9.

Known facts about a circuit are stored as a collection of assertions. Circuit analysis rules are implemented as demons. When an assertion is added to the database, one or more demons will match it and fire. When a demon fires, it will either make a new assertion or detect a contradiction. New assertions are generated using a technique of linear constraint propagation. Contradictions arise because non-linear devices are represented as piece-wise linear operators, and the system must sometimes guess the operating region of a device. An incorrect guess will ultimately lead to a contradiction. At that time, the system examines the set of constraints that led to the contradiction and uses a scheme called dependency directed backtracking to return only to those assertions which

Xsel            [McD81]    Application Specific Component Selection for R1
                           - *application specific rules and quantity formulae*
                           - *emphasis on user interaction and explanation*

DAA             [Kow85a]   Computer Architecture from Algorithmic Description
                           - *multiple OPS5 rule sets*
                           - *algorithmic cost functions*
                           - *independently assessed as "good" designer*

DSS             [Aco86]    Digital Synthesis from Behavioral Specification
                           - *abstracts and stores previous design plans*
                           - *tackles new designs by analogy*
                           - *plans stored as partially ordered sets of rules*

LSS             [Yos86]    Logic Synthesis from Functional Description
                           - *rule-based transformations*
                           - *search space limited by cost bounding*

Talib           [Kim84]    nMOS Leaf Cell Layout
                           - *multiple OPS5 rule sets*
                           - *prioritized control rules*
                           - *least commitment strategy to avoid backtracking*

TOPOLOGIZER     [Kol85]    CMOS Leaf Cell Layout
                           - *originally developed in OPS5; recoded in Lisp*
                           - *initial guess followed by iterative improvement*

Floyd           [Dic86]    Custom IC Floorplanning
                           - *domain knowledge in object-based data structures*
                           - *five algorithmic subsystems*
                           - *OPS5 control rules*

Flute           [Wat86]    Custom IC Floorplanning
                           - *multiple rule sets in Lisp*
                           - *procedural state controller*
                           - *algorithmic pattern match operators*

Blorec          [Jos85]    Solve Blockages in Printed Wiring Boards
                           - *multiple rule sets in Lisp*
                           - *procedural controller*

Weaver          [Joo85a]   Switch-box IC Router
                           - *multiple OPS5 rule-based experts*
                           - *communication via central blackboard*
                           - *control via "focus of attention" expert*

KB              [Oda85]    Printed Wiring Board Component Placement
                           - *empirical clustering knowledge stored as schemata*
                           - *procedural controller matches schemata to data*

Prosaic         [Bow85]    Analog Operational Amplifier Design
                           - *known solutions (schemata) stored in database*
                           - *solves stages (subproblems) independently*
                           - *iterates to merge stage solutions*

**Table 9.1**  Summary of Knowledge-Based Synthesis Systems

(1) If the voltage on one terminal of a voltage source is given, one can assign the voltage on the other terminal.

(2) If the voltage on both terminals of a resistor is given, and the resistance is known, then the current through it can be assigned.

(3) If the current through a resistor, and the voltage on one of its terminals is known, along with the resistance of the resistor, then the voltage on the other terminal can be assigned.

(4) If all but one of the currents into a node are given, the remaining current can be assigned.

**Figure 9.9**  Example EL Rules

may have caused the problem. Once the incorrect assertion has been identified, a new operating point can be selected and the forward analysis resumed.

Table 9.2 lists some other examples of knowledge-based analysis tools.

## Management Systems

Management of the design process is an area in which knowledge-based techniques show great promise. *Palladio* [Bro83] was one of the first attempts to build an integrated knowledge-based environment that would support interactive and automatic design tools. The Palladio approach proposed that design should proceed as a parallel refinement of structural and behavioral perspectives. *Perspectives* are descriptions of a particular level of abstraction in terms of primitive components and a set of rules that define their legal interconnection. Structural perspectives are used to specify logical interconnection. A structural perspective, for example, could define a circuit at the transistor level. Another could be used to represent it as a collection of clocked storage registers. Behavioral perspectives are used to specify changes of circuit state over time. A behavioral perspective could describe the behavior of a circuit in terms of 3-valued logic. Behavioral perspectives are expressed as rules relating signal behaviors.

Behavioral and structural perspectives are linked in Palladio in two ways. First, design procedures can be used to generate structure from behavior (for example, PLA generation). Second, a rule-based logical simulator allows structure to be simulated and checked against its supposed behavior. The authors of Palladio used a number of different programming techniques to implement different portions of the system. These are summarized in Table 9.3. Object-oriented, data-oriented, and rule-based operation is supported by Loops, an object-oriented programming system. Logic programming is provided by the logic language MRS.

| Critter | [Kel84] | Functional and Timing Constraint Propagation |
| | | *- circuit and signal represented as expressions* |
| | | *- rules simplify and transform expressions* |
| | | *- used as a component in a number of other systems* |
| Dialog | [DeM84] | Verify Logic Levels and Timing in CMOS Circuits |
| | | *- rules written in Lextoc (Prolog-like)* |
| | | *- backward chaining and unification* |
| Debugger | [Kld85] | Verify Correct nMOS Transistor Connectivity |
| | | *- rules written in Prolog-like language* |
| VeryFun | [Woo85] | Functional Correctness of IC Logic |
| | [Woo86] | *- circuit and data stored as Prolog clauses* |
| | | *- performs logical (algebraic) simulation* |
| | | *- user-tailored primitives for incomplete circuits* |
| Prove | [Sri86] | Verify Logic against Functional Specification |
| | | *- logic network and function converted to C-Prolog* |
| | | *- rules determine equivalence of two forms* |
| TLTS | [Sim86] | Analysis and Redesign of Transmission Line Circuits |
| | | *- objects store deep knowledge of circuit behavior* |
| | | *- demons check for constraint consistency* |
| | | *- redesign by production rules in ORBS* |
| DFT | [Hor84] | Verify LSSD Design for Testability |
| | | *- circuit described by Prolog clauses* |
| | | *- rules check for controllability and observability* |
| | | *- algorithmic operators to speed operation* |
| PLA-ESS | [Bre85] | Select PLA Test Strategy |
| | | *- knowledge stored as empirical costs* |
| | | *- relies on user to make value judgements* |
| | | *- rule-driven menu-based interface* |
| FP | [Del85] | Functional Partitioning for Test Generation |
| | | *- object-based classification hierarchy* |
| | | *- forward chained rule-based clustering* |
| Amber | [How86] | Area Estimation Assistant for Custom Layout |
| | | *- estimation knowledge stored in module objects* |
| | | *- rule-based approximations for unknown modules* |

**Table 9.2**  Summary of Knowledge-Based Analysis Systems

| Paradigm | Application |
|---|---|
| Object-Oriented | Structural specification |
| Logical Language | Behavioral specification and simulation |
| Rule-Based | Incrementally constructed expert design aids |
| Data-Oriented | Constraint propagation |

**Table 9.3**  Programming Techniques Used in Palladio

Table 9.4 lists some other examples of knowledge-based tools used to manage the design process.

Schema        [Zip83]      Circuit Design Assistant
                                  *- object-based using Flavors*
                                  *- viewpoints focus attention on relevant data*
                                  *- user interactive redesign loop*

Class         [Bir85]      Custom IC Layout Assistant
                                  *- invokes other tools (design agents)*
                                  *- rule-based constraint propagation*
                                  *- control by user selection from alternatives*

Vexed         [Mit84]      Digital Circuit Design Consultant
                                  *- implementation methods stored as schemata*
                                  *- control knowledge provided by user*
                                  *- constraint propagation provided by Critter*

Cadre         [Ack85]      Custom IC Layout
                                  *- multiple agents under control of central manager*
                                  *- mix of rule-based and procedural subsystems*

Ulysses       [Bus85]      Custom IC Layout
                                  *- multiple agents communicate through blackboard*
                                  *- rules written in Scripts language*
                                  *- rules compile into Lisp, OPS5 and objects*

Hephaestus    [Sim85]      Interactive IC Layout
                                  *- rules interpret high level editing commands*
                                  *- rule-based layout checker*
                                  *- procedural layout agents (generators)*

Synapse       [Sub86]      Circuit Design from High Level Specification
                                  *- all design data represented as algebraic expressions*
                                  *- design by provably correct transformations*

**Table 9.4**  Summary of Design Management Systems

## Miscellaneous Systems

There are a number of knowledge-based systems that do not fit neatly into the three categories listed above. One of these is the Redesign system [Ste84] developed at Rutgers. Redesign is a rule-based interactive system that aids the user in modifying existing digital circuits to meet new functional or behavioral specifications. It requires, along with a description of the circuit to be modified, a copy of the design plan which describes how and why the original design decisions were made. This assumes that the original design was prepared or analyzed automatically by a system with an excellent explanation facility. Such a plan could, of course, be prepared manually, but it would no doubt take considerably longer than the proposed redesign. Rule-based knowledge of design tactics together with the ability to analyze and propagate signal constraints (using Critter) allow the system to suggest design modifications that will meet the new constraints. The system can also evaluate any changes suggested by the user. Table 9.5 lists a few other miscellaneous systems.

| | | |
|---|---|---|
| Assistant | [Dro85] | Datapath/Control Graph Trade-offs<br>*- circuit and control graph stored as Prolog clauses*<br>*- rules propose changes and maintain consistency* |
| Hitest | [Ben84] | Test Generation for Digital Circuits<br>*- object-based classification hierarchy*<br>*- heuristic knowledge of circuit function* |
| TDS | [Kar86] | Interface Timing Diagrams from Component Specs.<br>*- manufacturer's timing specs. as Prolog clauses*<br>*- timing analysis by forward chaining production rules*<br>*- interprets groups of Prolog clauses as rules* |
| Librarian | [How85] | IC Standard Cell Librarian<br>*- object-oriented database*<br>*- rules select cells that best match user specs.* |
| Neptune | [Foo86] | Select Modules from VLSI Library<br>*- objects store module attributes*<br>*- rule-based inheritance*<br>*- algorithmic selection written in C* |

**Table 9.5** Miscellaneous Knowledge-Based Tools

## 9.6  Application of Knowledge-Based Techniques

In this section, we take a closer look at six knowledge-based design systems, the design task they perform, and how they have been implemented. Each uses a different approach and illustrates a particular set of techniques.

### Weaver – Blackboard Systems

*Weaver* [Joo85a] is a channel/switchbox knowledge-based routing program. The problem it tackles is how best to route together a set of terminals placed on the boundary of a rectangular region using a specified number of routing layers. This is an important task in VLSI layout, and a number of algorithmic solutions have been proposed over the years. Finding the optimal solution is an NP-complete problem, and so heuristics are generally employed to limit the search space. These heuristic algorithms are frequently unable to complete 100% of the routes. Some of the reasons for this are (1) they use simple metrics, (2) they use an oversimplified model of the routing process, and (3) they do not know how to deal with special cases.

Weaver tries to directly capture the knowledge used by a human routing expert. Humans make routing decisions based upon a number of different metrics such as area, completion rate, and path length. At any one point in the process, these metrics may suggest different actions which must be resolved according to priorities and global goals. This leads to the architecture shown in Figure 9.10. It consists of a number of knowledge sources, each of which know how to plan or criticize according to some metric or simple rule. They communicate by posting results and suggestions onto a global working memory known as a blackboard. Control is performed by a "focus of attention" expert based on a set of priorities. Table 9.6 shows the priority assigned to each knowledge source and lists the number of OPS5 rules required by each.

The blackboard in Weaver is divided into three major partitions. The "problem partition" contains the design data objects (pins, nets, wires, *etc.*). The "decision partition" lists suggestions made by the various experts (for example, which net should be routed next). Both these partitions are accessed by all knowledge sources. The "scratch-pad partition" provides private areas for each knowledge source to perform internal bookkeeping.

An interesting aspect of Weaver is that, unlike human experts, it uses no backtracking. The authors felt that human backtracking decisions were too poorly understood to be coded directly into rules and that structured backtracking techniques were inadequate and inefficient. They discovered, however, that by combining enough predictive knowledge into the planning experts, they could arrive at good solutions (better than those achievable by algorithmic techniques) with no backtracking.

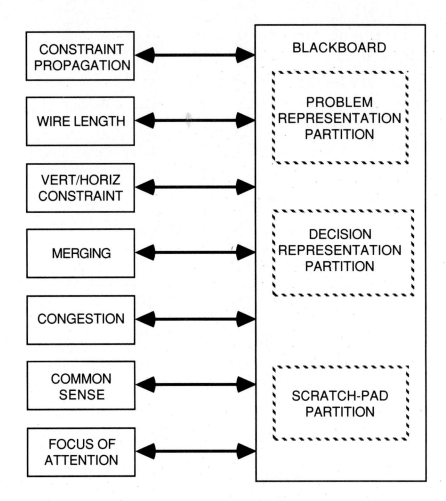

**Figure 9.10** Weaver Architecture

## Cadre − Cooperating agents

*Cadre* [Ack85] is a synthesis tool for converting hierarchical structural circuit descriptions into full custom VLSI layout. It is modeled on the human design process and consists of a number of expert agents connected through a central manager as shown in Figure 9.11. Each agent is expert in one aspect of VLSI layout (for example, floorplanning). The manager allocates tasks to the various agents and interprets their responses. The manager also provides an interactive

| Knowledge Source | No. of Rules |
|---|---|
| Constraint Propagation | 106 |
| Wire Length | 30 |
| Vert./Horiz. Constraints | 68 |
| Merging | 70 |
| Congestion | 10 |
| Common Sense | 31 |
| Focus of Attention | 26 |
| (Other) | 95 |

**Table 9.6**  Weaver Rule Priorities

user interface. Most importantly, the manager is responsible for ensuring constructive cooperation between the agents in order to arrive at a good global solution. In other words, the manager provides the overall design strategy.

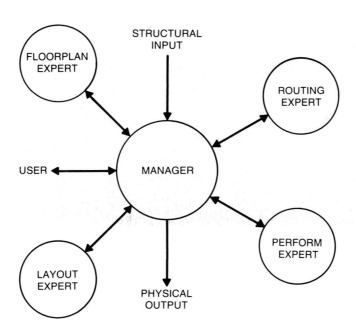

**Figure 9.11**  Cooperating Expert Agents

One advantage of this architecture is that it allows the various experts to be constructed independently, each agent concentrating on its own area of expertise. Another is that agents may be implemented in different ways. Some, for example, may be implemented as rule-based expert systems, while others use an algorithmic approach. For more difficult tasks, a human designer may even be called upon to play the role of an expert agent. There are initially three agents − a floorplanner [Wat86], a leaf cell layout expert, and a third agent known as the evaluator which is used to analyze the design and help the manager make backtracking decisions.

Like the human designer, Cadre relies on hierarchical subdivision to manage the design problem. The manager begins the design process using a top-down strategy in which a cell is designed according to restrictions placed on it by higher level cells already processed. It then analyzes and measures the result. If the manager judges the result to be poor, it uses information extracted from the analysis to redesign the cell's parent. This backtracking to the parent may cause part of the design to be discarded. It may also cause the parent's design to be re-analyzed. This bottom-up reaction to lower level design decisions can proceed as far up the tree as the manager deems cost effective. Design proceeds therefore as a series of top-down and bottom-up passes over the design hierarchy.

The crucial element in this system is, of course, the manager. It is the only part of the system that understands the overall design problem. Its responsibilities include maintaining the design database, assigning tasks to agents, maintaining consistency in the various design representations, supporting user interactions, maintaining a design history (for backtracking and explanation), and making design trade-offs.

Conceptually, the manager is organized into four pieces as shown in Figure 9.12. Modules are the objects that contain the problem specific design data (specifications, results and design history). Operators are design agents that are internal to the manager. Some of these act as an interface to the external agents shown in Figure 9.11. Operators are attached to modules according to a classification hierarchy. The tasker provides meta-control − design strategies based on the capabilities of the agents and user directives. The kernel provides the top level strategy. Its role is to drive Cadre from the initial (input) specifications, through various design refinements, towards the final design goal.

### Critter − Constraint Propagation

*Critter* [Kel84] is a tool used for verifying the functional correctness, speed, and timing robustness of a digital circuit. It takes, as input, algebraic expressions that describe (1) the hierarchical structure of the circuit, (2) the input/output behavior of each module in the circuit, and (3) the external interface (*i.e.,* the type and timing relationships of all the I/O signals). Based on this input, Critter algebraically analyzes the behavior of the circuit. It then produces a critique of the circuit highlighting (1) any errors that will cause the circuit not to work, (2) any signals that are close to their specification limit, and (3) the likely cause of these problems.

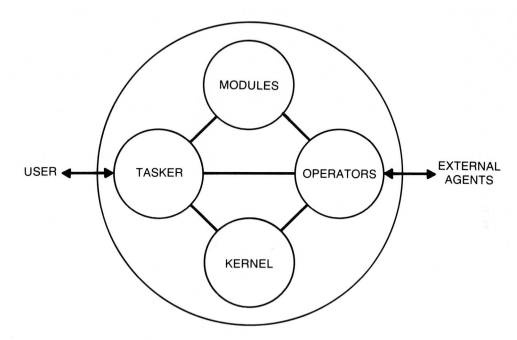

**Figure 9.12**  Internal Structure of Cadre Manager

The system is built around a central database that describes a circuit and its behavior in terms of modules and signals. Signals are represented as data streams that describe the signal as a semi-infinite sequence of data elements. Figure 9.13 describes a 7-bit parallel ASCII stream called ASCII-IN. Modules are described in terms of their operating requirements (for example, set up and hold times) and their input/output mappings. Figure 9.14 shows the behavior of a 7-bit module that is being used to latch the signal ASCII-IN.

Using these representations, Critter can (1) derive a module's output in terms of its input, (2) describe what inputs would be required to produce a specified output, and (3) determine how well any signals meet their specifications. Output signals are derived from inputs (and *vice versa*) using a type of algebraic simulation. Input/output mappings are applied as functions to the inputs. This generates a correct but usually very complex representation of output behavior. A set of rules is then applied to simplify the output expression. These rules implement techniques such as addition/subtraction of polynomials and Boolean equivalences. By repeated application to each submodule, Critter is able to evaluate behavior for all the signals in a circuit.

```
NAME of signal  =        ASCII-IN
INDEX-STRUCTURE  =       [(i (from 1 to + ∞ ))]
DATA-TYPE  =             CHARACTER
DATA-VALUE(i)  =         {not given}
ENCODING  =             (ASCII-ENCODED vector of 7 PARALLEL bits mapped
                            to voltage by POSITIVE-LOGIC conventions)
START-TIME(i)  =        (560*i + [130...147]) ns after T0
END-TIME(i)  =          (560*i + [230...247]) ns after T0
DURATION(i)  =          [normal distribution: μ = 100, σ = 4] ns
```

**Figure 9.13**  Example of Critter Data Stream

```
NAME  =                      LATCHED ASCII-IN
INDEX-STRUCTURE  =       [(i (from 1 to + ∞ ))]
DATA-TYPE  =             DATATYPE[ASCII-IN(i)]
DATA-VALUE(i)  =         DATAVALUE[ASCII-IN(i)]
ENCODING  =             ENCODING[ASCII-IN(i)]
START-TIME(i)  =        (the time of the NEXT (RISE of CCK) after
                            START-TIME[ASCII-IN(i)] + [18...35] ns
END-TIME(i)  =          (the time of the NEXT (RISE of CCK) after
                            END-TIME[ASCII-IN(i)] + [3...10] ns
DURATION(i)  =          PERIOD[CCK] - [15...25] ns
```

**Figure 9.14**  Example of Critter Module

Once all signal behaviors have been determined, they are compared to the user's specifications. This is achieved by substituting the behavior expressions into the specifications and algebraically reducing the result to a simple true or false proposition. Based on these comparisons a critique is prepared. An example is shown in Figure 9.15. More detailed information on any reported failure can be gained by interactively browsing the signal database.

## Design Automation Assistant – Knowledge Acquisition

The Design Automation Assistant (DAA) [Kow85a, Kow85b] is a system that generates a technology independent list of operators, registers, datapaths, and control signals from an algorithmic (ISPS) description. Figure 9.16 shows a sample ISPS input (taken from an instruction decoder). Figure 9.17 shows the datapath design output by DAA.

The system splits the design up into a number of subtasks as shown in Figure 9.18. Global allocation is the process of assigning base variable storage elements (registers, memories, controllers). It also sets up ports and assigns constraints. Dataflow allocation assigns operators to control phases, allocates

\# Signal behaviors computed: 3
\# Specification enumerated: 21, of which 3 proved redundant
\# Specifications proven satisfied: 17, of which 14 involved timing,
     and were satisfied by margins between 15 and 482 ns.

{end report here if no specs. violated}

The following *1* specification was proven:
1.   "set-up of PAR-EBCDIC w.r.t. clock BIT-CK"
    at submodule SHIFT,REGISTER missed by a margin of 5 ns
    (on total path delay of 535 ns.)

LIKELY CAUSE: "pessimistic delay approx."
SYMPTOM: "timing error $< 3\%$ total path using discrete time estimate"
RECOVERY: "Re-evaluate specification probabilistically"

**Figure 9.15**  A Critter Critique

Example:-
**Begin**

\*\* Storage.Declaration \*\*

cpage\current.page$<0:4>$,
i\instruction$<0:11>$,
    pb\page.0.bit$<>$ := i$<4>$,
    pa\page.address$<0:6>$ := i$<5:11>$,

\*\*Address.Calculation\*\*

Global eadd\effective.address$<0:11>$ :=
    **Begin**
    Decode pb $=>$
        **Begin**
        0 := eadd = '00000 @ pa,
        1 := eadd = cpage @ pa
        **End**
    **End**
**End**

**Figure 9.16**  DAA Input Description (ISPS)

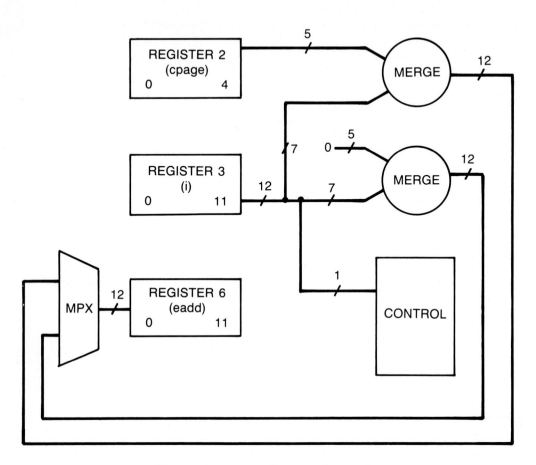

**Figure 9.17** DAA Output (datapath)

temporary registers, and then either assigns operator function to registers (for example, shift-register) or creates ALUs to perform the functions. Module allocation performs local improvements like removing unnecessary registers and redundant modules and merging ALUs. Global improvement removes unused modules and ports, uses fanout to replace duplicated outputs, and allocates bus structures.

Each of these subtasks is implemented as a set of rules coded in OPS5 (there are over 300 rules total). Figure 9.19 shows an example of one of the rules from the global improvement subtask. In addition there are some algorithmic operators written in $C$. These are used for estimating the gain (or loss) in terms of number of components and interconnection complexity that results from merging two modules.

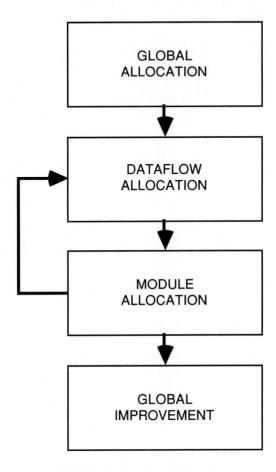

**Figure 9.18**  DAA Subtasks

An interesting aspect of DAA is the fact that knowledge was acquired using the classical expert system technique of having a knowledge engineer (who in this case had some knowledge of the field) interview a number of human experts. Kowalski gives an excellent account of this in his thesis [Kow85b]. The knowledge acquisition process began with the encoding of some straight book knowledge. This was followed by a series of case-study interviews in which designers discussed their views of the design process, including partitioning, component selection, allocation, *etc.* Four designers were interviewed: a novice, two who were moderately experienced, and an expert. Non-expert designers were included because it was felt that they might be better able to explain the reasoning behind their decisions.

**if:**     the most current active context is bus allocation
and there is a module that is a multiplexer
and there is another module that is also a multiplexer
and there is a link from a non-bus module to the first multiplexer
and there is a link from that module to the second multiplexer
and there is a link from another non-bus module to the first multiplexer
and there is a link from that module to the second multiplexer
**then:**   place these connections on an idle bus

**Figure 9.19**  Example of DAA Rule

From the information gained from these interviews, a prototype system was constructed. This system contained about 70 rules. The next step was to use this prototype to design a simple microprocessor (MC6502). This was followed by a number of acquisition interviews in which designers were asked to critique the design. Only expert designers were used during this phase. Based on these critiques, old rules were modified and new rules were added. This process continued until the system contained over 300 rules. Kowalski points out that this process will never be complete. There will always be the need to add new knowledge to improve the performance of the system.

The system that had been developed using the MC6502 example was then tested on a much more ambitious problem — the IBM 370. Only one design of this architecture was generated and then presented to a number of experts at IBM to evaluate the performance of the system. Their evaluation was that DAA had produced a design whose quality was what they would expect from "one of their better designers." This part of the experiment was significant in that it demonstrated the generality of the knowledge captured during earlier phases.

Thomas *et al.* [Tho83] wrote an interesting paper in which they compared DAA to another tool, *EMUCS*, which performed approximately the same function using a more traditional procedural approach. At that time, it was concluded that EMUCS achieved better register allocation and ran in a shorter time. DAA, however, provided greater flexibility and extensibility as was evidenced by the IBM 370 experiment. Since that time DAA has been improved to produce much better results than EMUCS.

More recently, DAA has been augmented to include an algorithmic preprocessor called *BUD* [McF86]. In describing this new version, Kowalski notes that the original DAA was limited in its ability to deal with global design issues. In addition, its top-down refinement strategy (with no backtracking) made it difficult to include any low level physical costs in making high level design decisions. BUD performs a global analysis of the design, predicting the throughput/area trade-offs of different resource allocations by actually evaluating trial floorplans. It uses its knowledge of the cost of layout, wiring, *etc.*, to suggest an initial assignment and clustering of modules. DAA then uses this new information to improve its module and bus allocation phases.

### Logic Verifier – Backward Chaining

Woo [Woo85] describes a Prolog-based tool for verifying the functional and behavioral characteristics of logic circuits. The verifier consists of two modules, an evaluator and a checker, as shown in Figure 9.20. The evaluator builds an expression for each output of the circuit based on the input specifications and the logic circuit description. The checker compares these outputs to the required output specifications and produces a report listing any discrepancies. The function of the logic verifier is quite similar, therefore, to that of Critter. It is implemented, however, using a quite different technique.

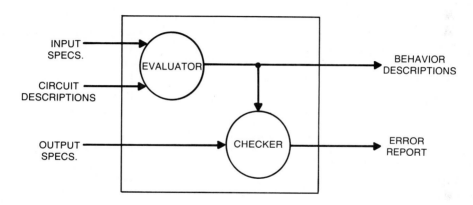

**Figure 9.20**  Logic Verifier

The circuit structure, input signals, and desired output signals are entered by the user as Prolog clauses. Signals are represented as a series of logic-value/time pairs. For example,

$$\text{sel}([f,0],[t,10],[a,T2]])$$

describes a signal *sel* whose initial value at time *0* is false, becomes true at time *10*, and finally takes on the value *a* at time *T2*. Output signals may also take on the value *don't-care* to suppress checking. Primitive logic components are described as sets of Prolog clauses that operate on signals. Figure 9.21, for example, shows the five clauses that specify the operation of a zero delay inverter. Note that *TA* and *TB* are signals (sequences of logic-value/time pairs) and that "|" is used as a concatenation operator. More complex circuits are then built out of primitive components as shown in the example of Figure 9.22.

1.    invg([[*t*,*Time*] | *TA*],[[*f*,*Time*] | *TB*]) :- invg(*TA*,*TB*).
2.    invg([[*f*,*Time*] | *TA*],[[*t*,*Time*] | *TB*]) :- invg(*TA*,*TB*).
3.    invg([[*X*,*Time*] | *TA*],[[not(*X*),*Time*] | *TB*]) :- invg(*TA*,*TB*).
4.    invg([[not(*X*),*Time*] | *TA*],[[*X*,*Time*] | *TB*]) :- invg(*TA*,*TB*).
5.    invg([],[]).

**Figure 9.21**  Prolog Description of Inverter

mux(*ENBar*, *Sel*, *I*0, *I*1, *Y*) :-
    inv(*Sel*, P1),                  % for inverter
    andg(*Sel*, *I*1, P2),       % for AND gate *A*1
    andg(P1, *I*0, P3),        % for AND gate *A*2
    org(*ENBar*, P2, P3, *Y*).   % for OR gate

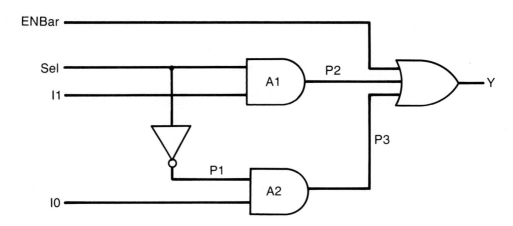

**Figure 9.22**  Combining Primitive Components into Complex Structures

Verification is initiated by generating a Prolog query. An example is given in Figure 9.23. This query is asking the Prolog interpreter if all six formulae can be simultaneously true. The first three bind input signals (*Data*, *CP* and *RBB*) to some predefined sequences (*sdin*, *cpin* and *rbbin*). The fourth applies these inputs to a module, *stp*8, and binds a variable *ReadyGen* to the *ready* output of that module (the other outputs are ignored). The last formula calls upon the checker to compare the output *ReadyGen* to the specified output *ReadyGiven* and produce a report called *ReadyReport*. Each term in these formulae is a consequent of a Prolog clause describing circuit structure of signal behavior.

$$: \quad \text{sdin}(Data), \text{cpin}(CP), \text{rbbin}(RBB),$$
$$\text{stp8}(Data, CP, RBB, ReadyGen, ,, ,, ,, ,, ,, ,, ,, ),$$
$$\text{ready}(ReadyGiven),$$
$$\text{checker}(ReadyGen, ReadyGiven, ReadyReport).$$

**Figure 9.23**  Prolog Verification Query

The Prolog interpreter searches for a proof of the query by recursively replacing each of the terms in these formulae with its antecedents. As it does so, it is constrained to bind variables in a consistent fashion. A signal cannot, for example, at some point in the search process have two different values. This process is known as *unification*. The process terminates when either (1) the query is proven true or (2) the search space is exhausted (which implies the query is false). In the logic verifier, the query always returns true but assigns, in the process, a series of values to the output of the checker which indicate whether or not the circuit is behaving as specified. An example of an error report is shown in Figure 9.24.

$$ReadyGiven = [[f,0],[t,77],[f,110]]$$
$$ReadyGen = [[f,0]]$$
$$ReadyReport = [[succ,0],[error,77],[succ,110]]$$

**Figure 9.24**  Logic Verifier Error Report

## Synapse − Transformation Schemata

*Synapse* [Sub86] is an expert system architecture for supporting provably correct VLSI design transformations. The primary goal of the system is to allow very high level functional and behavioral specifications to be mapped into custom VLSI layout.

A major factor that complicates the building of integrated design systems is that each representation of a circuit (for example, logic gates, truth tables, SLA patterns, algorithmic expressions) has its own set of concepts (primitives and semantics). In Synapse, all representations are built around a basic set of concepts that can be manipulated by formal reasoning techniques. Systems are described, in all representations and at all levels of abstraction, as algebraic expressions built on these basic concepts. The process of design then becomes a series of provably correct algebraic transformations that begin with the specification expressions and ultimately produce expressions describing mask layout.

The architecture of the system is shown in Figure 9.25. The working memory contains (1) the design data in the form of specification and result expressions and (2) a set of axioms that define the semantics (primitive components and their legal combinations) for each representation. The inference rules are formal techniques for manipulating and reasoning about the basic concepts. Examples are demodulation, paramodulation, and resolution. The strategy is a set of rules that guide the system in finding transformations that will lead toward a solution. This block also controls user interaction. The overall design environment includes an object-based representation [Gab87] that is intended to support the evolution of design paradigms, system software, and domain knowledge.

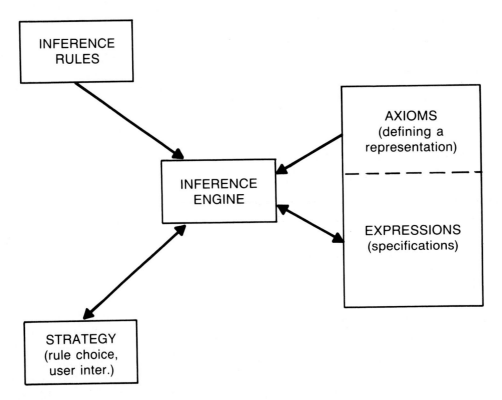

**Figure 9.25** Synapse Architecture

Transformation knowledge is also stored in the form of schemata. A schema is a plan that describes how a particular pattern in one representation can be transformed into a new pattern in another representation. A schema consists of (1) an input behavior expression (the pattern to be matched), (2) an output behavior expression (the implementation), (3) a set of properties that describe in general terms the effects of applying this transformation (for example, area/time trade-offs), (4) preconditions that must be satisfied by the input pattern, and (5) postconditions that will hold after the application of the schema. Individual schemata are proven correct before being added to the system. This eliminates the need to verify correctness every time the schema is used. Figure 9.26 shows the result of applying a schema that knows how to transform recursive behavior expressions into linear arrays. It has been applied to the list operator *deleteMax* to produce a simple array implementation of that function.

## 9.7  Future Directions

Finally, it is interesting to take a look at where the field is heading. One of the most pressing tasks is to find good models for capturing, representing and manipulating the many different forms of design knowledge. Just as the the production system model has been successfully used to solve many quite different diagnostic problems, so we need a general model for dealing with design problems. Such a model would represent control knowledge as well as domain knowledge and efficiently compose them together with procedural solutions to algorithmic subtasks.

### Learning

Another exciting area is the possibility of incorporating machine learning into these systems. Knowledge-based symbolic representations provide a solid foundation on which to conduct learning experiments. Tools that act as design assistants have direct access to human design techniques coded in a form that is understood by the system. Learning offers the possibility of short circuiting the long, expensive knowledge acquisition phase of building expert systems.

Constellation [Lat86] and Leap [Mit85] are two experimental systems that have been set up to explore learning possibilities. Leap has been developed at Rutgers as part of the Vexed project. Leap notes when the designer chooses not to select one of the options suggested by Vexed but instead proposes a new solution. The system takes the problem and its new solution as a training example. Leap then tries to analytically prove the transformation expressed in the training example. If a proof is found, the steps taken in the proof are used to expand or generalize the training example into a new transformation rule.

Acosta *et al.* [Aco86] describe a tool for designing digital systems that abstracts design plans and stores them as a partially ordered set of Lisp rules. When a new design is requested, the system searches its set of plans for an old problem that most closely matches the new design specification. If a suitable match is found, it proceeds by analogically mapping the old plan onto the new problem.

*Behavioral Input:*

DeleteMax($l$) is defined as:

DeleteMax(*NewList*) =   *NewList*
DeleteMax(Add($l,x$)) =   $l$ **if** x > Max($l$)
                          Add(DeleteMax($l$),$x$) otherwise

*Structural Output:*

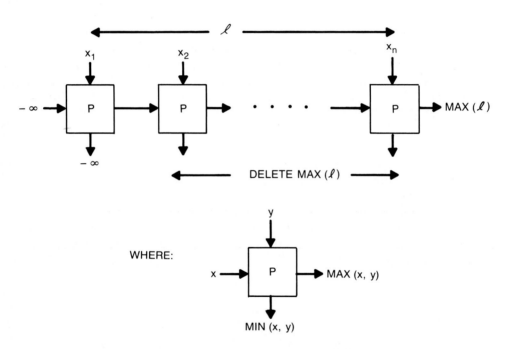

**Figure 9.26**  Transformation by Schema

## Multiprocessing

Another development will likely be the use of multiprocessing engines to speed up the execution of knowledge-based tools. Systems that are configured around multiple agents or knowledge sources would seem especially suited to this style of implementation. This raises all sorts of interesting problems, however, in data consistency and constraint propagation. Transaction processing [Ens86], a database technique for ensuring reliable multiple user access to common data, may then become an integral feature of large CAD systems.

**Summary**

Physical design is a knowledge intensive process. New programming techniques based on explicit knowledge representation offer the hope of extending design automation into new areas of the design process. Many prototype knowledge-based design systems have been built, a number of which have been described in this chapter. The results obtained by these systems are very encouraging; they demonstrate viable techniques for dealing with difficult problems like uncertainty, incomplete knowledge, conflicting constraints, special cases, design trade-offs, and expert user interaction. Before jumping to the conclusion, however, that the future of CAD lies in expert systems, it would be valuable to put this work in perspective.

Most of the CAD tools in existence today were programmed using conventional procedural programming techniques. For those tasks which map well into algorithmic form, this remains the implementation of choice. It makes no sense to try to apply expert systems techniques to those problems for which there exists a good algorithmic solution, for the algorithmic solution will invariably be faster to implement and simpler to debug and execute more reliably in considerably less time. Knowledge-based systems are not a replacement for algorithmic CAD tools. Rather, they are a powerful new technique for exploring those portions of the design process which, up until now, have proved too difficult or expensive to automate.

Many of the systems described in this chapter are experimental prototypes. They have been built to demonstrate concepts and are neither complete nor robust enough to be used in a real design environment. This does not diminish their value as exploratory systems. It does, however, point out the enormous amount of work that remains to be done in this field. Developing a real-world rule-based expert system from initial concept through problem modeling, knowledge acquisition, prototyping and rule refinement to operation in the field is an enormous task usually covering many years of effort. In addition, we still do not have a good general model of how to represent all the various kinds of design knowledge in a single integrated system. Programming tools are evolving as systems are being built. While it is important to recognize the steps taken in these experimental systems, it is also important not to oversell the state of the art and the distance that remains to be traveled.

Finally, these caveats are not intended, in any way, to devalue the application of knowledge-based techniques to physical design automation. Rather, they are to recognize the fact that much good work has been done in this field and to protect that work from being lost in the frenzy created as the world rushes to embrace expert systems.

## Exercises

**9.1**  What is the difference between a rule-based production system and a *C* or *Pascal* program coded as a series of "if-then" statements? Could a production system be written in Pascal? What code would have to be written to support rule-based operation?

**9.2**  In what ways can knowledge-based systems be used to improve the (a) programmer interface and (b) user interface?

**9.3**  Why is the synthesis of a system usually more difficult than its analysis? How can the redesign model be used to simplify the synthesis process? Why is the redesign model not always applicable? Give examples.

**9.4**  Describe some of the ways in which human experts learn from experience. Why is learning so important to the future of expert systems?

**9.5**  How is backtracking used in design systems? Why is it sometimes very expensive? How can its cost be reduced?

**9.6**  Explain the difference between forward and backward chaining inference engines. What analogies can you draw to top-down/bottom-up design? In what ways do these two concepts differ?

**9.7**  Try to describe building a kite without using the words "then" or "next" (or any other implied sequence word).

**9.8†**  Choose a design task with which you are familiar. Describe (giving examples) the different types of knowledge you would use to perform this task. Try to encode some of that knowledge as rules.

**9.9†**  Generate a set of rules that can be used to determine the logic state (0, 1 or X) of any one terminal of an n- or pMOS transistor given the state of the other two. Expand the rule set to that required to statically analyze an arbitrary network of n- and pMOS transistors in which a few nodes (for example, inputs) have been set to 0 or 1 and all others have been initialized X. What control structures are needed to correctly apply the rule set to a circuit?

**9.10‡**  Locate a colleague who is expert in some field in which you are relatively unfamiliar. Interview the expert to gain some knowledge of how problem solving occurs in that particular field. Try to develop a simple model describing the objects in that field and encode some of the knowledge as rules operating on those objects. What type of knowledge is simple to encode? What type is difficult?

# References

[Ack85]  Ackland, B., A. Dickenson, R. Ensor, J. Gabbe, P. Kollaritsch, T. London, C. Poirier, P. Subrahmanyam, and H. Watanabe, "CADRE — a system of cooperating VLSI design experts," in *IEEE Intl. Conf. on Computer Design*, pp. 99-104, 1985.

[Aco86]  Acosta, R., M. Huhns, and S. Liuh, "Analogical reasoning for digital system synthesis," in *Digest Intl. Conf. on Computer-Aided Design*, pp. 173-176, 1986.

[Afs86]  Afsarmanesh, H., D. Knapp, D. McLeod, and A. Parker, "Information management for VLSI/CAD," in *Proc. Intl. Conf. on Computer Design*, pp. 476-481, 1986.

[Ben84]  Bending, M. "Hitest: a knowledge-based test generation system," *IEEE Design and Test*, pp. 83-92, May 1984.

[Bir85]  Birmingham, W., R. Joobbani, J. Kim, D. Siewiorek, and G. York, "CLASS: a chip layout assistant," in *Digest Intl. Conf. on Computer-Aided Design*, pp. 216-218, 1985.

[Bir86]  Birmingham, W., R. Joobbani, and J. Kim, "Knowledge-based expert systems and their application," in *Proc. 23rd Design Automation Conf.*, pp. 531-539, 1986.

[Bob82]  Bobrow, D., and M. Stefik, "The LOOPS manual," Memo 2KB-VLSI-81-13, Xerox Palo Alto Research Center, Palo Alto, CA, August 1982.

[Bow85]  Bowman, R., and D. Lane, "A knowledge-based system for analog integrated circuit design," in *Proc. Intl. Conf. on Computer Design*, pp. 210-212, 1985.

[Bre85]  Breuer, M., and X. Zhu, "A knowledge-based system for selecting a test methodology for a PLA," in *Proc. 22nd Design Automation Conf.*, pp. 259-265, 1985.

[Bro83]  Brown, H., C. Tong, and G. Foyster, "Palladio: an exploratory environment for circuit design," *IEEE Computer*, vol. 16, no. 12, pp. 41-58, December 1983.

[Buc78]  Buchanan, B., and E. Feigenbaum, "Dendral and meta-dendral: their applications dimensions," *Artificial Intelligence*, vol. 11, pp. 5-24, 1978.

[Bus85]  Bushnell, M., and S. Director, "ULYSSES — an expert-system based VLSI design environment," in *Proc. Intl. Symposium on Circuits and Systems*, pp. 893-896, 1985.

[Bus86]  Bushnell, M., and S. Director, "VLSI CAD tool integration using the ulysses environment," in *Proc. 23rd Design Automation Conf.*, pp. 55-61, 1986.

[Cla85]   Clark, G., and R. Zippel, "Schema — an architecture for knowledge-based CAD," in *Digest Intl. Conf. on Computer-Aided Design*, pp. 50-52, 1985.

[Dah66]   Dahl, O-J., and K. Nygaard, "Simula — an algol based simulation language," *Communications of the ACM*, vol. 9, no. 9, pp. 671-678, 1966.

[Del85]   Delorme, C., P. Roux, L. Demains D'Archimbaud, H. Giambiasi, R. L'Bath, B. Mac Gee, and R. Charroppin, "A functional partitioning expert system for test sequences generation," in *Proc. 22nd Design Automation Conf.*, pp. 820-824, 1985.

[DeM84]   De Man, H., I. Bolsens, and E. Vanden Meersch, "An expert system for logical and electrical debugging of MOSVLSI networks," in *Digest Intl. Conf. on Computer-Aided Design*, pp. 203-205, 1984.

[Den82]   Denyer, P., D. Renshaw, and N. Bergamnn "A silicon compiler for VLSI signal processors," *European Solid State Circuits Conf.*, 1982.

[Dic86]   Dickinson, A., "Floyd: a knowledge-based floor plan designer," in *Proc. Intl. Conf. on Computer Design*, pp. 176-179, 1986.

[Dix84]   Dixon, J., and M. Simmons, "An architecture for application of artificial intelligence to design," in *Proc. 21st Design Automation Conf.*, pp. 634-640, 1984.

[Dro85]   Drongowski, P., "A graphical rule-based assistant for control graph — datapath design," in *Proc. Intl. Conf. on Computer Design*, pp. 208-211, 1985.

[Dud79]   Duda, R., J. Gaschnig, and P. Hart, "Model design in the prospector consultant system for mineral exploration," in Michie, D. (editor), *Expert Systems in the Microelectronic Age*, Edinburgh University Press, Edinburgh, Scotland, pp. 153-167, 1979.

[Ens86]   Ensor, R., and J. Gabbe, "Transactional blackboards," *Intl. Journal for Artificial Intelligence in Engineering*, vol 1, no. 2, pp. 80-84, Oct. 1986.

[Foo86]   Foo, Y., and H. Kobayashi, "A knowledge-based system for VLSI module selection," in *Proc. Intl. Conf. on Computer Design*, pp. 184-187, 1986.

[For77]   Forgy, C., and J. McDermott, "OPS, a domain independent production system language," *Intl. Joint Conf. on Artificial Intelligence*, pp. 933-939, 1977.

[Gab87]   Gabbe, J., and P. Subrahmanyam, "A representation for the evolution of VLSI designs," to appear in *Knowledge-based Approaches to Design*, edited by D. Sriram and C. Tong, Addison-Wesley, Reading, MA, 1987.

[Gaj84]  Gajski, D., "Silicon compilers and expert systems for VLSI," in *Proc. 21st Design Automation Conf.*, pp. 86-87, 1984.

[Gev87]  Gevarter, W., "The nature and evaluation of commercial expert system building tools," *IEEE Computer*, pp. 24-41, May 1987.

[Hor84]  Horstmann, P., and E. Stabler, "Computer aided (CAD) design using logic programming," in *Proc. 21st Design Automation Conf.*, pp. 144-151, 1984.

[How85]  Ho, W., Y. Hu, and D. Yun, "An intelligent librarian for VLSI cell databases," in *Proc. Intl. Conf. on Computer Design*, pp. 78-81, 1985.

[How86]  How, M., and B. Pan, "AMBER — a knowledge-based area estimation assistant," in *Proc. Intl. Conf. on Computer Design*, pp. 180-183, 1986.

[Joo85a]  Joobbani, R., and D. Siewiorek, "Weaver: a knowledge-based routing expert," in *Proc. 22nd Design Automation Conf.*, pp. 266-272, 1985.

[Joo85b]  Joobbani, R., D. Siewiorek, and S. Talukdar, "Application of knowledge-based expert systems to detailed routing of VLSI chips," in *Proc. Intl. Conf. on Computer Design*, pp. 199-202, 1985.

[Jos85]  Joseph, R., "An expert systems approach to completing partially routed printed circuit boards," in *Proc. 22nd Design Automation Conf.*, pp. 523-528, 1985.

[Kar86]  Kara, A., R. Rastogi, and K. Kawamura, "TDS: an expert system to automate timing design for interfacing VLSI chips in microcomputer systems," in *Digest Intl. Conf. on Computer-Aided Design*, pp. 362-365, 1986.

[Kel84]  Kelly, V., "The CRITTER system — automated critiquing of digital circuit designs," in *Proc. 21st Design Automation Conf.*, pp. 419-425, 1984.

[Kim84]  Kim, J., J. McDermott, and D. Siewiorek, "Exploiting domain knowledge in IC cell layout," *IEEE Design and Test*, pp. 52-64, August 1984.

[Kld85]  Kolodny, A., R. Friedman, and T. Ben-Tzur, "Rule-based static debugger and simulation compiler for VLSI schematics," in *Digest Intl. Conf. on Computer-Aided Design*, pp. 150-152, 1985.

[Kna83]  Knapp, D., J. Granacki, and A. Parker, "An expert synthesis system," in *Digest Intl. Conf. on Computer-Aided Design*, pp. 164-165, 1983.

[Kol85]  Kollaritsch, P., and N. Weste, "TOPOLOGIZER: an expert system translator of transistor connectivity to symbolic cell layout," *IEEE Journal Solid State Circuits*, vol. 20, no. 3, pp. 799-804, 1985.

[Kow85a] Kowalski, T., and D. Thomas, "The VLSI design automation assistant: what's in a knowledge base," in *Proc. 22nd Design Automation Conf.*, pp. 252-258, 1985.

[Kow85b] Kowalski, T., *An Artificial Intelligence Approach to VLSI Design*, Kluwer Academic Press, Norwell, MA, 1985.

[Lat77] Latombe, J., "Artificial intelligence in computer-aided design: the TROPIC system," J. Allan III, editor, *CAD Systems*, North Holland, Amsterdam, the Netherlands, pp. 61-99, 1977.

[Lat86] Lathrop, R., and R. Kirk, "A system which uses examples to learn VLSI structure manipulation," *National Conf. of AAAI*, pp. 1024-1028, 1986.

[McD81] McDermott, J., "Domain knowledge and the design process," in *Proc. 18th Design Automation Conf.*, pp. 580-588, 1981.

[McF86] McFarland, M. and T. Kowalski, "Assisting DAA: the use of global analysis in an expert system," in *Proc. Intl. Conf. on Computer Design*, pp. 482-485, 1986.

[Mit84] Mitchell, T., L. Steinberg, and J. Shulman, "A knowledge-based approach to design," *IEEE Workshop on Principles of Knowledge-Based Systems*, pp. 27-34, December 1984.

[Mit85] Mitchell, T., S. Mahadevan, and L. Steinberg, "LEAP: a learning apprentice for VLSI design," *Intl. Joint Conf. on Artificial Intelligence*, pp. 575-580, 1985.

[Mit86] Mittal, S., and A. Araya, "A knowledge-based framework for design," *National Conf. of AAAI*, pp. 856-863, 1986.

[Nil80] Nilsson, N., *Principles of Artificial Intelligence*, Tioga Publishing Company, Palo Alto, CA, 1980.

[Oda85] Odawara, G., K. Iijima, and K. Wakabayashi, "Knowledge-based placement technique for printed wiring boards," *Proc. 22nd Design Automation Conf.*, pp. 616-622, 1985.

[Rub83] Rubin, S., "An integrated aid for top down electrical design," in *Proc. 2nd Intl. Conf. on VLSI (VLSI83)*, pp. 63-72, 1983.

[Sho76] Shortliffe, E., *Computer Based Medical Consultations: MYCIN*, Elsevier Scientific, New York, 1976.

[Shr83] Shrobe, H., "AI meets CAD," in *2nd Intl. Conf. on VLSI*, pp. 387-399, 1983.

[Sim85] Simoudis, E., and S. Fickas, "The application of knowledge-based design techniques to circuit design," in *Digest Intl. Conf. on Computer-Aided Design*, pp. 213-215, 1985.

[Sim86]   Simoudis, E., B. Bee, R. Evans, and D. Vonada, "An AI system for improving the performance of complex transmission line networks," in *Digest Intl. Conf. on Computer-Aided Design*, pp. 366-369, 1986.

[Sis82]   Siskind, J., J. Southard, and K. Crouch, "Generating custom high performance VLSI designs from succinct algorithmic descriptions," *Conf. on Advanced Research in VLSI*, pp. 28-39, January 1982.

[Spi85]   Spickelmier, R., and A. R. Newton, "Connectivity verification using a rule-based approach," in *Digest Intl. Conf. on Computer-Aided Design*, pp. 190-192, 1985.

[Sri86]   Srinivas, N., and V. Agrawal, "PROVE: prolog based verifier," in *Digest Intl. Conf. on Computer-Aided Design*, pp. 306-309, 1986.

[Sta77]   Stallman, R., and G. Sussman, "Forward reasoning and dependency-directed backtracking in a system for computer-aided circuit analysis," *Artificial Intelligence*, vol. 9, pp. 135-196, 1977.

[Ste82]   Stefik, M., and L. Conway, "Toward the principled engineering of knowledge," *The AI Magazine*, pp. 4-16, Summer 1982.

[Ste84]   Steinberg, L., and T. Mitchell, "A knowledge-based approach to VLSI CAD — the redesign system," in *Proc. 21st Design Automation Conf.*, pp. 412-418, 1984.

[Sub86]   Subrahmanyam, P., "Synapse: an expert system for VLSI design," *IEEE Computer*, vol. 19, no. 7, pp. 78-89, July 1986.

[Sus80]   Sussman, G., J. Holloway, and T. Knight, "Design aids for digital integrated systems, an artificial intelligence approach," *IEEE Conf. on Circuits and Computers*, pp. 612-615, 1980.

[Sym84]   Symbolics Inc. *Reference Guide to Symbolics Lisp*, Symbolics Inc., Cambridge, MA, 1984.

[Tho83]   Thomas, D., C. Hitchcock III, T. Kowalski, J. Rajan, and R. Walker, "Automatic datapath synthesis," *IEEE Computer*, vol. 16, no. 12, pp. 59-70, December 1983.

[Wal85]   Walker, R., and D. Thomas, "A model of design representation and synthesis," in *Proc. 22nd Design Automation Conf.*, pp. 453-459, 1985.

[Wat86]   Watanabe, H., and B. Ackland, "Flute — a floorplanning agent for full custom VLSI design," in *Proc. 23rd Design Automation Conf.*, pp. 601-607, 1986.

[Wei84]   Weiss, S., and C. Kulikowski, *A Practical Guide to Designing Expert Systems*, Rowman & Allenheld, Totowa, NJ, 1984.

[Wol86]   Wolf, W., "An object-oriented, procedural database for VLSI chip planning," in *Proc. 23rd Design Automation Conf.*, pp. 744-751, 1986.

[Woo85]   Woo, N., "A prolog based verifier for the functional correctness of logic circuits," in *Proc. Intl. Conf. on Computer Design*, pp. 203-207, 1985.

[Woo86]   Woo, N., "Validating the functional correctness of incomplete logic circuits with the aid of a verification system," in *Digest Intl. Conf. on Computer-Aided Design*, pp. 302-305, 1986.

[Yos86]   Yoshimura, T., and S. Goto, "A rule-base and algorithmic approach for logic synthesis," in *Digest Intl. Conf. on Computer-Aided Design*, pp. 162-165, 1986.

[Zip83]   Zipple, R., "An expert system for VLSI design," in *Intl. Symposium on Circuits and Systems*, pp. 191-193, 1983.

# Chapter 10

# The Combinatorial Complexity of Layout Problems

**Thomas Lengauer**

*University of Paderborn*
*Paderborn, West Germany*

## 10.1 Introduction

The circuit layout problem is a constrained optimization problem: we look for a layout that is optimal with respect to one or more cost measures (area, maximum wire length, *etc.*) and that meets certain layout requirements (sufficient distance between different wires, restricted number of wiring layers, *etc.*). In general the layout problem is so difficult that optimal layouts cannot be found in reasonable time. Thus, heuristic methods are used to approximate the optimal layouts by "good" layouts. One such heuristic consists of breaking the process of laying out a circuit into phases: partitioning, placement, global routing, and detailed routing. Each phase solves a subproblem that is not quite as hard as the total layout problem. On the other hand, breaking the layout process into phases precludes finding optimal layouts, because the early phases of optimization use cost measures that only approximate the demands of later phases. As an example, for placement optimization, the wire-length of a net may be approximated by the length of the half-perimeter of the rectangle circumscribing the net. This figure of merit, however, is only a lower bound on the real wire length of the net in Manhattan geometries. The exact wire length is not known at the time of placement. Thus, the result of the placement process may actually be suboptimal with respect to the following routing phase.

The separation of the layout process into phases introduces subproblems that still are hard. For instance, placement is hard to do; no fast way of finding optimal placements is known. Thus, placement is broken into phases again, for example, partitioning, orientation of the slicing tree, *etc.* Eventually the subproblems can be solved efficiently, or we resort to heuristics for finding good, if not optimal, solutions to the subproblems. It is evident that there is no precise definition of a "good" layout in this context. Usually the layout procedure is evaluated by comparing its performance with those of other competing layout procedures. The notion of performance used here entails both the computing resources expended by the layout procedure and the quality of the resulting layout. However, this approach yields a quite subjective measure of quality. A much more objective measure would be obtained by comparing the performance of the layout procedure to all *conceivable* layout procedures. Rather than discovering that the layout system under evaluation, on *typical* circuits, finds layouts that are about 15% smaller in area than corresponding layouts done on *some competing* layout system, one would show that the layout system on *all* circuits finds layouts that are, say, at most 5% larger in area than the *optimal* layout. Methods for making such absolute statements about the performance of layout algorithms actually exist. They will be discussed in this chapter. Unfortunately they cannot be applied to all layout procedures but only to algorithms that are analyzable in this sense. As of yet such algorithms often do not reach into the range of performance needed for economic layout. In other words, rather than ensuring at most 5% deviation from optimal area, the figures are more like several hundred percent — statements which are of no use for practical purposes. Also, such layouts can often be found only by procedures that expend a lot of computing resources. But this is not the point. There is every reason to believe that objectively analyzable layout methods can be heuristically refined in such a way as to be of practical service.

This chapter gives an introduction to the mathematical framework that is used to develop objectively analyzable algorithms. Section 10.2 makes precise what it means for an algorithmic problem to be efficiently solvable (feasible) and provides a framework for evaluating heuristics for solving infeasible problems. Section 10.3 considers the complexity of the layout problems as a whole. Section 10.4 reviews results on the complexity of subproblems occurring when layout is performed in phases such as placement and routing. Section 10.5 discusses low-level layout optimization and verification problems, and Section 10.6 gives conclusions. For previous discussions of the complexity of layout problems, the reader is referred to [Sah80, Ull84, Sah86, Shi86].

## 10.2  Computational Complexity

We will only consider sequential algorithms. (The mathematical model used here is the Turing machine, but for intuition it suffices to think of a von Neumann computer with large storage space.) The computing resources expended by a computation are runtime and storage space. Runtime is measured in terms of number of instructions. Storage space is measured by the number of storage cells used; each cell can hold an arbitrary large integer. (If the size of the numbers stored is an issue, then space is measured by the number of storage bits used, and instructions for manipulating large numbers use up more time than instructions for manipulating small numbers.) Of course, both the amount of runtime and storage space expended by an algorithm are heavily dependent on such parameters as programming style, programming languages, compilers, machine architecture, *etc.* In order to be able to abstract from these details, and make complexity a characteristic of the algorithm rather than its implementation, we only consider the *asymptotic* behavior of time and space.

Let $\Pi$ be an algorithmic problem. $\Pi$ is defined as a mapping $\Pi : I \rightarrow S$ from problem instances to solutions. Here the problem instance (for example, the netlist for a circuit) is given by some natural encoding as is the solution to the instance (for example, an optimal layout for the circuit). Each problem instance $p \in I$ is assigned a length $l(p)$. Usually $l(p)$ is the amount of storage needed to store the encoding of the instance. Now let $T_A(p)$ and $S_A(p)$ be the time and space, respectively, used by the algorithm $A$ solving problem $\Pi$ on instance $p$. Rather than being interested in the exact values of $T_A(p)$, $S_A(p)$ for each $p \in \Pi$, which are very hard to obtain, we are only interested in values that depend on the length of the instance, usually in the largest value for all instances whose length is $n$.

$$T_A(n) = \max_{l(p)=n} T_A(p)$$

$$S_A(n) = \max_{l(p)=n} S_A(p)$$

$T_A(n)$ $(S_A(n))$ is called the *worst case* time (space) for algorithm $A$ on instances $p \in I$ of length $n$. (In *average case* analysis $T_A(n) = \sum_{l(p)=n} \pi(p)T_A(p)$ where $\pi(p)$ is a given probability that instance $p$ occurs among all instances of length $n$. We will not discuss average case analysis in detail.) The exact value of $T_A(n)$ $(S_A(n))$ is dependent on the implementation, so we are only interested in the *asymptotic*, worst case complexity that specifies how fast $T_A(n)$ $(S_A(n))$ grows as $n$ increases. Specifically we use the following notation suggested by [Knu76]:

$$T_A(n) = O(f(n)) \quad \text{if there are constants } n_0 \in I\!N \text{ and } c > 0$$
$$\text{such that for all } n > n_0 \text{ we have } T_A(n) \leq cf(n)$$
$$T_A(n) = \Omega(f(n)) \quad \text{if there are constants } n_0 \in I\!N \text{ and } \varepsilon > 0$$
$$\text{such that for all } n > n_0 \text{ we have } T_A(n) \geq \varepsilon f(n)$$
$$T_A(n) = \Theta(f(n)) \quad \text{if } T_A(n) = O(f(n)) \text{ and } T_A(n) = \Omega(f(n))$$

$T_A(n) = O(f(n))$ means that as $n$ grows $T_A(n)$ always stays below $f(n)$ multiplied by some constant. $T_A(n) = \Omega(f(n))$ means that as $n$ grows $T_A(n)$ always stays above $f(n)$ multiplied by some constant. If $T_A(n) = \Theta(f(n))$ then $T_A(n)$ can be bounded from above and below by $f(n)$ multiplied by suitable constants. We say that $T_A(n)$ grows *asymptotically* like $f(n)$. The asymptotic behavior of $T_A(n)$ $(S_A(n))$ is called the *time (space) complexity* of algorithm $A$ and is a characteristic only of the algorithm and not of the implementation. The rest of this section considers only time complexity.

The complexity of problem $\Pi$ is obtained by minimizing over the complexity of all its algorithms, *i.e.*, problem $\Pi$ has time complexity $f(n)$ if for all algorithms $A$ solving $\Pi$ we have $T_A(n) = \Omega(f(n))$ and there is an algorithm $A$ solving $P$ such that $T_A(n) = O(f(n))$.

The advantage of being independent from details of the implementation with these definitions has been obtained at a cost. Since no further restrictions are placed on the constants $c$ and $\varepsilon$, they can take on values that are outside the range of practical interest. For example, the function $f_1(n) = 10^4 n$ grows more slowly asymptotically than the function $f_2(n) = 10^{-4} n^2$ even though $f_2(n) \leq f_1(n)$ for $n \leq 10^8$. Thus, it is important to enhance the analysis of the asymptotic complexity such that the factors $c$ and $\varepsilon$ are determined explicitly. The more reasonable the values of $c$ and $\varepsilon$ are, the more relevant the analysis will be for practical applications. Unfortunately in many cases $c$ and $\varepsilon$ are not known or do not have practical values. Here more work must be invested to make asymptotically efficient algorithms usable in practice.

We are now ready to define what we mean by a feasible problem.

Definition 10.1:

A *feasible* (or *polynomial time*) algorithm $A$ is an algorithm such that $T_A(n) = O(n^\alpha)$ for some $\alpha \geq 0$.

A problem $\Pi$ is called *feasible* (or *polynomial time*) if there is an algorithm $A$ solving $\Pi$ that is feasible. The class of feasible problems is denoted with P.

In Definition 10.1, the constant $\alpha$ is not specified further; this lack of specification causes the same problems as described above. Suffice it to say that most of the feasible problems we will discuss have algorithms $A$ with $T_A(n) = O(n^\alpha)$ where $\alpha$ is small ($\alpha \leq 3$).

It is extremely hard to show that a problem $\Pi$ is infeasible. Indeed there are only a handful of natural problems for which this is known, and we will not consider any of them. But there is a large class of problems for which infeasibility is extremely likely. These are the NP-complete problems.

Definition 10.2:

> An *optimization problem* $\Pi : I \rightarrow S$ is an algorithmic problem such that $S$ is a set of *configurations*. There is a predicate $\varphi : I \times S \rightarrow \{0, 1\}$ that determines for each instance $p \in I$ the set of *legal configurations* (namely those with $\varphi(p, s) = 1$) and a cost function $c : S \rightarrow I\!N$. For $p \in I$, $\Pi(p)$ is an *optimal configuration*, i. e., a configuration $s \in S$ such that $\varphi(p,s) = 1$ and $c(s)$ is minimum.

Most of the layout problems we will consider are optimization problems. Here is an example of a problem that occurs in placement.

*Graph Partition*

| | |
|---|---|
| Instance: | An undirected graph $G = (V, E)$, an integer $\alpha$ ($|V|/2 \leq \alpha \leq |V| - 1$). |
| Configurations: | All vertex partitions $V' \subset V$ of $G$. |
| Legal configurations: | All partitions $V' \subset V$ such that $|V'|, |V - V'| \leq \alpha$. |
| Cost function: | $c(V') = |\{\{v, w\} \in E \mid v \in V', w \in V - V'\}|$. |

*Graph Partition* asks for partitioning the vertex set of a graph $G$ into two approximately equal size subsets such that the number of edges between the two subsets (*i.e.* the *cutsize*) is minimized. (See also Chapter 4.)

Our goal is to analyze the complexity of optimization problems. However, for reasons that will become clear, NP-completeness is not defined on optimization problems but on *decision* problems.

Definition 10.3:

> A *decision problem* is an algorithmic problem $\Pi : I \rightarrow S$ such that $S = \{0, 1\}$.

A decision problem asks a binary question of an instance where 0 represents the answer No and 1 represents the answer Yes. Each optimization problem has a canonically corresponding decision problem, the so-called threshold problem.

Definition 10.4:

> Let $\Pi : I \rightarrow S$ be an optimization problem with legality predicate $\varphi$ and cost function $c$. The *threshold problem*, $\Pi_{TH} : I \times I\!N \rightarrow \{0, 1\}$, for $\Pi$ is the problem that, given the instance $(p, k) \in I \times I\!N$, asks whether there exists a configuration $s \in S$ with $\varphi(p, s) = 1$ and $c(s) \leq k$.

For example, the problem *GraphPartition$_{TH}$* asks if, given an instance $(G, \alpha)$ of *Graph Partition* and an integer $k$, there exists a partition with cutsize at most $k$.

We now introduce the concept of NP-completeness for decision problems.

Definition 10.5:

> Let $\Pi$ be a decision problem. A *non-deterministic* algorithm $A$ for $\Pi$ is an algorithm that, at each step, is allowed to make a binary guess as to how to proceed. The algorithm $A$ solves $\Pi$ if, given $p \in I$, there is a way for $A$ to guess such that 1 is computed exactly if $\Pi(p) = 1$. $T_A(p)$ is the shortest time for $A$ to produce a 1 under any choice of guesses. If $\Pi(p) = 0$ no choice of guesses will produce a 1 and $T_A(p) := 1$.
>
> NP denotes the class of all problems that have polynomial time, non-deterministic algorithms.

The definition of a non-deterministic algorithm at first seems unintuitive, and it certainly is not very practical. However, non-determinism is an extremely important concept for the complexity analysis of algorithmic problems. Intuitively if there is a fast non-deterministic algorithm for a decision problem $\Pi$, this means that for each $p \in I$, if $\Pi(p) = 1$ then there exists a short proof of this fact. Indeed, the fastest non-deterministic computation can be taken as such a proof. Going through its steps will convince everybody that $\Pi(p) = 1$. (This is why we restrict our attention to decision problems. A non-deterministic computation of the optimal solution of an optimization problem does not constitute a proof of optimality!)

A non-deterministic algorithm for solving a decision problem $\Pi$ is only burdened with checking the shortest proof of the fact that $\Pi(p) = 1$. (It has to do nothing if $\Pi(p) = 0$.) It does not spend any time on finding the proof. Thus, we are not surprised that NP contains a lot of problems that do not seem feasible deterministically. For instance, *GraphPartition$_{TH}$* is in NP. It is an easy task to guess an optimal partition and check whether its cutsize is at most $k$. Yet, no feasible method for finding optimal partitions is known, and, as we will see, none is likely to exist. Despite this intuitive feeling, the question P = NP? persistently resists being answered. However, as time proceeds the evidence that P $\neq$ NP grows overwhelmingly. This is because we know a large number of natural problems that are in NP and that cannot be solved in polynomial time so far. These are the NP-*complete* problems.

Definition 10.6:

A *polynomial transformation* $\Phi$ from a decision problem $\Pi_1 : I_1 \rightarrow \{0,1\}$ to a decision problem $\Pi_2 : I_2 \rightarrow \{0,1\}$ is a function $\Phi : I_1 \rightarrow I_2$ that is computable by a polynomial time deterministic algorithm such that for all $p \in I_1$

$$\Pi_1(p) = \Pi_2(\Phi(p)).$$

If such a transformation $\Phi$ from $\Pi_1$ to $\Pi_2$ exists, $\Pi_1$ is called *(many-to-one) reducible* to $\Pi_2$ (written $\Pi_1 \leq_m \Pi_2$).

A problem $\Pi_1 \in$ NP is called *NP-complete* if $\Pi \leq_m \Pi_1$ for all $\Pi \in$ NP.

A polynomial transformation quickly translates an instance of $\Pi_1$ into an equivalent instance of $\Pi_2$. An NP-complete problem is a problem in NP that each problem in NP can be reduced to. Thus, if any NP-complete problem $\Pi_1$ is feasible, all are. We only have to transform the instance of any problem $\Pi \in$ NP into an equivalent instance of $\Pi_1$ via $\Phi$ and then solve this instance of $\Pi_1$ using a feasible algorithm.

Several thousand natural NP-complete problems have been found in such diverse areas as logic, graph theory, algebra, optimization, and number theory. There is no known feasible algorithm for any of them. As the set of NP-complete problems grows, the conviction grows that P $\neq$ NP. Thus, while nondeterminism is not a practical concept *per se*, it helps in providing evidence for the fact that certain algorithmic problems are hard. A discussion of NP-completeness can be found in textbooks such as [Gar79, Meh84]. For this chapter we will accept that we cannot solve an NP-complete problem within reasonable time.

Let us now examine the consequences that the NP-completeness of $\Pi_{TH}$ has on $\Pi$. As an example, *GraphPartition$_{TH}$* is NP-complete (even when restricted to problem instances where $G$ has few edges and $\alpha = |V|/2$ [Bui87]). But this means that almost with certainty, *Graph Partition* is infeasible. Assume there were a polynomial time algorithm for *Graph Partition*. By running it and then testing whether the resulting partition has cutsize at most $k$, we get a polynomial time algorithm for *GraphPartition$_{TH}$*; this contradicts our conjecture. For this reason *Graph Partition* is called NP-*hard*. The same argument works for all optimization problems where the cost function $c(s)$ can be computed in polynomial time in the length of the encoding for $s$. This will be the case for all problems we consider. Thus, the NP-completeness of the threshold problem implies (almost with certainty) the infeasibility of the corresponding optimization problem.

Now assume that $\Pi : I \rightarrow S$ is an NP-hard optimization problem. How can we proceed solving $\Pi$, anyway, and how can we make absolute (asymptotic) performance guarantees? There are several ways of doing this, and they will be listed in order of decreasing attractiveness.

- It could be the case that we do not need to solve all instances in $I$, but that all instances we are interested in may come from a subset $I' \subset I$ such that $\Pi|_{I'} : I' \rightarrow S$ is feasible. For example, this is the case for *Graph Partition* if we restrict the problem to all instances with $\alpha = |V| - 1$. Then the problem can be solved in polynomial time with network flow methods.

- While it may be that the problem at hand is NP-hard for all interesting classes of instances, there may be a *pseudopolynomial time* algorithm. A pseudopolynomial time algorithm is an algorithm solving an optimization problem involving integers, that runs in polynomial time in the length of the instance *and in the size of the integers involved*. Thus, such an algorithm is feasible as long as the numbers involved are not too large, *i.e.*, they are polynomial in the length of the instance. An NP-complete problem that does have a pseudopolynomial time algorithm is called *weakly NP-complete*. Several problems occurring in layout are only weakly NP-complete. Since the numbers involved in layout problems are usually bounded by the area of the resulting layout, such problems still have polynomial time algorithms in the resulting layout area. Even though this area may be much larger than the length of the problem instance, such algorithms may still be economical in practical applications.

- If we cannot restrict the set of problem instances, we may still be able to approximate the optimal solution efficiently. The metric by which we measure the guaranteed maximum deviation of the cost of the solution $A(p)$ computed by algorithm $A$ on instance $p$ from the cost of an optimal solution $opt(p)$ is usually

$$ERROR(A) = \max_{p \in I} \frac{c(A(p))}{c(opt(p))}.$$

Such *approximation algorithms*, with bounded error, are not known for *Graph Partition*. But recently Rao [Rao87] presented a polynomial time approximation algorithm for *Graph Partition*, when restricted to planar graphs, that has an error of $O(\log|V|)$.

- In many cases we do not succeed in finding a good approximation algorithm. (There are even problems for which there is no feasible approximation algorithm $A$ with $ERROR(A) \leq c$, for any constant $c > 0$, unless P = NP.) In those cases we take one further step back in our quest for making absolute statements about the algorithm's performance. It is often possible to bound the cost of the solution for $p \in I$ computed by algorithm $A$ with respect to some parameter of $p$. For example, if the *Graph Partition* problem is restricted to planar graphs $G$, then we can find a partition $V'$ in linear time such that $|V'|$, $|V - V'| \leq 2|V|/3$ and the cutsize of $|V'|$ is at most $2\sqrt{2}\sqrt{|V|}$ [Lip79]. Thus, while this algorithm may have a large error in certain instances, we can at least bound the cost of its outcome; this makes it more predictable. Furthermore, for some planar graphs with $n$ vertices, all partitions $V'$ with $|V'|, |V - V'| \leq 2|V|/3$ have cutsize at least $7/6 \sqrt{|V|}$ [Phi80]. Thus, the algorithm comes close to the optimal solution for at least some graphs. Most of the layout algorithms, whose performance can be analyzed, are of this type.

- Another approach is to introduce a random element. We enable the algorithm to toss a coin and proceed depending on the outcome. This approach is a quite practical concept, and sometimes it helps a lot. Of course, now we cannot predict the algorithm's performance with certainty, but there are a number of problems in which some of the following statements can be made.

(1)   The algorithm always computes the optimal solution.

(2)   The algorithm computes the optimal solution with high probability (tending to 1 as $n$ grows).

(3)   If the algorithm computes the optimal solution, it produces a proof of optimality.

(4)   The algorithm's runtime is small in the worst case.

(5)   The algorithm's expected runtime is small on each instance $p \in I$.

Note that randomized algorithms with properties (1) and (5) are more powerful than deterministic algorithms with good average case behavior. Since there is randomization internal to the algorithm, every instance $p \in I$ has the chance of being solved quickly. For a deterministic algorithm with good average case, but bad worst case runtime, there are inherently bad examples that always need a long time to be solved.

Algorithms that use random coin tossing are slowly entering the field of layout. Most of them are not analyzable yet, i.e., they actually belong in the next class. The future is bound to present more sophisticated randomized algorithms that are both analyzable and efficient in practice.

- The last resort is to proceed heuristically. Promising tricks (maybe also involving randomization) may get close to the optimal solution, but we have no means of finding out, in absolute terms, where this method puts us. Even though we may be quite happy with the solution obtained for some problem instance, we do not know how much better we could have done, or how good we will do on the next instance. This is by far the most widespread method of solution today. All of the iterative improvement techniques (deterministic and randomized) fall in this category.

We have now defined what we mean by a hard problem and reviewed some algorithmic techniques for solving hard problems. The following sections cover all phases of the layout process. In each section we define the combinatorial problems that are the basis of the corresponding layout subproblems. Then we summarize the known results and their complexity and survey existing methods for solution. We do not discuss the problems of floorplanning and module generation. There are complexity results for such problems, but many closely relate to results we present, and for others the definition of the optimization problem itself is somewhat involved; discussing such problems is beyond the scope of this chapter.

## 10.3  Layout of Graphs

In this section we consider the layout problem as a whole. Our model of a circuit is a hypergraph where vertices represent circuit elements and hyperedges represent wires. The layout problem is modeled as an embedding problem of such hypergraphs in a square grid. Here vertices of the hypergraph are embedded into nodes of the grid, and hyperedges are embedded into connected subgraphs of the grid (mostly trees) to which the vertices incident to the hyperedge belong. Different hyperedges are required to be embedded into edge-disjoint paths (knock-knee model). This model formalizes the essential boundary conditions on topological layout for a two-layer fabrication process that disallows running wires above each other (for example, to avoid cross talk). The asymptotic bounds for area requirements in this model generalize to models for fabrication processes of VLSI chips that use any bounded number of layers [Tho80]. Most of the results on this layout problem are restricted to graphs, *i.e.*, multi-terminal nets are excluded.

*Hypergraph Embedding*

| | |
|---|---|
| Instance: | A hypergraph $G$. Each vertex of $G$ belongs to at most four hyperedges. (Otherwise, no legal embedding into the grid exists.) |
| Configurations: | All embeddings of $G$ into a rectangular square grid. |
| Legal configurations: | All embeddings of $G$ into a square grid, such that different hyperedges are embedded into edge disjoint subgraphs of the grid. |
| Cost function: | Number of nodes of the grid. |

*Hypergraph Embedding* is NP-complete, even if $G$ is only a graph [Kra84]. This result is not surprising. As shown in later sections, even strongly restricted subproblems of this problem are NP-complete. Nevertheless, approximation algorithms are known for interesting restrictions of the layout problem. In the following sections we consider only graphs.

Definition 10.7:

> Let $G$ be a graph with $N$ vertices. $G$ has an $\alpha$-*bifurcator* of size $F$ if $G$ can be decomposed into two disjoint subgraphs $G_1$, $G_2$ by removing at most $F$ edges and, if furthermore, $G_1$, $G_2$ can be decomposed likewise into disjoint subgraphs $G_{11}$, $G_{12}$ and $G_{21}$, $G_{22}$ respectively by removing at most $F/\alpha$ edges, and so on. In general, in the $i$-th step, removing $F/\alpha^{i-1}$ edges from a subgraph accomplishes the splitting. This process continues until each subgraph is either empty or an isolated vertex [Bha84].

Of particular interest for graph layout are $\sqrt{2}$-bifurcators. They are the parameter that determines the area-requirement for layouts of a graph.

Theorem 10.1:

> Let $F$ be the size of the smallest $\sqrt{2}$-bifurcator of graph $G$. Let $A$ be the minimum size of a grid into which $G$ can be embedded, $L$ be the smallest maximum edge length of $G$ in any embedding, and $C$ be the crossing number of $G$ (*i.e.*, the minimum number of edge crossings needed in any embedding for $G$).
>
> $$F^2 \leq A \leq O(F^2 \log^2(N/F))$$
>
> $$\Omega(F^2) \leq C + N \leq O(F^2 \log(N/F))$$
>
> $$\Omega(F^2/N) \leq L \leq O(F \log(N/F)/\log\log(N/F))$$
>
> All three of the above upper bounds can be realized by the same layout. If a $\sqrt{2}$-bifurcator of size $F$ is given, this layout can be computed in polynomial time [Bha84].

The upper bounds of Theorem 10.1 are derived in three steps. In the first step the $\sqrt{2}$-bifurcator is balanced, so each split results in two, approximately equal size, halves. This splitting increases the size of the bifurcator only by a constant factor. In the second step the graph $G$ is embedded into a special graph $T$, the so-called *tree of meshes*. (See Figure 10.1.) $T$ is sized according to the balanced bifurcator. The size of the meshes at level $i$ from the top is proportional to $F/(\sqrt{2})^i \times F/(\sqrt{2})^i$. All vertices of $G$ are embedded in the meshes at the leaves of $T$. In the third step $T$ is embedded into a rectangular square grid.

Theorem 10.1 is a very powerful theoretical result: It presents a universally applicable layout strategy that guarantees small error on *any* graph.

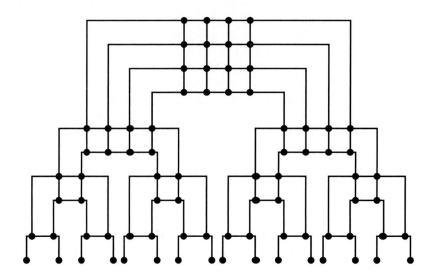

**Figure 10.1**  A Tree of Meshes

Indeed, if a $\sqrt{2}$-bifurcator of nearly minimum size is given, then Theorem 10.1 gives an approximation algorithm $A$ with

$$ERROR(A) = O(\log^2(N/F))$$

$$ERROR(C+N) = O(\log(N/F))$$

$$ERROR(L) = O((N/F)\log(N/F)/\log\log(N/F)).$$

On a number of practically interesting classes of graphs, the errors can be reduced further (for example, to $O(1)$ for area, for any graph in the class) [Lei83a, Bha84].

Nevertheless, several problems pose obstacles towards making this layout method more practical.

- The constants involved are large. At present these constants are chosen to simplify the presentation of the algorithms and proofs. Here there is room for heuristic improvement.

- Theorem 10.1 does not consider hypergraphs. Extensions of the method to hypergraphs may be possible.

- Finding bifurcators is difficult. This is the most critical problem with the method. The problem of finding bifurcators rests on the NP-hard *Graph Partition* problem. One can use heuristics (see Chapters 3 and 4), but using a non-analyzable heuristic destroys the analyzability of the whole procedure. Note, however, that many of the practical placement algorithms used today are based on partitioning of graphs. It seems that we cannot make progress on graph layout without making progress on graph partitioning.

The universality of the tree of meshes as a host graph, in which arbitrary graphs can be embedded efficiently, has led to the development of parallel computer architectures based on this graph [Lei85a].

## 10.4  Layout Subproblems

This section discusses the complexity of subproblems that have to be solved if the layout is done in phases.

### Partitioning

At the basis of all partitioning problems are variations of the following combinatorial problem.

> *Hypergraph Partitioning*
>
> Instance:            An undirected hypergraph $G = (V, E)$ with vertex weights $w : V \rightarrow I\!N$, edge weights $l : E \rightarrow I\!N$, and a maximum cluster size $B \in I\!N$.
>
> Configurations:      All partitions of $V$ into subsets $V_1, ..., V_m$, $m \geq 2$.
>
> Legal configurations: All partitions such that
> $$\sum_{v \in V_i} w(v) \leq B \text{ for all } i = 1, ..., m.$$
>
> Cost function:       $c(V_1, ..., V_m) =$
> $$\sum_{e \in E} (|\{i \in \{1, ..., m\} \mid V_i \cap e \neq \varnothing\}| - 1) l(e).$$

The *Hypergraph Partitioning* problem is an equivalent restatement of Problem 3.2 in Chapter 3. The legal configurations are the partitions in which each cluster $V_i$ has a total vertex weight not exceeding $B$. The weights of the vertices represent block sizes, and the weights on the edges represent connection costs. The maximum cluster size $B$ is a parameter that controls the balance of the partitions.

The *Hypergraph Partitioning* problem is NP-complete, even if $B \geq 3$ is fixed and $w \equiv 1, l \equiv 1$ [Hya73]. The problem is only weakly NP-complete if $G$ is restricted to be a tree [Ber82]. In this case there is a pseudopolynomial time algorithm that solves the problem in time $O(nB^2)$ [Luk74]. If $G$ is a tree and all edge weights are identical [Had74], or if $G$ is a tree and all vertex weights are identical [Gar79], then the problem is in P.

Currently none of the results cited above has practical impact on VLSI layout procedures; the instances of the *Hypergraph Partitioning* problem that occur in circuit layout go beyond the graph classes considered in the restrictions of the problem.

In many applications, such as min-cut placement, we choose $B$ around $W/2$ where $W = \sum\limits_{v \in V} w(v)$, and consider only *bipartitions*, *i.e.*, partitions such that $m=2$. Then the problem becomes the weighted hypergraph version of the *Graph Partition* problem discussed in Section 10.2. [Bui87] study deterministic and randomized bipartitioning algorithms that perform well on some classes of random graphs. These algorithms are efficient, find the optimal bipartition almost always, and provide a proof of optimality if they find the optimal bipartition. While these results are not immediately applicable to circuit layout, they may present a starting point for the development of better partitioning heuristics, and possibly even practical approximation algorithms.

## Placement and Assignment

The placement problem can be stated in terms of several non-equivalent combinatorial problems, depending on what cost measure should be minimized. Each one of these problems is NP-complete.

As described in Chapter 4, many placement techniques are based on hypergraph partitioning. This approach is aimed at minimizing wire congestion. The corresponding combinatorial problems were described in the previous section.

Methods that try to minimize a cost measure, representing interconnection cost, look for an assignment of coordinates in the plane, say for the centers of blocks, such that a cost measure estimating the total amount of wiring is minimized.

*Two-Dimensional Placement*

| | |
|---|---|
| Instance: | A hypergraph $G=(V, E)$ and an interconnection cost function $c : E \rightarrow I\!N$, a set $V' \subset V$ of fixed vertices and positions $s_i, t_i \in I\!N$ for all vertices $v_i \in V'$. |
| Legal configurations: | All placements $((x_i, y_i))$ $i=1,...,|V|$ with $x_i, y_i \in I\!N$, such that $x_i = s_i$ and $y_i = t_i$ for all $v_i \in V'$. |
| Cost function: | $\sum\limits_{e \in E} c(e) \max\limits_{v_i, v_j \in e} d(x_i, y_i, x_j, y_j)$. |

The cost function $c(e)$ determines the cost per unit length of wiring net $e$. The total wiring cost for net $e$ is approximated by the maximum $L_1$-distance between any two blocks belonging to net $e$.

$$d_1(x_i, y_i, x_j, y_j) = |x_i - x_j| + |y_i - y_j|.$$

The *Two-Dimensional Placement* problem is in P. In fact, it can be stated in terms of two identically structured linear programs, one for each dimension. This characteristic remains true even if the *Two-Dimensional Placement* problem is generalized to incorporate block sizes, pin positions in blocks, *etc.* [Wei87]. The corresponding linear programs are dual (in the linear programming sense) to the mincost flow problem [Eic86], for which efficient algorithms have recently been developed by [Gal86, Gol87]. (See also the discussion on compaction in Section 10.5, and Exercise 10.11.)

If the $L_1$-metric is substituted by a quadratic distance measure

$$d_2(x_i, y_i, x_j, y_j) = (x_i - x_j)^2 + (y_i - y_j)^2$$

similar results hold for the two-dimensional placement of graphs. The solution method rests on finding appropriate eigenvectors of a matrix which is a modification of the cost matrix of the graph [Hal70]. In this case the placement coordinates may become irrational, so coordinates are only approximated. If the quadratic distance measure is used for hypergraph placement, the interconnection costs of the hypergraph are approximated by representing each hyperedge by a complete graph on all of its terminals. The edges of this graph receive suitably reduced interconnection costs.

In general the solution of the *Two-Dimensional Placement* problem only gives a first approximation of the placement; the blocks have finite dimensions and may still overlap in the placement. Insisting that blocks be non-overlapping amounts to requiring that the coordinates of the placement be integer. Assume, for example, that all blocks are unit squares and that they should be placed along a line, as done in standard cell placement.

*Optimal Linear Arrangement*

| | |
|---|---|
| Instance: | A hypergraph $G = (V, E)$ and a cost function $c : E \to IN$. |
| Legal configurations: | All permutations $\pi : \{1,...,n\} \to \{1,...,n\}$. |
| Cost function: | $\sum_{e \in E} c(e) \max_{v_i, v_j \in e} |\pi(i) - \pi(j)|$. |

This problem is NP-hard, even if $G$ is only a graph and $c \equiv 1$ [Gar76]. It can be solved in polynomial time $O(n^{2.2})$ if $G$ is a tree and $c \equiv 1$ [Shi79]. If the quadratic distance measure $(\pi(i) - \pi(j))^2$ is used in the cost measure, then the problem is still NP-hard for graphs with $c \equiv 1$ (the proof of [Gar76] carries over), but no polynomial time algorithm for trees is known.

A more interesting cost measure for placement problems in standard cell layout is the *cutwidth*:

$$\max_{2 \leq i \leq |V|} \sum \{c(e)| \ \exists \ (v_j, v_k) \in e \ \ \pi(j) < i \leq \pi(k)\}$$

The cutwidth correlates with the channel density in standard cell placement. With this cost measure the one-dimensional placement problem is called the *Min-cut Linear Arrangement* problem. The *Min-cut Linear Arrangement* problem also occurs in module generators that are based on Weinberger arrays. (See Chapter 7.) It is NP-hard even for planar graphs with maximum vertex degree 3 and $c \equiv 1$, and also for trees with arbitrary cost function [Mon86]. On the other hand, a polynomial time algorithm with runtime $O(n\log(n))$ is known for trees with $c \equiv 1$ [Yan85].

Furthermore, if the problem is restricted to hypergraphs with cutwidth at most $k$, then an $O(n^{p(k)})$-time dynamic programming algorithm exists, where $p(k)$ is a quadratic polynomial in $k$ [Mil86]. Obviously the runtime of this algorithm becomes impractical even for small values of $k$. Nevertheless, these results show that some progress has been made on interesting restrictions of linear layout problems. (Recently, the existence of $O(n^2)$ algorithms for finding the cutwidth of a graph with cutwidth $\leq k$ was proved [Fel88]. But the constant factor rises extremely quickly as $k$ grows, and the proof does not give an explicit construction of the algorithm.)

The two-dimensional arrangement problem corresponding to the *Optimal Linear Arrangement* problem is usually considered only on graphs:

*Quadratic Assignment*

Instance:    A graph $G=(V, E)$, $|V|=n$, a cost function $c : E \rightarrow I\!N$, an $n \times n$ integer distance matrix $D=((d_{ij}))$.

Legal configurations:    All permutations $\pi : \{1,...,n\} \rightarrow \{1,...,n\}$.

Cost function:    $\displaystyle \sum_{e=(v_i, v_j) \in E} c(e)\ d_{\pi(i),\ \pi(j)}.$

The distance matrix $D$ defines the mutual distances between $n$ "slots" into which the $n$ blocks are to be placed. Rather than arranging the slots along a line, they can now be placed arbitrarily. In fact, any distance matrix is allowed. The *Quadratic Assignment* problem is a generalization of the *Optimal Linear Arrangement* problem, which it becomes when $d_{ij}=|i-j|$. This is one of the central problems studied in operations research. [Bur84] gives an overview of the knowledge about this problem. There are no approximation algorithms. However, it can be shown that if $n$ is large and the instances are chosen at random, with high probability a problem instance is chosen for which the costs of the legal configurations do not differ by very much [Bur85]. This means that heuristics are likely to find good solutions. In fact, most often there are few bad solutions.

For layout the distance matrix that corresponds to the square of the $L_2$-metric is most interesting. New heuristic techniques for this restriction of the *Quadratic Assignment* problem have been reported by [Fra86, Fra87].

In order to get around the difficulty of dealing with NP-hard problems explicitly, [Ake81] bases placement on the simpler linear assignment problem.

*Linear Assignment*

| | |
|---|---|
| Instance: | An $n \times n$ integer cost matrix $C = ((c_{ij}))$. |
| Legal configurations: | All permutations $\pi : \{1,...,n\} \to \{1,...,n\}$. |
| Cost function: | $\displaystyle\sum_{1 \le i \le n} c_{i,\,\pi(i)}.$ |

Here the placement is not evaluated by assigning costs to each pair of blocks. Rather, the $c_{ij}$ give the cost of placing block $i$ into slot $j$ of $n$ possible slots $1,...,n$. The linear assignment problem can be solved in time $O(n^3)$ [Pap82].

Clearly the problem with this approach lies in determining a suitable cost matrix that will reflect the interconnection cost of the placement. [Ake81] gives heuristic methods for doing so.

## Global Routing

The input to the global routing problem is a floorplan that can be represented as a planar graph $F = (V, E)$, whose vertices represent the blocks and whose edges represent possible routing channels. Each edge $e$ has two values associated with it, a capacity $c(e)$ and a length $l(e)$. The capacity is a measure of how many wires fit into the corresponding routing channel. The length defines the length of the routing channel. In addition a multi-set $N$ of nets is given − each net $n$ being a subset of the vertices of $F$. ($N$ is a multi-set, because the same subset of vertices can occur several times in $N$.) The nets have to be connected through the routing channels.

*Global Routing*

| | |
|---|---|
| Instance: | A planar graph $F = (V, E)$ a multi-set $N$ of nets on $V$, a length function $l : E \to I\!N$, a capacity function $c : E \to I\!N$. Let $k_n$ be the multiplicity of $n$ in $N$. |
| Legal configurations: | A set $N' \subset N$ of routable nets, and a set of subtrees $T_{n,i}$ of $F$, $n \in N'$, $i \in 1,...,k_n$. Here $T_{n,i}$ must connect all terminals of net $n$. $T_{n,i}$ is called a *Steiner tree* for net $n$. For $e \in E$, let $U(e)$ denote the set of pairs $(n, i)$ such that $e$ is an edge in $T_{n,i}$. Then for all $e \in E$, $|U(e)| \le c(e)$ has to hold. |
| Cost function: | $\displaystyle\sum_{e \in E} l(e)|U(e)| \;+\; \lambda|N - N'|$. Here $\lambda$ is some large constant. |

The constraints $|U(e)| \le c(e)$ ensure that no edge is used above its capacity. The large constant $\lambda$ in the cost function ensures that the number of routable nets is maximized with priority. (The edge capacities $c(e)$ are estimates of the expected channel density. If they are not desired to be strict upper bounds, they can be removed from the constraints and put into the cost function to punish routings exceeding edge capacities.) The *Global Routing* problem is NP-hard even if $F$ is a grid; all nets are two-terminal nets and $c \equiv 1$ [Kra84].

If $N$ has only one net $n$, and $k_n = 1$, this problem becomes the *Minimum Steiner Tree* problem. The *Minimum Steiner Tree* problem is also NP-hard, but various approximation algorithms exist that have an error of $2(1-1/l)$ where $l$ is the minimum number of leaves in a minimum cost Steiner tree [Kou81, Wu86]. The fastest version has recently been developed [Meh88a].

The *Minimum Steiner Tree* problem remains NP-hard if the graph is restricted to belonging to any of the following classes of graphs, even if all edge lengths are 1: graphs with maximum vertex degree $k$, planar graphs, and grid graphs. On the other hand, there are a lot of graph classes that admit polynomial (even linear) time solutions of the *Minimum Steiner Tree* problem. These include outerplanar graphs [Wal82], partial 2-trees [Wal83], and others. [Ber87] give a general method for constructing linear time algorithm for the *Minimum Steiner Tree* problem on special graph classes, including series-parallel graphs. If the *Minimum Steiner Tree* problem is restricted to the case that $n$ is a two-terminal net, it becomes the shortest path problem, which is efficiently solvable with Dijkstra's algorithm (see for example, [Tar83].) All maze routers and several Steiner tree heuristics are based on this observation. (See also Chapter 5.) [Wid87] gives an excellent survey on the *Minimum Steiner Tree* problem. [Joh84a, Joh85] contain more results on special cases of the *Minimum Steiner Tree* problem.

Thus, there is a large body of knowledge about the *Minimum Steiner Tree* problem. The difficulty with solving this problem for global routing is that each instance of the *Minimum Steiner Tree* problem considers only one net. Some order has to be chosen in which the nets are processed in order to achieve global routing. The outcome depends highly on the order in which the nets are processed since nets that are routed earlier can obstruct nets that are routed later. Sometimes nets routed early are ripped up and rerouted later in order to even the advantages different nets get during the routing. The order in which nets are routed has to be chosen heuristically, and finding good heuristics is a major stumbling block for finding good routings. Recently investigations have been carried out to determine how all nets can be routed simultaneously.

Most approaches to simultaneous routing of all wires reformulate the global routing problem as an integer linear program. Each possible Steiner tree for each net $n \in N$ receives a variable. Let $I_n$ be the number of Steiner trees for net $n$. The variable $x_{n,i}$, $n \in N$, $i \in 1,..., I_n$ denotes the number of instances of net $n$ in $N$ that are routed with the $i$-th Steiner tree. Furthermore, we provide a variable $x_n$ denoting the number of unroutable instances of net $n$. The integer program then looks as follows:

$x_{n,i} \geq 0,\ x_{n,i}$ integer  for all $n \in N,\ i \in 1, ..., I_n$

$x_n \geq 0,\ x_n$ integer      for all $n \in N$

$x_n + \sum_{1 \leq i \leq I_n} x_{n,i} = k_n$  for all $n \in N$  (All net instances have to be accounted for.)

$\sum_{(n,i) \in U(e)} x_{n,i} \leq c(e)$     for all $e \in E$  (No edge can be used above its capacity.)

With these constraints we want to minimize the cost function

$$\sum_{e \in E} \left[ l(e) \sum_{(n,i) \in U(e)} x_{n,i} \right] + \lambda \sum_{n \in N} x_n$$

Note that if all possible Steiner trees for net $n$ are included, even the size of this integer linear program is exponential in the length of the problem instance. Such a large integer linear program cannot be generated explicitly. In order to solve it (or to approximate its solution), only significant parts of the instance are generated. Sometimes the variety of Steiner trees for a net is restricted in order to reduce the size of the integer linear program. [Gar72] give an introduction into techniques for solving integer linear programs. However, global routing problem instances tend to be too large to be solvable with such techniques.

Recently several people proposed to combine the ideas of global routing by integer linear programming and global routing by successive Steiner tree construction to get efficient high performance routers. [Bur83] consider a hierarchical top-down approach for gridlike floorplans (for example, gate arrays). [Luk87] extend this method to arbitrary floorplans. The idea is to do the routing on a very coarse abstraction of the floorplan first. This abstraction is obtained by lumping blocks of the floorplan together such that only a small number of blocks remain. The corresponding routing problem is so small that its integer linear program can be solved explicitly. This amounts to finding a coarse routing for all nets simultaneously. The floorplan is then refined in several steps by breaking up the lumped blocks, and the coarse routing is detailed on the refined floorplan. For each refinement step, this again involves solving small integer linear programs for the portions of the routing that are confined to each lumped block. These portions of the routing are then connected to a global routing of the whole floorplan using Steiner tree techniques. The advantage is that the *Minimum Steiner Tree* problem only arises on graphs on which it is efficiently solvable. [Hu85] present a different bottom-up approach in which global routings achieved for parts of the floorplan are pasted together by solving integer linear programs. The hierarchical global routing algorithms are good heuristics, but not approximation algorithms.

[Kar87, Rag85, Ng87, Rag86, Rag87] consider a modification of the routing problem that occurs in gate array layout. Here $F$ is a grid graph, and the cost function is not the total wire length but the maximum "width" of an edge

$$W \equiv \max_{e \in E}(|U(e)| - c(e))$$

The NP-hardness of this version of the global routing problem follows from [Kra84]. [Kar87] prove lower and upper bounds on $W$. They also present a polynomial time approximation algorithm whose error is $O(\log(n/W))$ where $n$ is the number of nets. (All nets are two-terminal nets.) [Rag85, Ng87] present a randomized technique for minimizing $W$. The routing problem is stated as an integer linear program, whose linear relaxation is solved first. (The linear relaxation is obtained by dropping the constraint that the variables be integer. It is a linear program and can thus be solved efficiently.) Then the fractional parts of the solutions of the linear program relaxation are used as a bias for a random experiment that determines in which way the variables are rounded to get the integer solution. This is a method that does not work well for general integer programs [Pap82], but it can be applied successfully to routing problems. In fact, [Rag85] give a randomized routing algorithm whose error is close to 1 with high certainty. [Rag86] makes this method deterministic and achieves polynomial approximation algorithms for this version of the global routing problem whose error is bounded by a constant. [Rag87] extend this method to multi-terminal nets. The method of [Rag86, Rag87] should be extendible to other versions of the global routing problem.

**Detailed Routing**

Detailed routing follows global routing. A detailed routing problem instance consists of a planar routing graph $F$ and a set of nets, just as in global routing. The routing graph $F$ can be regarded as a refinement of the floorplan graph considered in global routing. In the floorplan graph each module is represented by a vertex, and routing regions are represented by edges. Here modules are represented by holes in the graph $F$, and routing regions are described by explicating the routing tracks they consist of. In addition to the nets, there is a set of approximate routings that are the result of the preceding global routing phase. (See Figure 10.2.) Detailed routing problems differ with respect to the shape of the graph $F$ and the routing model used. Three routing models have been investigated in detail.

- *River Routing:* Different nets are required to be wired as *vertex disjoint* subgraphs of $F$.

- *Knock-knee Routing:* Different nets must be wired as *edge disjoint* subgraphs of $F$.

- *Manhattan Routing:* The routings for different nets must be edge disjoint, and any vertex that is shared must be a *cross-over*. "Knock-knees" (*i.e.*, routings where two opposite wire bends share a grid point) are disallowed. (The Manhattan model is only used when $F$ is a grid graph.)

The river routing model allows only planar one-layer routings. Both of the other models allow cross-overs. The Manhattan model induces a simple assignment of wire segments to two routing layers. Horizontal wire segments are wired in one; vertical segments are wired in the other layer. Layer assignment for knock-knee routing is more difficult.

The most popular routing graph is a *channel*. A channel is a rectangular grid graph that has $w$ rows and arbitrarily many columns. Terminals of nets are located on the top and bottom rows. $w$ is called the channel width. A *switchbox* is a channel that has a specified number of columns; the terminals can also appear on the leftmost and the rightmost columns. Recently results have been obtained on routing in more general grid graphs, including switchboxes with ragged edges (all terminals on the outside of the grid graph) and with holes representing blocks (all terminals on the outside or on the boundary of a hole; see Figure 10.2).

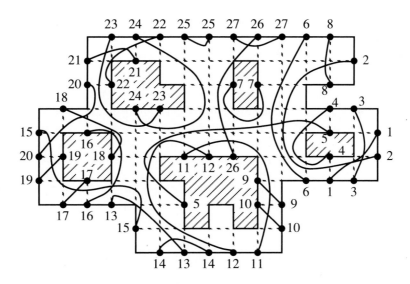

**Figure 10.2**  A Detailed Routing Problem with Two-terminal Nets and a General Routing Graph with Holes

The notion of a global routing that is specified with the problem instance is only significant if the routing graph has holes. In this case the global routing determines how the wire maneuvers around the holes in the routing region. Any routing determined for the wire during the solution of the detailed routing problem has to be transformable into the global routing without pulling any wires across holes in the routing graph.

We now give a survey of results known on each routing model.

**River Routing.** We consider two-terminal nets only. Generalizations to multi-terminal nets are possible. Optimizing the width of a channel in the river routing model can be done by a practical, linear time algorithm. Even a placement problem that arranges blocks on the outside of a channel, such as to minimize the channel width, can be done in linear time [Mir87]. These results may be generalized to channels with a ragged boundary. [Bak84a] present a quadratic algorithm for river routing within a rectangular routing graph with one hole. All nets have one terminal on the outside of the routing graph and one terminal on the hole. (This models the routing problem that occurs when connecting a circuit to a ring of pads.) The algorithm even places the circuit to make a routing possible. [Col84, Lei85b] present efficient algorithms for detailed river routing in arbitrary routing graphs with holes. They also reduce the test of whether a river routing problem is solvable to the solution of a set of simple linear constraints.

**Knock-Knee Routing.** Feasible algorithms exist for routing two-terminal nets in a channel [Meh86b], in a switchbox [Meh86a], in a convex switchbox with ragged edges [Nis85], in an arbitrary switchbox with ragged edges [Kau86], in an arbitrary planar graph with all terminals on the outside of the graph [Bec86], and in a grid graph with holes, given a global routing [Kau87]. In the case of general routing graphs, a detailed routing can be found only if a technical "evenness" condition is satisfied [Bec86, Kau87]. As in the river routing model, testing a routing problem for solvability amounts to solving a set of simple linear constraints. The channel routing algorithm extends to multi-terminal nets where it turns into an approximation algorithm with an error of 2. [Meh86b] give an approximation algorithm with error 3/2 for routing two- and three-terminal nets in a channel. [Gao87] extend this result to arbitrary multi-terminal nets. Their algorithm can be modified to apply to the Manhattan model as well. [Sar87] shows that with multi-terminal nets, even testing whether a channel is routable is NP-complete.

Several results exist for layer assignment of knock-knee routings. An efficient channel routing algorithm for two-terminal nets that minimizes channel width and uses three layers is presented in [Pre84]. If only two layers are available, the channel width may have to be doubled to achieve the routing [Lei81], but doubling the channel width always suffices [Riv81]. [Lip84] shows that determining whether three layers are necessary for a given channel routing is NP-complete. In all these results on layer assignment, contacts may only be placed on grid nodes. The models for multi-layer wiring allow several contacts on a grid node as long as no shorts are introduced. [Bra84] consider general knock-knee routings (including multi-terminal nets). The authors prove that each such routing can be routed in four layers. Difficulties with layer assignment have restricted the use of the knock-knee model in practical layout procedures.

**Manhattan Routing.** Here even the most restricted subproblems are NP-hard. [Szy85] shows that minimizing channel width is NP-hard, even for two-terminal nets. Even though very good heuristics have been known for quite some time, the first efficient approximation algorithm was presented by [Bak84b]. This is because, in the Manhattan model, the channel width does not only depend on the intuitive density of a channel routing problem (see Chapter 5), but also on the more elusive measure of *flux* that was first defined in [Bak84b]. The reason flux is so elusive is that the flux of most routing problems is very small. This is also why so many heuristics that do not consider flux can route within 1 or 2 of density on typical routing problems. However, there are also natural channel routing problems whose flux exceeds their density. One example is the shift-right-by-1 problem. (The shift-right-by-$k$ problem has $n$ two-terminal nets with one terminal on the top and one on the bottom side of the channel. For each net the bottom terminal is $k$ columns to the right of the top terminal. On each side of the channel, the terminals occupy adjacent columns.) Many practical heuristics solve a restriction of the channel routing problem in which doglegs (*i.e.*, track changes of a net) are forbidden. Without doglegs, a channel routing problem may not be solvable because of cyclic vertical constraints. (See Chapter 5.) Whether a routing problem is solvable without doglegs can be determined quickly. However, minimizing the channel width is NP-hard even if doglegs are disallowed [LaP80]. No good approximation algorithm is known for this restriction of the Manhattan model.

Recently other routing models have been investigated that allow overlap of wires. These models are interesting for both on-chip and printed circuit board routing. Heuristics and approximation algorithms for the channel routing problem in these models are given in [Gao86, Ham85, Ber86, Gao87].

## 10.5 Low-Level Layout Problems

In this section we discuss the complexity of problems occurring when optimizing and validating mask layouts.

### Compaction

From a combinatorial point of view, the compaction problem essentially reduces to the following packing problem:

### Two-Dimensional Packing

Instance:           A set $R = \{R_1,...,R_n\}$ of rectangles that lie parallel
                    to the coordinate axes. Each rectangle $R_i$ is
                    defined by a quadruple $(x_i, y_i, h_i, w_i)$, where $(x_i, y_i)$
                    are the coordinates of the lower left corner; $h_i$ and
                    $w_i$ are height and width, respectively.

Legal configurations:   A layout $((u_i, v_i))$ $i=1,...,n$ that assigns new coordi-
                    nates to the lower left corners of the rectangles
                    such that no two rectangles overlap.

Cost function:       The area of the smallest rectangle enclosing the
                    layout.

The rectangles represent blocks and straight wire sections. The no-overlap constraint models the minimum distance requirements imposed by the fabrication
design rules. There are several useful restrictions of this problem. If $v_i = y_i$ for
all $i$ is required, the problem is called *One-Dimensional Packing*. If $v_i = y_i$ for all
$i$, and $x_i \le x_j = > u_i \le u_j$ is required for all pairs $R_i, R_j$ of rectangles that "overlap" when projected on the y-axis, then the problem is called *Ordered One-
Dimensional Packing*.

   *Two-Dimensional Packing* and *One-Dimensional Packing* are both NP-hard
[Doe87]. The *Ordered One-Dimensional Packing* problem can be solved
efficiently by methods based on shortest path computations (see Chapter 6).
These methods allow the incorporation of additional constraints into the compaction, such as minimum or maximum distances between pairs of rectangles in the
compacted layout. This is why most compactors are one-dimensional compactors. (Besides being NP-hard, compaction on the basis of the *One-Dimensional
Packing* problem has the disadvantage that, after the compaction, some wires
may have to be rerouted between the newly placed blocks. Thus the *One-
Dimensional Packing* problem is not used in compaction.)

   Both the *One-Dimensional Packing* problem and the *Two-Dimensional
Packing* problem gain their complexity through the variety of choices for ensuring that a pair $R_i$ and $R_j$ of rectangles do not overlap. In the *One-Dimensional
Packing* problem we can choose to put $R_i$ sufficiently far to the left or to the
right of $R_j$. In the *Two-Dimensional Packing* problem $R_i$ and $R_j$ can also be
separated by putting $R_i$ above or below $R_j$. Thus, there are four choices for each
pair of rectangles. We can assign to each pair $R_i$, $R_j$ a decision variable $\alpha_{ij}$ taking on values in the set $\{0, 1\}$ (for *One-Dimensional Packing*) or $\{0, 1, 2, 3\}$ (for
*Two-Dimensional Packing*) to denote the choice. Each assignment of all
$\alpha_{ij}$ $i, j = 1,...,n$ then leads to one or two independent *Ordered One-Dimensional
Packing* problems that can be solved efficiently. The right assignment of choices
can be found with branch-and-bound methods. [Ked84, Sch83b] propose two
different versions of this packing strategy. [Ked84] enhance their version with

compaction specific heuristics that reduce the number of choices to be made. This approach can be regarded as the branch-and-bound solution of a special mixed integer, linear program. The method guarantees finding the best layout if sufficient (*i.e.*, in the worst case exponential) runtime is given. If the method is stopped early, then only weaker guarantees can be given, even though in many cases the optimal layout is actually found early; most of the runtime is spent excluding other layout alternatives. Several other heuristic ways of solving the *Two-Dimensional Packing* problems are mentioned in Chapter 6.

Compaction that involves only shortest path computations to solve the *Ordered One-Dimensional Packing* subproblems suffers from an imbalanced placement of the circuit modules. The modules tend to be heaped on one side of the layout. To balance the placements, a minimization of the total wire length is incorporated [Sch83a, Eic86]. This problem is a special linear programming problem that has the same structure as the linear programs occurring in the *Two-Dimensional Placement* problem. [Eic86] exhibits the duality of this problem to the mincost flow problem and applies special versions of the simplex algorithm to this problem. [Sch83a] presents graph-based heuristics for solving this problem.

Traditionally compaction follows the detailed routing phase. The recent progress made in the development of a theory of detailed routing has led to developing one-dimensional compaction algorithms that precede the detailed routing phase. The idea is to push the blocks together as long as the detailed routability of the corresponding placement and global routing is guaranteed. Fortunately [Lei85b, Kau87] provide criteria for testing routability in terms of linear constraints. Thus, the one-dimensional compaction problem, which minimizes layout area while preserving detailed routability, is again a special linear program, which can be constructed and solved efficiently [Mal85, Meh88b]. This kind of compaction is one-dimensional compaction with completely general jog introduction. In fact only the way in which a wire maneuvers between the blocks is kept invariant during the compaction. This compaction algorithm applies to compaction problems with two-terminal nets in the river routing model and the knock-knee model.

**Layout Verification**

Design rule checking and circuit extraction problems reduce to the following decision problem.

> *Polygon Intersection*

> Instance:    A set of polygons in the plane having $n$ corners in total.

> Question:    Do some of the polygons intersect?

This problem can be stated statically (as above) or dynamically (where adding and deleting polygons is possible, and new intersections should be reported each time the polygon set is changed). These problems are all efficiently solvable; most solutions are based on the scan-line paradigm. For details see Chapter 8.

Often the polygons are defined hierarchically. Whole sets of polygons are grouped together and represented by a special polygon (usually the convex hull or the smallest enclosing rectangle). This approach reduces the length of the description of a polygon file dramatically. If the input is given in such a succinct fashion, the *Polygon Intersection* problem becomes (weakly) NP-complete, even if we restrict our attention to rectangles that lie parallel to the coordinate axes [Ben83]. Basically all the schemes for hierarchical design rule checking are aimed at finding methods for solving this infeasible problem that are fast on typical problem instances. For details on the corresponding heuristics, see Chapter 8.

The netlist comparison problem is based on the following decision problem:

*Graph Isomorphism*

| Instance: | Two hypergraphs $G_1 = (V_1, E_1)$, $G_2 = (V_2, E_2)$. A bijective mapping $\Phi : V'_1 \to V'_2$ where $V'_1 \subset V_2$, $V'_2 \subset V_2$. |
|---|---|
| Question: | Are there bijective mappings $\Phi_V : V_1 \to V_2$ and $\Phi_E : E_1 \to E_2$ such that $\Phi_V$ is an extension of $\Phi$ and $v \in e \ <=> \ \Phi_V(v) \in \Phi_E(e)$ for all $v \in V$, $e \in E$? |

The two hypergraphs represent the two netlists to be compared. The mapping $\Phi$ represents known correspondences among certain circuit elements, for example, I/O pins. The pair $(\Phi_V, \Phi_E)$ is called a *hypergraph isomorphism*. Note that this problem does not become easier when we restrict $G_1$, $G_2$ to be graphs. This is because we can turn a hypergraph $G$ into a bipartite graph $G'$ by making each net $e$ into a net-vertex $v_e$ such that $v$ is adjacent to $v_e$ in $G'$ exactly if $v \in e$ in $G$. Then we add an additional vertex $x$ that is adjacent to all vertices in $V$. With this definition two hypergraphs $G_1$, $G_2$ are isomorphic exactly if $G'_1$ and $G'_2$ are graph-isomorphic such that $x_1$ turns into $x_2$. (The additional vertex $x$ makes sure that net-vertices in $G'_1$ are turned into net-vertices in $G'_2$ by any isomorphism.) Thus, we restrict $G_1$, $G_2$ to be graphs in the following.

The complexity of the *Graph Isomorphism* problem is not known. It is not known to be NP-complete, and there is quite some evidence that it may not be. On the other hand, no feasible algorithm for this problem is known. The best known isomorphism test has a complexity of $O(\exp(n^{1/2 + \varepsilon}))$ where $\varepsilon > 0$ is an arbitrarily small constant [Zem85]. The problem is in P for planar graphs [Hop74] and for graphs with bounded degree [Gal82, Luk82, Fur83]. So far the concept of randomization has only improved the expected runtime of deterministic polynomial time algorithms [Joh84b]. Furthermore, since fast graph isomorphism tests neglect the problem of diagnosis, none of the theoretical results in this field seems to have practical relevance to layout verification at this point.

## 10.6  Conclusion

The complexity of the physical design process has two aspects: a systems aspect and a combinatorial aspect. The systems aspect deals with the problems of providing the designer with a manageable view of the design process, supporting the designer's activities reliably and consistently, and guiding him or her along the way. The combinatorial aspect, on which we concentrated in this chapter, concerns finding good solutions to precisely stated combinatorial problems that have to be solved during the design process using reasonable computing resources.

Both aspects of the physical design process are equally important. How well we can solve the combinatorial problems has a great impact on the physical design process. If we can compact floorplans reliably and generate high quality, basic cells, layout verification will have reduced importance in the design process. Advanced techniques that integrate floorplanning, placement, and global routing free the designer from having to guide the layout system explicitly through these phases of the layout generation.

As the preceding chapters witness, most combinatorial layout problems are solved by heuristics that are based on (sometimes ingenious) experiential insights but not on a thorough understanding of the problem. As a result, the only way in which these heuristics can be evaluated is by a competitive comparison with other methods of solution. Developing high performance combinatorial algorithms that can be evaluated in absolute terms presupposes a thorough understanding of the underlying layout problem. As an example, the search for an approximation algorithm for channel routing in the Manhattan model (Section 10.4) produced a new structural parameter, flux, and it could be shown that flux and density are exactly the two parameters that determine the channel width (up to a constant factor). By the same token, the fundamental work on graph layout (Section 10.3) exhibited the strong interrelationship between layout and graph partitioning. In fact, up to a $\log^2$-factor the $\sqrt{2}$-bifurcator size determines the area requirement to lay out the graph. It is insights of this sort that are among the most valuable returns of theoretical research. Even if the immediate layout procedures obtained do not yet reach the range of practical performance, the structural insights gained bear high promise of triggering the development of high performance layout procedures that yield good layouts in absolute terms.

## Appendix: Definitions from Graph Theory

A *hypergraph* $G = (V, E)$ consists of a finite set $V = 1,...,n$ of vertices and a (multi-) set $E \subset 2^V$ of *hyperedge*s or *net*s. We denote the number of hyperedges with $|E| = m$. The vertices in a hyperedge $e \in E$ are called *terminal*s of $e$. $|e|$

denotes the cardinality of the hyperedge $e$. $p = \sum_{e \in E} |e|$ is the total number of terminals in $G$. The *degree* of a vertex $v \in V$ is the number of hyperedges to which $v$ belongs.

   If all $e \in E$ satisfy $|e| = 2$, then $G$ is called an (undirected) *graph*. The *crossing number* of a graph $G$ is the minimum number of edge crossings that is needed to draw $G$ in the plane. A *planar graph* is a graph whose crossing number is 0. A *path* $p$ in $G$ is a sequence $v_0, e_1, v_1, e_1, ..., v_{k-1}, e_k, v_k$ such that $v_i \in V$ for $i = 0, ..., k$, $e_i \in E$, $e_i = (v_{i-1}, v_i)$ for $i = 1, ..., k$. If $v_0 = v_k$ then $p$ is called a *cycle*. A *connected graph* is a graph that has a path $v = v_0, ..., v_k = w$ for any two vertices $v, w \in V$. An *acyclic graph* is a graph that contains no cycle. An acyclic graph is also called a *forest*. An acyclic connected graph is also called a *tree*. A *bipartite graph* is a graph such that $V$ can be partitioned into two vertex sets $V_1, V_2$ such that edges only run between vertices in $V_1$ and nodes in $V_2$.

   A *directed graph* is a graph $G = (V, E)$ such that $E \subset V \times V$, i.e., each edge is directed. A directed edge $e = (v, w) \in E$ is said to *leave* $v$ and *enter* $w$. The *indegree* (respectively *outdegree*) of a vertex $v$ is the number of edges entering (respectively leaving) $v$. A *directed path* $p$ in $G$ is a sequence $v_0, e_1, ..., e_{k-1}, v_k$ such that $v_i \in V$ for $i = 0, ..., k$ and $e_i = (v_{i-1}, v_i)$, $e_i \in E$ for $i = 1, ..., k$. If $v_0 = v_k$ then $p$ is called a *circuit*. In a *strongly connected graph* each pair $v, w \in V$ of vertices lies on a common circuit. A *directed acyclic graph (dag)* has no circuit. If $G$ is a dag and each vertex except for one vertex $r$ has indegree 1, then $G$ is called a *rooted tree*. $r$ is the *root* of $G$.

## Exercises

**10.1**   Show that

   **a)**   any polynomial of degree $d$ is $O(n^d)$

   **b)**   $\log n = O(n^\varepsilon)$ for all $\varepsilon > 0$

**10.2**   Which of the following placement problems is (are) in P?

   **a)**   *Quadratic Assignment*

   **b)**   *Linear Assignment*

   **c)**   *Optimal Linear Arrangement*

**10.3†**   Construct a linear arrangement of the complete binary tree of height $n$ that minimizes the cutwidth.

**10.4**    Which of the following routing problems is (are) in P?

    **a)**    Global routing of a single multi-terminal net in a grid-like floor-plan

    **b)**    Global routing of a set of two-terminal nets in a grid-like floorplan

    **c)**    Global routing of a single two-terminal net in a grid-like floorplan

    **d)**    Detailed routing of a set of two-terminal nets in a routing graph with holes

        **1)**    in the river routing model

        **2)**    in the Manhattan model

**10.5†**    Develop a linear time algorithm for the *Minimum Steiner Tree* problem on $2 \times n$ point grid graphs. (This problem is solved in hierarchical global routing of gate arrays.)

**10.6†**    Develop a linear time algorithm for routing a set of two-terminal nets in a channel with a minimum number of tracks in the river routing model. (This involves finding an algorithm, finding a lower bound on the number of necessary tracks, and proving that the algorithm achieves the lower bound.)

**10.7†**    Repeat Exercise 10.6 for the knock-knee model.

**10.8†**    Route the shift-right-by-2 channel routing problem with as few tracks as you can

    **a)**    in the river routing model

    **b)**    in the knock-knee model

    **c)**    in the Manhattan model

**10.9‡**    Develop a channel routing algorithm in the Manhattan model that has the asymptotic worst case performance of the algorithm by [Bak84b] but whose performance compares favorably with current channel routers on typical instances of the channel routing problem.

**10.10**    Construct an instance of the channel routing problem in the knock-knee model that cannot be routed with two wiring layers if contacts can only be placed on grid points.

**10.11†** Formulate a suitable version of the *Two-Dimensional Placement* problem that incorporates block sizes and pin locations in blocks as a set of two linear programs. Define the *Wire Length Minimization* problem for compaction formally, develop its corresponding linear program, and show that it has the same structure as the linear programs occurring in the *Two-Dimensional Placement* problem. Show that these linear programs have solution vectors whose coordinates are all integer, as long as the input only consists of integers. Prove that their dual (in the linear programming sense) is the mincost flow problem.

**10.12‡** Take steps towards developing efficient approximation algorithms for bipartitioning graphs and hypergraphs.

**10.13‡** It is known that an $N$-vertex planar graph has an $O(\sqrt{N})$ size $\sqrt{2}$-bifurcator. Thus, by [Bak84b] it can be laid out in area

$$A = O(N\log^2 N)$$

[Lei83b] proves that there are $N$-vertex graphs for which $A = \Omega(N\log N)$ is necessary. Tighten the remaining gap of a $\log N$-factor for planar graphs, *i.e.*, either prove that less than $O(N\log^2 N)$ area is possible for all planar graphs or exhibit $N$-vertex graphs that require more than $\Omega(N\log N)$ area or both.

# References

[Ake81]  Akers, S. B., "On the use of the linear assignment algorithm in module placement," in *Proc. 18th Design Automation Conf.*, pp. 137-144, 1981.

[Bak84a] Baker, B. S., and R. Y. Pinter, "An algorithm for the optimal placement and routing of a circuit within a ring of pads," in *Proc. 24th Annual IEEE Symposium on Foundations of Computer Science*, pp. 360-370, 1984.

[Bak84b] Baker, B., S., N. Bhatt, and T. Leighton, "An approximation algorithm for Manhattan routing," in *Advances in Computing Research*, vol. 2, VLSI Theory, F. P. Preparata, ed., Jai Press Inc., Greenwich, CT, pp. 205-229, 1984.

[Bec86]  Becker, M., and K. Mehlhorn, "Algorithms for routing in planar graphs," *Acta Informatica*, vol. 23, pp. 163-176, 1986.

[Ben83]  Bentley, J. L., T. Ottmann, and P. Widmayer, "The complexity of manipulating hierarchically defined sets of rectangles." *Advances in Computing Research*, Jai Press Inc., Greenwich, CT, vol. 1, pp. 127-158, 1983.

[Ber82]   Bertolazzi, P. M. Lucertini, and A. M. Spaccamela, "Analysis of a class of graph partitioning problems," *R.A.I.R.O. Theoretical Informatics*, vol. 16, pp. 255-261, 1982.

[Ber87]   Bern, M. W., E. L. Lawler, and A. L. Wong, "Linear-time computation of optimal subgraphs of decomposable graphs," in *J. Algorithms*, vol. 8, pp. 216-235, 1987.

[Ber86]   Berger, B., M. Brady, D. J. Brown, and F. T. Leighton, *Nearly Optimal Bounds and Algorithms for Multilayer Channel Routing*, typescript, 1986.

[Bha84]   Bhatt, S. N., and F. T. Leighton, "A framework for solving VLSI graph layout problems," *Journal of Computer Systems and Science*, vol. 28, pp. 300-343, 1984.

[Bui87]   Bui, T. N., S. Chauduri, F. T. Leighton, and M. Sipser, "Graph bisection algorithms with good average case behaviour," *Combinatorica*, vol. 7, pp. 1818-192, 1987.

[Bra84]   Brady, M. L., and D. J. Brown, "VLSI routing: four layers are enough," *Advances in Computing*, vol. 2, VLSI Theory, F. P. Preparata, ed., Jai Press, Inc., Greenwich, CT, pp. 245-257, 1984.

[Bur83]   Burstein, M., and R. Pelavin, "Hierarchical wire routing," *IEEE Trans. CAD of Integrated Circuits and Systems*, vol. CAD-2, pp. 223-234, 1983.

[Bur84]   Burkard, R. E., "Quadratic assignment problems," *European Journal of Operational Research*, vol. 15, pp. 283-289, 1984.

[Bur85]   Burkard, R. E., and U. Fincke, "Probabilistic asymptotic properties of optimization problems," *Discrete Applied Mathematics*, vol. 12, pp. 21-29. 1985,

[Col84]   Cole, R., and A. Siegel, "River routing, every which way but loose," in *Proc. 25th Annual IEEE Symposium on Foundations of Computer Science*, pp. 65-73, 1984.

[Doe87]   Doenhardt, J., and T. Lengauer, "Algorithmic aspects of one-dimensional layout compaction," *IEEE Trans. CAD of Integrated Circuits and Systems*, vol.CAD-6, pp. 863-878, 1987.

[Eic86]   Eichenberger, P. A., *Fast Symbolic Layout Translation for Custom VLSI Integrated Circuits*, Dissertation, Department of Electrical Engineering, TR. 86-295, Stanford University, Stanford, CA, 1986.

[Fel88]   Fellows, M. R., and A. Langston, "Layout permutation problems and well-partially-ordered sets," in *Proc. of Fifth MIT Conference on Advanced Research in VLSI*, J. Allen and F. T. Leighton, eds., MIT Press, Cambridge, MA, 1988.

[Fra86]  Frankle, J., and R. M. Karp, "Circuit placements and cost bounds by Eigenvector decompositions," *Digest Intl. Conf. on Computer-Aided Design*, pp. 414-417, 1986.

[Fra87]  Frankle, J., *Circuit Placement Methods Using Multiple Eigenvectors and Linear Probe Techniques*, Dissertation, TR. UCB/ERL M87/32, Department of Electrical Engineering and Computer Science, University of California, Berkeley, CA, 1987.

[Fur83]  Fürer, M., W. Schnyder, and E. Specker, "Normal forms for trivalent graphs and graphs of bounded valence," in *Proc. 15th Annual ACM Symposium on Theory of Computing*, pp. 161-170, 1983.

[Gal82]  Galil, Z., C. M. Hoffmann, E. M. Luks, C. P. Schnorr, and A. Weber, "An $O(n^3 \log n)$ deterministic and an $O(n^3)$ probabilistic isomorphism test for trivalent graphs," in *Proc. 23rd Annual IEEE Symposium on Foundations of Computer Science*, pp. 118-125, 1982.

[Gal86]  Galil, Z., and E. Tardos, "An $O(n^2 \log n(m + n \log n))$ min-cost flow algorithm," in *Proc. 27th Annual IEEE Symposium on Foundations of Computer Science*, pp. 1-9, 1986.

[Gao86]  Gao, S., and S. Hambrusch, "Two-layer channel routing with vertical unit-length overlap," *Algorithmica*, pp. 223-232, 1986.

[Gao87]  Gao, S., and M. Kaufmann, "Channel routing of multiterminal nets," in *Proc. 28th Annual IEEE Symposium on Foundations of Computer Science*, pp. 316-325, 1987.

[Gar72]  Garfinkel, R. S., and G. L. Nemhauser, *Integer Programming*, John Wiley & Sons, New York, NY, 1972.

[Gar76]  Garey, M. R., D. S. Johnson, and L. Stockmeyer, "Some simplified NP-complete graph problems," *Theoretical Computer Science*, vol. 1, pp. 237-267, 1976.

[Gar79]  Garey, M. R., and D. S. Johnson, *Computers and Intractability: A Guide to the Theory of NP-Completeness*, San Francisco, CA, W. H. Freeman & Co., 1979.

[Gol87]  Goldberg, A., and R. Tarjan, "Solving minimum-cost flow problems by successive approximation," in *Proc. 19th Annual ACM Symposium on Theory of Computing*, pp. 7-18, 1987.

[Had74]  Hadlock, F. O., "Minimum spanning forests of bounded trees," in *Proc. Fifth Southeastern Conference on Combinatorics, Graph Theory and Computing*, Utilitas Mathematica Publishing, Winnipeg, Canada, pp. 449-460, 1974.

[Hal70]  Hall, K. M., "An r-dimensional quadratic placement algorithm," *Management Science*, vol. 17, pp. 219-229, 1970.

[Ham85] Hambrusch, S., "Channel routing algorithms for overlap models," *IEEE Trans. CAD of Integrated Circuits and Systems*, vol. CAD-4, pp. 23-30, 1985.

[Hop74] Hopcroft, J. E., and J. K. Wong, "Linear time algorithm for isomorphisms of planar graphs," in *Proc. 16th Annual ACM Symposium on Theory of Computing*, pp. 172-184, 1974.

[Hu85] Hu, T. C., and M. T. Shing, "A decomposition algorithm for circuit routing," *VLSI Layout: Theory and Design*, T. C. Hu, E. S. Kuh, eds., IEEE Press, pp. 144-152, 1985.

[Hya73] Hyafil, L., and R. L. Rivest, *Graph Partitioning and Constructing Optimal Decision Trees Are Polynomially Complete Problems*, Report No. 33, IRIA-Laboria, Rocquencourt, France, 1973.

[Joh84a] Johnson, D. S., "The NP-completeness column: an ongoing guide," *J. Algorithms*, vol. 5, pp. 147-160, 1984.

[Joh84b] Johnson, D. S., "The NP-completeness column: an ongoing guide," *J. Algorithms*, vol. 5, pp. 433-447, 1984.

[Joh85] Johnson, D. S., "The NP-completeness column: an ongoing guide," *J. Algorithms*, vol. 6, pp. 434-451, 1985.

[Kar87] Karp, R. M., F. T. Leighton, R. L. Rivest, C. D. Thompson, U. Vazirani, and V. Vazirani, "Global wire routing in two-dimensional arrays," *Algorithmica*, vol. 2, pp. 113-129, 1987.

[Kau86] Kaufmann, M., and K. Mehlhorn, "Routing through a generalized switchbox," *J. Algorithms*, vol. 7, pp. 510-531, 1986.

[Kau87] Kaufmann, M., and K. Mehlhorn, "On local routing of two-terminal nets," in *Proc. 4th Annual Symposium on Theoretical Aspects of Computer Science*, F. J. Brandenburg et al., eds., Springer Lecture Notes in Computer Science No. 247, Springer-Verlag, New York, NY, pp. 40-52, 1987.

[Ked84] Kedem, G., and H. Watanabe, "Graph optimization techniques for IC layout and compaction," *IEEE Trans. CAD of Integrated Circuits and Systems*, vol. CAD-3, pp. 12-20, 1984.

[Knu76] Knuth, D. E., "Big omicron and big omega and big theta," SIGACT News, pp 18-24, April-June, 1976.

[Kou81] Kou, L., G. Markowsky, and L. Berman, " A fast algorithm for Steiner trees," *Acta Informatica*, vol. 15, pp. 141-145, 1981.

[Kra84] Kramer, M. R., and J. van Leeuwen, "The complexity of wire routing and finding minimum area layouts for arbitrary VLSI circuits," *Advances in Computing Research*, vol. 2, VLSI Theory, F. P. Preparata, ed., Jai Press Inc., Greenwich, CT, pp. 129-146, 1984.

[LaP80]   LaPaugh, A. S., *Algorithms for Integrated Circuit Layout: An Analytic Approach*, Ph.D. Thesis, MIT, 1980.

[Lei81]   Leighton, F. T., *New Lower Bounds on Channel Routing*, MIT VLSI Memo 82-71, 1981.

[Lei83a]  Leiserson, C. E., *Area-efficient VLSI Computation*, MIT Press, Cambridge, MA, 1983.

[Lei83b]  Leighton, F. T., *Complexity Issues in VLSI*, MIT Press, Cambridge, MA, 1983.

[Lei85a]  Leiserson, C. E., "Fat trees: universal networks for hardware-efficient supercomputing," in *Proc. Intl. Conf. on Parallel Processing*, pp. 393-402, 1985.

[Lei85b]  Leiserson, C., and F. M. Maley, "Algorithms for routing and testing routability of planar VLSI layouts," in *Proc. 17th Annual ACM Symposium on Theory of Computing*, pp. 69-78, 1985.

[Lip79]   Lipton, R. J., and R. E. Tarjan, "A separator theorem on planar graphs," *SIAM Journal Applied Mathematics*, vol. 36, pp. 177-189, 1979.

[Lip84]   Lipski, W., "On the structure of three-layer wireable layouts," *Advances in Computing Research*, vol. 2, VLSI Theory, F. P. Preparata, ed., Jai Press Inc., Greenwich, CT, pp. 231-244, 1984.

[Luk74]   Lukes, J. A., "Efficient algorithm for the partitioning of trees," *IBM Journal of Research Development*, vol. 18, pp. 217-224, 1974.

[Luk82]   Luks, E. M., "Isomorphism of graphs of bounded valence can be tested in polynomial time," *Journal of Computer and Systems Science*, vol. 25, pp. 42-65, 1982.

[Luk87]   Luk, W. K., P. Sipala, M. Tamminen, D. Tang, L. S. Woo, and C. K. Wong, "A hierarchical global wiring algorithm for custom chip design," *IEEE Trans. CAD of Integrated Circuits and Systems*, vol. CAD-6, pp. 518-533, 1987.

[Mal85]   Maley, F. M., "Compaction with automatic jog introduction," *1985 Chapel Hill Conference on VLSI*, H. Fuchs, ed., Computer Science Press, Rockville, MD, pp. 261-283, 1985.

[Meh84]   Mehlhorn, K., *Data Structures and Algorithms, Volume 2: Graph Algorithms and NP-completeness*, Springer-Verlag, New York, NY, 1984.

[Meh86a]  Mehlhorn, K., and F. Preparata, "Routing through a rectangle," *Journal of the ACM*, vol. 33, pp. 60-85, 1986.

[Meh86b]  Mehlhorn, K., F. P. Preparata, and M. Sarrafzadeh, "Channel routing in knock-knee mode: Simplified algorithms and proofs," *Algorithmica*, vol. 1, pp. 213-221, 1986.

[Meh88a] Mehlhorn, K., "A faster approximation algorithm for the Steiner problem in graphs," to appear in *Inf. Proc. Let.*

[Meh88b] Mehlhorn, K., and S. Näher, "A faster compaction algorithm with automatic jog introduction," in *Proc. of the Fifth MIT Conference on Advanced Research in VLSI*, J. Allen, F. T. Leighton, eds., 1988.

[Mil86] Miller, Z., and I. H. Sudborough, "A polynomial algorithm for recognizing small cutwidth in hypergraphs," *VLSI Algorithms and Architectures*, F. Makedon et al., eds., Springer Lecture Notes in Computer Science No. 227, Springer-Verlag, New York, NY, pp. 252-260, 1986.

[Mir87] Mirzaian, A., "River routing in VLSI," *Journal of Computer Systems and Science*, vol. 34, pp. 43-54, 1987.

[Mon86] Monien, B., and I. H. Sudborough, "Min-cut is NP-complete for edge weighted trees," in *Proc. 13th Intl. Conf. on Automata, Languages and Programming*, L. Kott, ed., Springer Lecture Notes in Computer Science No. 226, Springer-Verlag, New York, NY, pp. 265-273, 1986.

[Ng87] Ng, A. P. C., P. Raghavan, and C. D. Thompson, "Experimental results for a linear program global router," *Computers and Artificial Intelligence*, vol. 6, pp. 229-242, 1987.

[Nis85] Nishizeki, T., N. Saito, and K. Suzuki, "A linear-time routing algorithm for convex grids," *IEEE Trans. CAD of Integrated Circuits and Systems*, vol. CAD-4, pp. 68-76, 1985.

[Pap82] Papadimitriou, C. H., and K. Steiglitz, *Combinatorial Optimization: Algorithms and Complexity*, Prentice-Hall, Englewood Cliffs, NJ, 1982.

[Phi80] Philipp, R., and E.-J. Prauss, "Separators in planar graphs," *Acta Informatica*, vol. 14, pp. 87-106, 1980.

[Pre84] Preparata, F. P., and W. Lipski, "Optimal three-layer channel routing," *IEEE Trans. Computers*, vol. C-33, pp. 427-437, 1984.

[Rag85] Raghavan, P., and C. D. Thompson, "Provably good routing in graphs: regular arrays," in *Proc. 17th Annual ACM Symposium on Theory of Computing*, pp. 79-87, 1985.

[Rag86] Raghavan, P., "Probabilistic construction of deterministic algorithms: approximating packing integer programs," in *Proc. 27th Annual IEEE Symposium on Foundations of Computer Science*, pp. 10-18, 1986.

[Rag87] Raghavan, P., and C. D. Thompson, *Multiterminal Global Routing: A Deterministic Approximation Scheme*, TR-12806, IBM T. J. Watson Research Laboratory, Yorktown Heights, NY, 1987.

[Rao87] Rao, S., "Finding nearly optimal separators in planar graphs," in *Proc. 28th Annual IEEE Symposium on Foundations of Computer Science*, pp. 225-237, 1987.

[Riv81]   Rivest, R. L., A. Baratz, and G. Miller, "Provably good channel routing algorithms," 1981 *CMU Conf. on VLSI Systems and Computations*, H. T. Kung et al., eds., Computer Science Press, Rockville, MD, pp. 153-159, 1981.

[Sah80]   Sahni, S., and A. Bhatt, "The complexity of design automation problems," in *Proc. 17th Design Automation Conf.*, pp. 402-411, 1980.

[Sah86]   Sahni, S., A. Bhatt, and R. Raghavan, "The complexity of design automation problems," in *Proc. of Automation 86 High Technology Computer Conf.*, van Nostrand Reinhold, New York, NY, pp. 82-98, 1986.

[Sar87]   Sarrafzadeh, M., "Channel-routing problem in the knock-knee mode is NP-complete," *IEEE Trans. CAD of Integrated Circuits and Systems*, vol. CAD-6, pp. 503-506, 1987.

[Sch83a]  Schiele, W., "Improved compaction by minimized length of wires," in *Proc. 20th Design Automation Conf.*, pp. 121-127, 1983.

[Sch83b]  Schlag, M., Y. Z. Liao, and C. K. Wong, "An algorithm for optimal two-dimensional compaction of VLSI layouts," *INTEGRATION*, vol. 1, pp. 179-209, 1983.

[Shi79]   Shiloach, Y., "A minimum linear arrangement algorithm for undirected trees," *SIAM Journal of Computing*, vol. 8, pp. 15-32, 1979.

[Shi86]   Shing, M. T., and T. C. Hu, "Computational complexity of layout problems," in *Layout Design and Verification*, T. Ohtsuki, ed., North Holland, New York, NY, Chapter 8, pp. 267-294, 1986.

[Szy85]   Szymanski, T. G., "Dogleg channel routing is NP-complete," *IEEE Trans. CAD of Integrated Circuits and Systems*, vol. CAD-4, pp. 31-41, 1985.

[Tar83]   Tarjan, R. E., *Data Structures and Network Algorithms*, SIAM Monograph No. 44, 1983.

[Tho80]   Thompson, C. D., *A Complexity Theory for VLSI*, TR 80-140, Department of Computer Science, Carnegie-Mellon University, 1980.

[Ull84]   Ullman, J. D., *Computational Aspects of VLSI*, Computer Science Press, Rockville, MD, 1984.

[Wal82]   Wald, J. A., and C. J. Colbourn, "Steiner trees in outerplanar graphs," in *Proc. 13th Southeastern Conference on Combinatorics Graph Theory and Computing*, pp. 15-22, 1982.

[Wal83]   Wald, J. A., and C. J. Colbourn, "Steiner trees, partial 2-trees, and minimum IFI networks," *Networks*, vol. 13, pp. 159-167, 1983.

[Wei87]   Weis, B. X., and D. A. Mlynski, "A new relative placement procedure based on MSST and linear programming," in *Proc. of International Symposium on Circuits and Systems*, pp. 564-567, 1987.

[Wid87]   Widmayer, P., *Fast Approximation Algorithms for Steiner's Problem in Graphs,* Habilitation, University of Karlsruhe, 1987.

[Wu86]    Wu, Y. F., P. Widmayer, and C. K. Wong, "A faster approximation algorithm for the Steiner problem in graphs," *Acta Informatica*, vol. 23, pp. 223-229, 1986.

[Yan85]   Yannakakis, M., "A polynomial algorithm for the min-cut linear arrangement of trees," *Journal ACM 32*, pp. 950-988, 1985.

[Zem85]   Zemlyachenko, V. N., N. M. Korneeko, and R. I. Tyshkevich, "Graph isomorphism problem," *Journal of Soviet Mathematics*, vol. 29, pp. 1426-1481, 1985.

# Index

**Bold** *numbers indicate page numbers on which terms are defined. In the text, terms being defined appear in italics.*